SPANNING WASHINGTON

A computer-enhanced image showing how the new Tacoma Narrows Bridge (left) will appear when completed in 2007.

Tacoma Narrows Constructors

S·P·A·N·N·I·N·G WASHINGTON

Historic Highway Bridges of the Evergreen State

Craig Holstine and Richard Hobbs

Afterword by Eric DeLony

Washington State University Press
Pullman, Washington

Washington State University Press
PO Box 645910
Pullman, Washington 99164-5910
Phone: 800-354-7360
Fax: 509-335-8568
E-mail: wsupress@wsu.edu
Web site: wsupress.wsu.edu

First printing 2005

Library of Congress Cataloging-in-Publication Data

Holstine, Craig E.
Spanning Washington : historic highway bridges of the Evergreen State / Craig Holstine and Richard Hobbs.
p. cm.
Includes bibliographical references.
ISBN 0-87422-281-8 (alk. paper)
1. Bridges--Washington (State)--History. I. Hobbs, Richard, 1947- II. Title.
TG24.W3H65 2005
624.2'09797--dc22

ON THE COVER:
"Into the Fog, Deception Pass Bridge," by Dan Karvasek, www.dankarvasek.com

Table of Contents

Foreword vii
Douglas B. MacDonald, Washington Secretary of Transportation
Acknowledgments ix
Dedication xi
Introduction 1

PART I: HISTORICAL PERSPECTIVES
Chapter 1 150 Years of Spanning Washington 29
Chapter 2 Failure and Progress 51
Chapter 3 Loss and Preservation 67
Chapter 4 Designers of Dreams: Bridge Engineers 83

PART II: PREMIER BRIDGES
Chapter 5 Bridges of the Columbia River 97
Chapter 6 Bridges of Spokane and Northeast Washington 123
Chapter 7 Bridges of Central and Southeast Washington 139
Chapter 8 Bridges of Seattle 151
Chapter 9 Floating Bridges 167
Chapter 10 Bridges of Northwest Washington 187
Chapter 11 Bridges of Southwest Washington 197
Chapter 12 Bridges of the Kitsap and Olympic Peninsulas 215

Afterword 227
Eric DeLony, Retired Chief, Historic American Engineering Record
Appendix Washington State Historic Highway Bridges National Register of Historic Places/Historic American Engineering Record 239
Glossary 253
Bibliography 257
Index 265
The Authors 268

The Monroe Street Bridge (1911), Spokane.

Eric DeLony

Foreword

"Technology, like art, is a soaring exercise of the human imagination."
—Daniel Bell

Washington's most distinctive treasure is its landscape. Bridges are the structures we Washingtonians have built to tie our communities together over the valleys, through the mountains, and across the innumerable waterways of our small but glorious place on earth. And bridges—each one stretched across space by the sheer force of human will—in their course themselves become a part of the very landscapes for which they have been drawn.

Spanning Washington is about Washington's historic bridges. It seeks to preserve and commemorate the strength and elegance of these structures, and the ingenuity and physical efforts of their builders. It also celebrates the ways in which visions of transportation have been realized in the works of civil engineering, and then, through the decades, have fundamentally shaped our daily lives across the length and breadth of this state.

The covered Pe Ell Bridge (1934) crosses the Chehalis River.
Craig Holstine

Some of Washington's bridges are famous around the world. Some stand as major monuments in the history of bridge design and construction. Joining these examples of supremely successful bridges, however, are a few whose spectacular failures remind us that fallibility, as well as victory, can accompany the journey of human inspiration.

But this book is also about bridges we might regard as common and everyday—the small or unpretentious spans across a local creek or a steep and narrow ravine. Look closely—bridges, like people, reveal their true merit in their individuality and the success with which they meet their own purposes in their unique times and places.

Seek in these pages the special structures so wondrous and beautiful as to be not-so-humble human contributions to the scenery itself. Choose your favorite; I know mine is here. We hope the bridges in this book can take you on your own special adventure along the path of imagination through history and into our treasured realm of the earthly and human landscape.

—Douglas B. MacDonald
Secretary of Transportation

The Rainbow Bridge (1957) near La Conner.
WSDOT

Acknowledgments

The authors' work stands on the shoulders of many others who contributed to this book. First and foremost, *Spanning Washington* would not have been published had it not been for the tireless efforts and unwavering support of Sandie Turner in WSDOT's Environmental Services Office. Her former supervisor Rick Singer envisioned a book that would be more than a published list of Washington's bridge inventory. Although we take full responsibility for its content, the authors have attempted to realize Rick's objective. Visually the book has benefited from the map-making skills of Tanya Johnson in ESO's GIS Program, and from the scanning skills of Diana Martinez at ESO and John Darin in WSDOT's Print Services.

We benefited greatly from the work of other writers, and depended heavily on Historic American Engineering Record (HAER) reports and National Register of Historic Places (NRHP) nominations, rather than on the primary sources used by the authors of those documents. The experts who authored most of those studies include Oscar R. "Bob" George and Robert Krier, retired WSDOT bridge engineers with acute sensitivities for historic structures; Lisa Soderberg, who conducted the state's first bridge inventory and prepared a thematic NRHP nomination; and Robert Hadlow, Jonathan Clarke, and Wm. Michael Lawrence, bridge historians. The latter, along with their colleagues under the supervision of HAER chief Eric DeLony, created a written, graphic, and photographic record that is surely the envy of other states in the nation. As evidenced by his Afterword in this book, Eric's continuing interest in Washington bridges has gratified and inspired us, and we thank him for putting his retirement aside on our behalf. Our friend and historian Jeff Creighton contributed substantially to the information in Chapter 6 on Spokane's historic bridges, and gathered materials relating to other bridges as well. Glenn Hartmann of Western Shore Heritage Services provided administrative support and thoughtful guidance. We gratefully extend our appreciation to them. We thank WSU Press staff Glen Lindeman, Mary Read, Nancy Grunewald, Kerry Darnall, Nella Letizia, and Jean Taylor for their diligent efforts in seeing this project through to publication. And we thank our families, especially our wives, Marsha and Lynette, for their help, patience, and understanding.

Vital to our efforts was assistance provided by the individuals and institutions listed below. Their contributions were instrumental in bringing this book to completion.

Washington State Department of Transportation (present and past employees):

Sandie Turner, Diana Martinez, Tanya Johnson, Pam Conley, Ken Stone, Heather Zolzer, Jef Lucero, Trent de Boer, Pam Trautman, and Brenda Kent, Environmental Services Office

Rick Singer, Business Manager, Tacoma Narrows Bridge Project

Sharan Linzy, Doris Zahn, Larry Veden, and Harvey Coffman, Bridge Preservation Office

Patrick Clarke and DeWayne Wilson, Bridge and Structures Office

Rich Gleckler, Contracts Review

Randy Powell and John Johnson, Engineering Records

Grant Griffin and Greg Kolle, Highways and Local Programs

Jennifer Boteler and Rebecca Christie, Library

Leroy Slemmer, Tacoma Project Engineer's Office

Elizabeth Robbins, Policy Development and Regional Coordination

Battelle Corporation, Pacific Northwest Laboratory: David Harvey

City of Olympia: Jay Burney

City of Spokane, Historic Preservation Office: Teresa Brum

Douglas County Transportation and Land Services Engineering Department: Carol Hardie

Ferry County Public Works: Keith Muggoch

Gifford Pinchot National Forest: Rick McClure

Hood River Historical Museum, Hood River, Oregon

King County Office of Cultural Resources: Charlie Sundberg

King County Department of Transportation: Fennelle Miller

Lewis County Historical Museum: Margaret Shields

Museum of History and Industry, Seattle: Carolyn Marr

North Central Washington Museum, Wenatchee: Mark Behler, Keith Williams

Northwest Museum of Arts and Culture, Spokane: Karen DeSeve, Rose Krause

Okanogan County Museum: Richard Ries

Port of Hood River, Oregon: Melissa Child and Linda Shames

Seattle Municipal Archives: Scott Cline

Seattle Public Library

Skagit County Public Works Department: Barbara Hathaway

Snohomish County Planning Department: Louise Lindgren

Spokane Public Library, Northwest Room: Nancy Compau, Rayette Sterling

University of Washington, Alumni Office

University of Washington Libraries, Manuscripts, Archives, and Special Collections: Gary Lundell and other staff

University of Washington, Registrar's Office: Virjean Edwards

Washington State Archives: David Hastings, Patricia Hopkins, Mary Hammer

Washington State Archives, Eastern Regional Office: Susan Beamer

Washington State Archives, Puget Sound Regional Office: Mike Saunders

Washington State Board of Registration for Professional Engineers and Land Surveyors: John Pettainen, George Twiss

Washington State Department of Natural Resources: Eric Carlsen, Brian Davis, Lee Stilson

Washington State Historical Society: Ed Nolan, Joy Werlink, Elaine Miller

Washington State Library: Shirley Lewis, Leah Cushman

Washington State Office of Archaeology and Historic Preservation: Susan Goff, Sara Steel, Russell Holter, Michael Houser, Greg Griffith, Rick Anderson

Washington State University, Alumni Office

Washington State University, Manuscripts, Archives, and Special Collections

Wenatchee Valley Museum and Culture Center: Mark Behler

Whatcom County Public Works: Steve Dillon

Whatcom Museum of History and Art: Jeff Jewell

Individual contributors include: C. W. Eldridge, Eleanor Hadley, and Steve Hauff

To our friends and mentors for their exceptional contributions to
Pacific Northwest history and bridge preservation

David H. Stratton
Washington State University Professor Emeritus of History

Oscar R. "Bob" George and Robert H. Krier
Retired WSDOT Bridge Engineers

❧

Washington State University Press gratefully
acknowledges the support and assistance from the
Washington State Department of Transportation
that made the publication of this book possible.

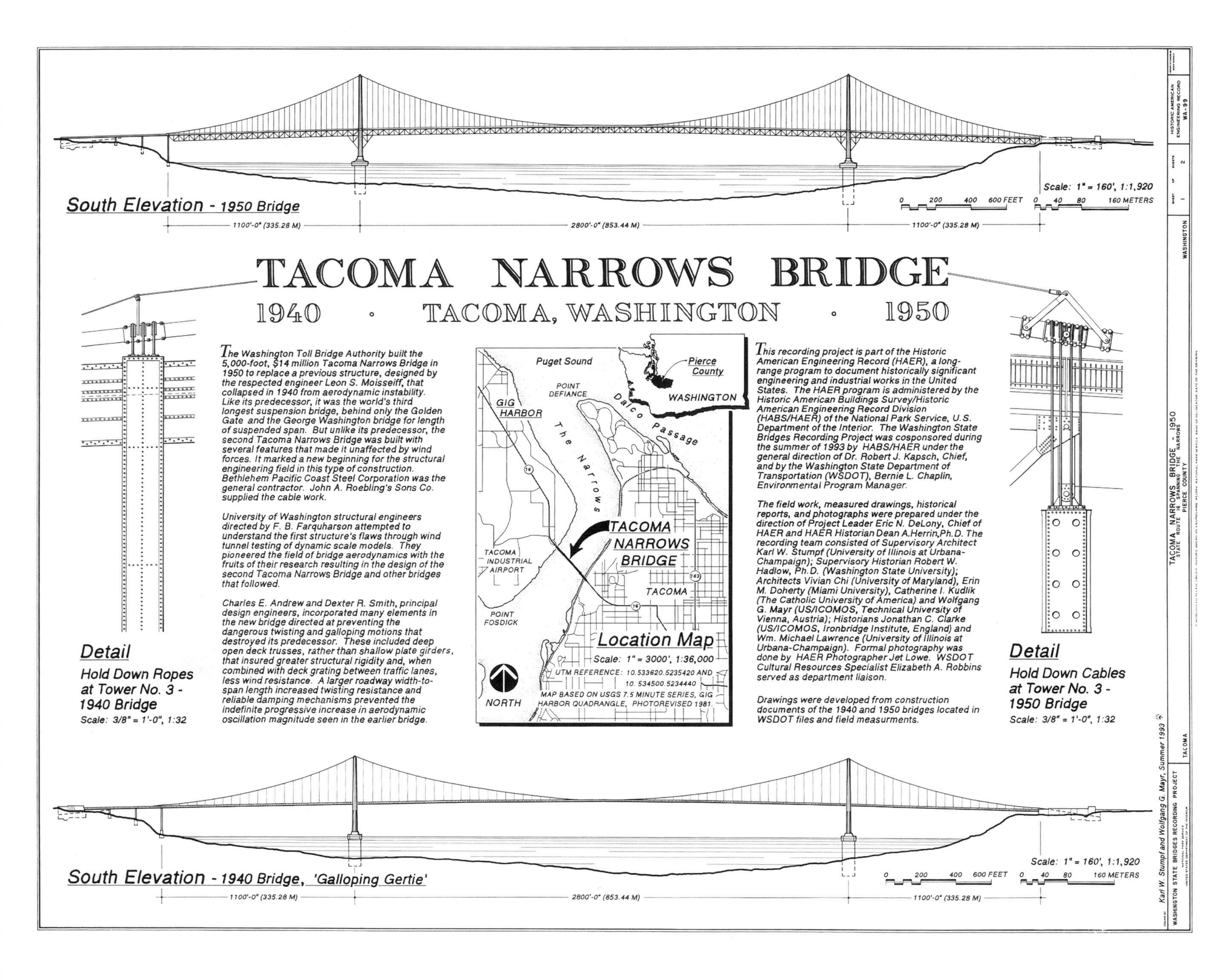

South Elevation - 1950 Bridge
1100'-0" (335.28 M)
2800'-0" (853.44 M)
1100'-0" (335.28 M)
Scale: 1" = 160', 1:1,920
0 200 400 600 FEET
0 40 80 160 METERS
TACOMA NARROWS BRIDGE
1940 · TACOMA, WASHINGTON · 1950
The Washington Toll Bridge Authority built the 5,000-foot, $14 million Tacoma Narrows Bridge in 1950 to replace a previous structure, designed by the respected engineer Leon S. Moisseiff, that collapsed in 1940 from aerodynamic instability. Like its predecessor, it was the world's third longest suspension bridge, behind only the Golden Gate and the George Washington bridge for length of suspended span. But unlike its predecessor, the second Tacoma Narrows Bridge was built with several features that made it unaffected by wind forces. It marked a new beginning for the structural engineering field in this type of construction. Bethlehem Pacific Coast Steel Corporation was the general contractor. John A. Roebling's Sons Co. supplied the cable work.
University of Washington structural engineers directed by F. B. Farquharson attempted to understand the first structure's flaws through wind tunnel testing of dynamic scale models. They pioneered the field of bridge aerodynamics with the fruits of their research resulting in the design of the second Tacoma Narrows Bridge and other bridges that followed.
Charles E. Andrew and Dexter R. Smith, principal design engineers, incorporated many elements in the new bridge directed at preventing the dangerous twisting and galloping motions that destroyed its predecessor. These included deep open deck trusses, rather than shallow plate girders, that insured greater structural rigidity and, when combined with deck grating between traffic lanes, less wind resistance. A larger roadway width-to-span length increased twisting resistance and reliable damping mechanisms prevented the indefinite progressive increase in aerodynamic oscillation magnitude seen in the earlier bridge.
Puget Sound
Pierce County
WASHINGTON
POINT DEFIANCE
Dalco Passage
GIG HARBOR
The Narrows
TACOMA NARROWS BRIDGE
TACOMA INDUSTRIAL AIRPORT
TACOMA
POINT FOSDICK
Location Map
Scale: 1" = 3000', 1:36,000
UTM REFERENCE: 10.533620.5235420 AND 10. 534500.5234440
NORTH
MAP BASED ON USGS 7.5 MINUTE SERIES, GIG HARBOR QUADRANGLE, PHOTOREVISED 1981.
This recording project is part of the Historic American Engineering Record (HAER), a long-range program to document historically significant engineering and industrial works in the United States. The HAER program is administered by the Historic American Buildings Survey/Historic American Engineering Record Division (HABS/HAER) of the National Park Service, U.S. Department of the Interior. The Washington State Bridges Recording Project was cosponsored during the summer of 1993 by HABS/HAER under the general direction of Dr. Robert J. Kapsch, Chief, and by the Washington State Department of Transportation (WSDOT), Bernie L. Chaplin, Environmental Program Manager.
The field work, measured drawings, historical reports, and photographs were prepared under the direction of Project Leader Eric N. DeLony, Chief of HAER and HAER Historian Dean A.Herrin,Ph.D. The recording team consisted of Supervisory Architect Karl W. Stumpf (University of Illinois at Urbana-Champaign); Supervisory Historian Robert W. Hadlow, Ph.D. (Washington State University); Architects Vivian Chi (University of Maryland), Erin M. Doherty (Miami University), Catherine I. Kudlik (The Catholic University of America) and Wolfgang G. Mayr (US/ICOMOS, Technical University of Vienna, Austria); Historians Jonathan C. Clarke (US/ICOMOS, Ironbridge Institute, England) and Wm. Michael Lawrence (University of Illinois at Urbana-Champaign). Formal photography was done by HAER Photographer Jet Lowe. WSDOT Cultural Resources Specialist Elizabeth A. Robbins served as department liaison.
Drawings were developed from construction documents of the 1940 and 1950 bridges located in WSDOT files and field measurments.
Detail
Hold Down Ropes at Tower No. 3 - 1940 Bridge
Scale: 3/8" = 1'-0", 1:32
Detail
Hold Down Cables at Tower No. 3 - 1950 Bridge
Scale: 3/8" = 1'-0", 1:32
South Elevation - 1940 Bridge, 'Galloping Gertie'
1100'-0" (335.28 M)
2800'-0" (853.44 M)
1100'-0" (335.28 M)
Scale: 1" = 160', 1:1,920
0 200 400 600 FEET
0 40 80 160 METERS
Karl W. Stumpf and Wolfgang G. Mayr, Summer 1993
WASHINGTON STATE BRIDGES RECORDING PROJECT
TACOMA NARROWS BRIDGE - 1950
STATE ROUTE 16 SPANNING THE NARROWS
PIERCE COUNTY
TACOMA
WASHINGTON
HISTORIC AMERICAN ENGINEERING RECORD
WA-99

Introduction

They sweep gracefully over streams, rivers, and gorges. They solidly straddle sloughs, bays, and canals. They span railroads, highways, and busy streets. Some have garnered headlines around the globe. Many have received prestigious awards. Some are charming and complex, while others are unadorned and subtle. We see and use them every day. And we usually take them for granted. Marvelous and mundane, they connect us to the past, the future, and to one another in ways too numerous to mention.

Washington State's historic highway bridges are extraordinary by any measure. Yet their stories are largely unknown. Built by visionary engineers and anonymous workmen, our bridges are amazing pieces of engineering, and they have played no small part in our state's history. Crossing the many natural barriers in Washington's varied landscape has been possible only with the advancement of bridge-building technology. Where challenges arose, innovative engineers stepped forward with creative, and sometimes surprising, solutions.

Every bridge that motorists, bicyclists, and pedestrians use today is much more than a mass of steel, wood, or concrete. Some bridges hold a unique beauty, derived from the artful combination of aesthetic and practical considerations. Others are simple, hard-working structures built for function, not form. All embody important aspects of history.

Our bridge heritage is closely linked with the growth of the automobile and highways. In 1910 only some 9,300 state residents owned a car. By 1920, the number was twenty times larger, and automobiles were a part of everyday life. The rapid expansion of city roads and country highways helped to knit Washington's once distant regions closer together. The bridges devised to cross meandering creeks, raging rapids, deep ravines, wide lakes, and other obstacles became an integral part of our political, economic, and social progress. These structures tell us a story about who we were and are as Washingtonians. And they tell us much about the evolution of bridge-building technology in regional, national, and international contexts.

The Historic American Engineeering Record (HAER) project produced drawings of some of Washington's most significant historic bridges, including both Tacoma Narrows Bridges, completed in 1940 and 1950. Drawing 1993.

Karl W. Stumpf and Wolfgang G. Mayr, HAER

Our historic bridges are an elemental part of the landscape, bringing a unique sense of place and time to Washington's scenery and culture. For example, a glance at the Rainbow Bridge across the Swinomish Slough reveals how deftly it mirrors the brilliant rainbows of color in nearby tulip fields. We value such spans as important historical landmarks, as cornerstones of our identity, and as environmental and artistic statements.

Everson-Goshen Trestle, Whatcom County, ca. 1900. Relatively easy to design and construct, using nature's bounty, timber trestle bridges once flourished in Washington. Image # 2002.44.2.

Whatcom Museum of History and Art, Bellingham

Imagine blue Lake Washington without its floating bridges, Deception Pass without its green cantilever span stretching dramatically over Puget Sound's swirling tides, the Columbia River without the soaring "Blue Bridge" and graceful Ed Hendler cable-stayed bridge between Pasco and Kennewick, the Spokane River without its majestic Monroe Street concrete arch, or Seattle without the ornate

The Lyons Ferry Bridge, moved from its original Columbia River crossing at Vantage and rebuilt over the Snake River in 1968. Photo 1993.

Jet Lowe, HAER.

Montlake Bridge. It is not surprising that both Seattle and Spokane claim the title "City of Bridges." Their bridges shaped them and have become integral parts of their character. Other cities share that experience.

Washington is home to some of America's greatest bridges. Several attracted worldwide attention and the praise of professional engineers at the time they were completed. Others have influenced the course of bridge engineering.

A few examples will make the point. In 1911 the Monroe Street Bridge in Spokane was the nation's largest concrete arch span. The Longview Bridge, completed in 1930, was for a time North America's longest and tallest cantilever bridge. The Lake Washington Floating Bridge opened in 1940 as the world's first floating-concrete-pontoon highway bridge, catapulting Washington to international leadership in floating-bridge technology. Two decades later, the state again claimed fame when the Hood Canal Floating Bridge became the first of its type built on tidal waters. The most famous failure is the first Tacoma Narrows Bridge, known as "Galloping Gertie," whose astounding collapse in November 1940 changed the course of suspension-bridge design. These and the state's other premier bridges are featured in the pages that follow.

What is a historic bridge?

The National Bridge Inventory defines a bridge as a structure with a clear span of at least twenty feet. Washington's public road system includes more than 7,200 bridges owned by the Washington State Department of Transportation (WSDOT), cities, and counties. Of those, approximately 700 are "historic," that is, fifty years old by criteria established by the National Register of Historic Places (NRHP). Many more were lost, leaving no trace on the landscape or in the historical record. Most bridges stand where other structures stood previously. Photographs survive of some; drawings and engineering specifications were preserved for precious few.

This book portrays many of Washington's historically significant highway bridges in geographic and historic contexts. The focus here is on those in use today, with no intent to document all of the state's historic bridges. Recognition of "historical significance" has come relatively recently, and too late for most of our bridges. Not until the National Historic Preservation Act of 1966, which created the NRHP, were "historic properties" afforded consideration in the planning process. The act did not create the register specifically for bridges, but their inclusion in it has preserved histories that otherwise would have been forgotten.

In general, we find very little documentation for bridges lost in the last century and a half in Washington. Bridges listed in the NRHP have been documented to varying degrees. Other bridges have been "determined eligible" for inclusion in the register as the result of individual evaluations. In addition, many of our finest bridges were the subject of Historic American Engineering Record (HAER) studies.[1] Established within the National Park Service in 1969 as a forum for documenting properties of engineering significance, including bridges, HAER has been crucial in preserving detailed records of historic bridges. A 1993 effort by HAER architects and historians, led by HAER Chief Eric DeLony, produced reports and photographs for thirty-three outstanding structures. (See DeLony's Afterword following Chapter 12.) Other bridges were documented prior to their alteration or demolition. Bridges that are still standing—the NRHP-listed spans, the NRHP "eligible" bridges, and the HAER-documented structures—appear in the Appendix. Many of the most outstanding or unique, or simply good representatives of bridge types, are featured in the main chapters as premier bridges.

The WSDOT, together with the Washington State Office of Archaeology and Historic Preservation (OAHP), sponsored the state's first inventory of historic bridges in 1979–80. Specialists evaluated more than 1,400 bridges constructed in the state prior to 1941 for their historical and engineering significance. Included were bridges on state highways, county roads, and city streets, as well as privately owned structures, including railroad bridges. The result was the addition of eighty-nine bridges and tunnels to the NRHP in 1982. This became the first Multiple Property Documentation for Bridges and Tunnels in Washington State.

The 1979–80 inventory set important precedents in the state's on-going bridge management strategy. The study expedited WSDOT's legal compliance obligations under the National Historic Preservation Act by evaluating many bridges of similar design statewide. The broader context afforded historical and engineering perspectives lacking in a piecemeal, and more costly, process of judging individual structures in isolated contexts. Reliable conclusions were reached and significant bridge engineering was recognized and recorded. Thus, valuable evaluation methods and criteria were established for future studies. Equally important, the statewide approach saved taxpayer dollars.

In 1991 WSDOT conducted an inventory of 335 bridges built between 1941 and 1950 on state, county, and city transportation systems to identify additional bridges of engineering and/or historical significance. That evaluation resulted in the nomination and listing of sixteen more historic bridges in the NRHP in 1995. In 2001, under the continuing federal mandate to update bridge inventories, WSDOT began a third phase of historic bridge management. The department evaluated 812 bridges constructed between 1951 and 1960 on state, county, and city road systems. NRHP nominations for thirty-one structures were prepared. To date, twenty-two of those bridges have been recommended for listing in the NRHP.

So what makes a bridge "historic," that is, worthy of NRHP listing? NRHP Criteria for Evaluation (36 CFR Part 60.4) are starting points, but the need for refinement is always on the minds of bridge engineers and historians. In assessing Washington's bridges, WSDOT staff crafted the following "Evaluation Factors" (each factor is scored 0 to 10):

1. Unique or unusual bridge type in the total bridge inventory.
2. Innovative structural design features.
3. Representative of a specific type of bridge.
4. One of a few remaining examples of a specific type of bridge.
5. Creative or economical use of construction materials in the bridge's design.
6. Innovative or unusual construction methods.
7. Importance of the bridge to the social, economic, or industrial development of the locality, region, or state.
8. Association of the bridge with a historic individual or event or with a significant public effort to encourage its construction.
9. Structural integrity: Have alterations compromised the bridge's original appearance, design, or function? Have structural elements responded well to the "test of time"?
10. Aesthetics of the bridge to its site.

No specific numerical score was established for NRHP qualification prior to the evaluations. Rather, bridges tended to fall into groups according to score. Those in the highest-scoring groups received additional research, until finally NRHP eligibility was determined under Criteria A and C. Criterion A eligibility was based upon a bridge's association with the broad themes of bridge building and transportation in Washington, enhanced by its importance to local or regional transportation and to economic or social development, or by its association with significant events or federal projects, such as dams on the Columbia River. Possession of high degrees of integrity in design and materials, especially in main spans, contributed to bridge eligibility under Criterion C. Some degree of modification to eligible bridges was judged acceptable so long as the significant engineering design remained intact and the bridge's historic character preserved. Replacement or alteration of approach spans did not, in most instances, deprive bridges of their engineering significance or NRHP eligibility.[2]

A spectacular failure, filmed and distributed for the world to see, can also make a bridge instantly historic, as we learned in 1992 when the ruins of "Galloping Gertie," the first Tacoma Narrows Bridge (1940) that now lie on the bottom of Puget Sound, were listed in the NRHP. Few, if any, other states can claim such a distinction. For a time, there was talk of listing the ruins of two of Washington's sunken floating bridges. Stunning film footage of their demises does not exist, however, and the relative significance of those ruins has yet to be determined.

Columbia River Vancouver-Portland I-5 Bridges (1917 and 1958), Clark County. Examples of vertical lift structures. Photo 2002.
Craig Holstine, WSDOT

Why this book?

In July 2000, recognizing the continued loss of the state's historic bridges, WSDOT, the Federal Highway Administration, the State Historic Preservation Officer, and the federal Advisory Council on Historic Preservation concluded a programmatic agreement with provisions for documenting historic bridges. The agreement stipulated that WSDOT prepare a "book for the general public on Washington's historic bridges" as an expedient, cost-effective method for the state to meet some of its federal legal obligations to evaluate and document historic bridges. This book is the result of that agreement.

In this volume, readers will find Washington's most outstanding historic bridges. Other notable bridges, too, have been included in the interest of maintaining the book's focus on Washington's significant bridges, regardless of official "historical" recognition.

Chapter 1 provides an overview of bridge building in Washington, setting a general chronological framework for bridges in the development of the state's transportation system. Early bridge designs, materials, and failures provided the background for later achievements that include our monumental historic spans, many of which are still in service.

The stories of bridge failures and progress, loss and preservation, take center stage in Chapters 2 and 3. If bridges were never lost, perhaps a book like this would not be needed. Unfortunately, fate, time, and the forces of nature have claimed, and will continue to claim, bridges. While some are replaced out of necessity, others are preserved through the innovative use of traditional and new materials and technology, or simply as the result of public will.

Every bridge has a human face. Chapter 4 presents profiles of several of our most outstanding bridge engineers, outlining the life stories of these "designers of dreams" (a term borrowed from Henry Petroski, a noted historian of engineering history) whose ingenuity left us a legacy of commanding architecture. Not all of the state's engineers who deserve mention appear here. We selected bridge designers whose achievements were extraordinary and for whom documentation could be found. In today's world, with computer-assisted design and other automated conveniences, it is hard to remember that most of Washington's historic spans were built with little more than slide rules, crude calculators, and innumerable quantities of drafting paper and pencils.

Lower Custer Way Bridge (foreground, Luten arch, built 1915), and the Upper Custer Way Bridge (rear, steel deck-arch, built 1956), Tumwater. Photo 2002.
Craig Holstine, WSDOT

The hard, physical work of constructing bridges was accomplished by uncounted thousands of workmen. The companies that built bridges often were widely known in reports and newspaper accounts. But the craftsmen and laborers whose sweat and toil erected Washington's marvelous historic spans received little recognition. Their names are virtually unrecorded, unless they fell victim during the construction process. Some did, for the work often was dangerous.

Thereafter, this book follows a geographical organization, highlighting the state's premier bridges by region. It is hoped that this will provide readers with a convenient reference for planning to visit bridges or simply for armchair traveling. Travelers beware—some bridges featured here have been, or are scheduled to be, replaced.

The mundane utility of our historic bridges masks both their significance as cultural resources and their vulnerability to the ravages of regular use. Many of our oldest spans became the victims of the constant demand to build and rebuild highways. Old age, pollution, heavy traffic, and neglect have taken a serious toll as well. Hopefully, this story of Washington's historic highway bridges will enhance appreciation of their exceptional contributions to our past and present. As elemental parts of our physical and cultural heritage, they deserve our best efforts for their healthy maintenance and preservation.

Bridge Basics

Bridges derive their strength from the careful resolution of two opposing stress forces—tension and compression. Tension is the force that pulls or elongates a bridge component. Compression pushes the components together. How well a bridge can support its own weight, plus traffic loads, is partly determined by its design and partly by the materials of which it is made. By the time Washington became a state in 1889, bridge engineering and technology were developing rapidly, thanks to advancements made by the transcontinental railroads in the United States and road building worldwide. Basic bridge types in use for centuries were given new design adaptations, as new materials and new demands appeared on the scene.[3] As a result, the oldest log and timber bridges have disappeared from our roadways.

The most prevalent bridge types found in Washington are discussed below, and examples of each type are provided. Annotated lists of all Washington bridges listed in or determined eligible for listing in the NRHP can be found in the Appendix. For basic bridge terminology, see the Glossary, also in the back of this book.

Trestle Bridges

Trestles are simple structures consisting of spans supported by frame bents. They are often used to support approaches to other, more sophisticated types of bridge structures, such as trusses. However, some bridges in the state are entirely trestles.

Notable historic wood trestle bridges in Washington include the Outlet Creek Bridge (1935; demolished 1994) and the Valley Creek and Tumwater Creek bridges (1936) on 8th Street in Port Angeles, Clallam County.

Truss Bridges

The truss derives its strength from the triangular arrangement of three members, combining tension and compression forces. The most popular truss designs were patented in the 1840s by William Howe, Thomas and Caleb Pratt, and James Warren. Compared to the East Coast, Washington has few examples of these early truss forms. Our earliest examples were made of wood. Washington's abundance of forests led to the proliferation of timber trusses even into the twentieth century. Pacific Northwest weather, of course, accelerates the deterioration of exposed wood; not surprisingly, most historic wooden bridges have been replaced. Some were covered, although all but three of those are now gone.

Iron, and especially steel, perform very well under both tension and compression. Steel's superiority as a bridge-building material and the steel industry's growth prompted widespread use of the steel truss in the late nineteenth

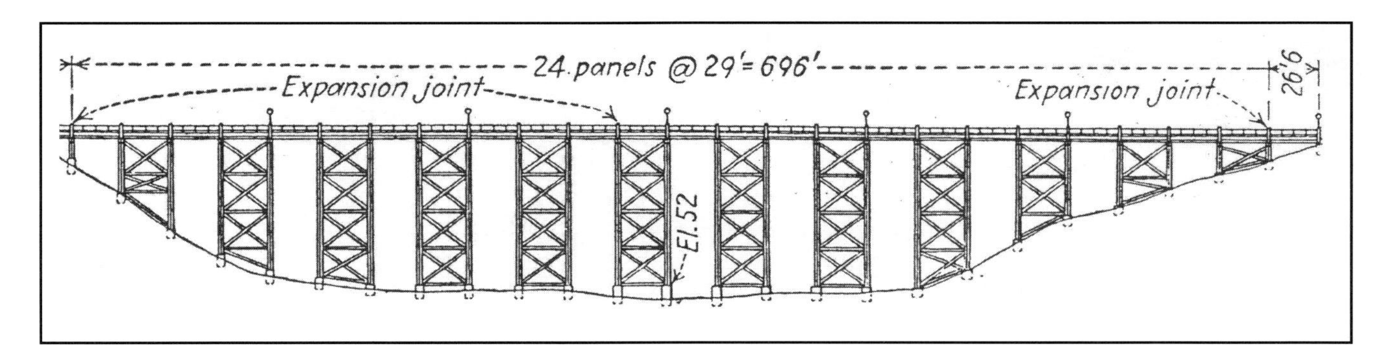

The 8th Street/Tumwater Creek Timber Trestle Bridge, constructed in Port Angeles in 1936. From the *Engineering News-Record*, August 6, 1936.

McGraw-Hill Companies, New York, NY

century. Initially, steel trusses employed pin connections. By the 1890s, however, the development of pneumatic technology for rivet guns led to their application for steel truss spans. Riveted-joint bridges began replacing pin-connected spans after 1910, since they provided a more rigid structure with a greater load capability, fewer maintenance problems, and a longer life expectancy. By the 1930s, concrete became a more economical choice, but the steel truss was typically preferred for longer bridges.

The use of truss spans declined considerably following the 1950s. Concrete-and-steel girder bridges became more popular, in part because they could be widened to accommodate traffic growth.

Truss bridges are of three basic groupings. A "through" truss carries its traffic load level with its bottom chords. A through-truss with no lateral (or sway) bracing between the top chords is a "pony" truss. A "deck" truss carries its traffic load level with its top chords.

Pony Truss

Although extremely popular for more than a century, pony trusses have largely disappeared. They are limited to relatively short spans, making them particularly susceptible to replacement by prestressed concrete girders. The South Fork Newaukum River Bridge (1930) in Lewis County and the Foss River Bridge (1951) in King County are notable examples of the pony truss type.

Foss River Bridge, a pony truss built in King County in 1951. Photo 2002.

Craig Holstine, WSDOT

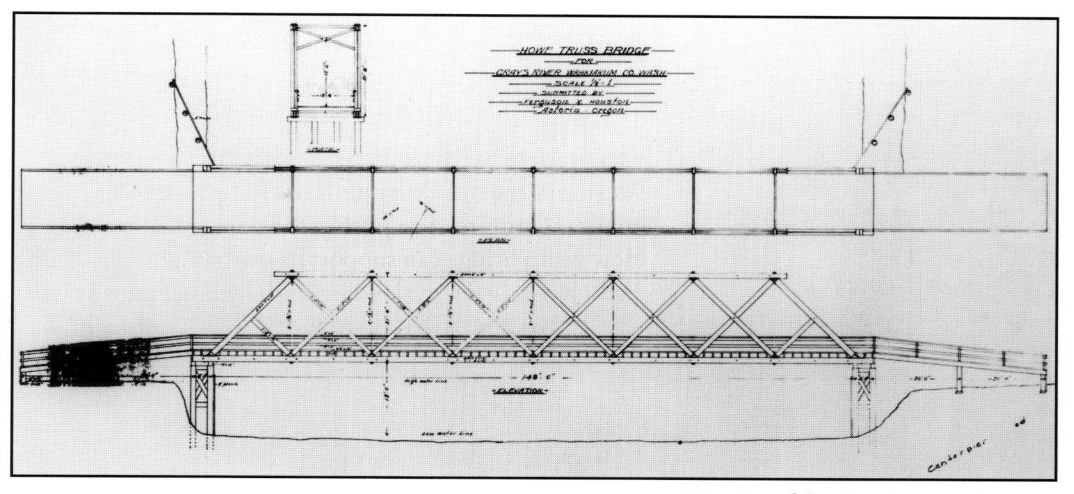

Drawing of the Grays River Bridge, ca. 1905. The Howe trusses were later covered with wood siding and metal roofing.

WSDOT

Howe Truss

In the Howe truss form, the bridge's vertical members are in tension and diagonals are in compression. This type of bridge truss has parallel chords, vertical (tension) members at the panel points, and diagonals forming an X pattern. Patented in 1840, the Howe truss became the most widely used design for timber railroad bridges.

Notable examples of timber Howe trusses in Washington include the Grays River Bridge (covered, 1905), Wahkiakum County; Harpole Bridge (covered, 1922), Whitman County; Pe Ell Bridge (covered, 1934), Lewis County; and the Norman Bridge (1924; reconstructed 1950; demolished 2004), King County.

Norman Bridge over the Snoqualmie River, King County. A Howe timber truss built in 1924, reconstructed 1950, and demolished in 2004.

Mark Ruwedel, King County Parks Department

McMILLIN BRIDGE

PUYALLUP RIVER 1934 WASHINGTON

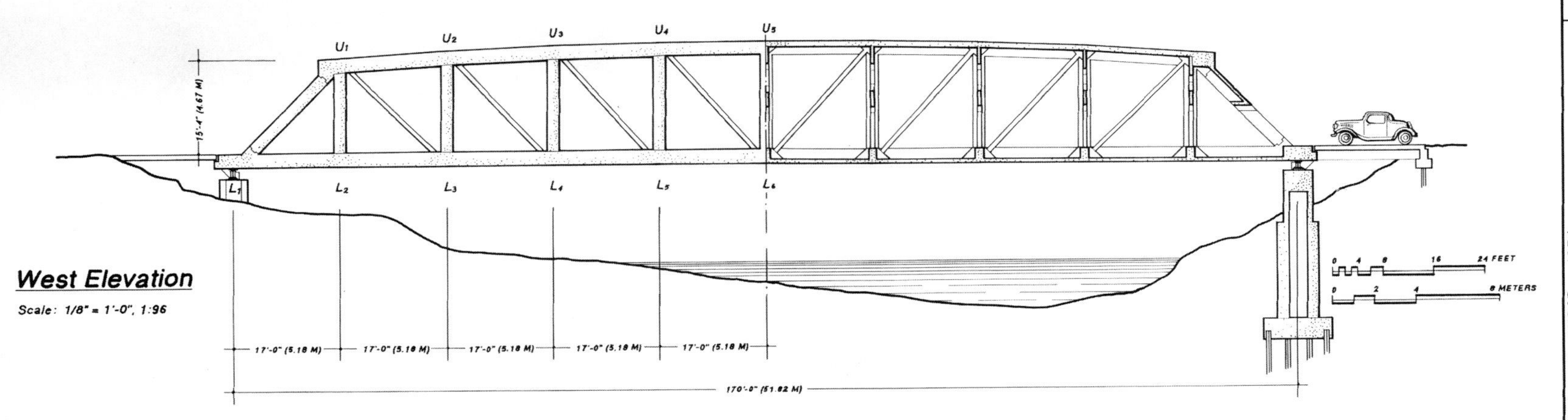

The McMillin Bridge was built by Pierce County beginning in 1934. It replaced a steel truss span of the same name and location, damaged by flooding of the Puyallup River. Similar to its predecessor, the new bridge employed a through-truss type, enabling it to accommodate the largest practical waterway. This required its width and length to increase.

This reinforced concrete design was chosen over alternate bids for structural steel trusses because of its inherent maintenance advantages and lower cost due in part to the simplicity of its construction. The resulting Pratt truss bridge consisted of a 170-foot span with two 20-foot approaches. "Double trusses" flanking the 24-foot wide road deck creates covered walkways for pedestrians. The breadth of these paired trusses affords the structure its great stiffness, therefore eliminating the need for lateral bracing over the roadway.

At the date of its completion the McMillin Bridge was recognized as the longest reinforced concrete span exclusive of arches in the country. It is significant regionally as well as nationally because the structural properties of concrete make it an unusual material choice for truss construction.

Homer M. Hadley, regional structural engineer of the Portland Cement Association conceived the Mc Millin's design. However, W.H. Witt Company of Seattle prepared the construction documents that were executed by general contractor Dolph Jones of Tacoma. The project was supervised by Pierce County engineer W.E. Berry and his successor Forest R. Easterday.

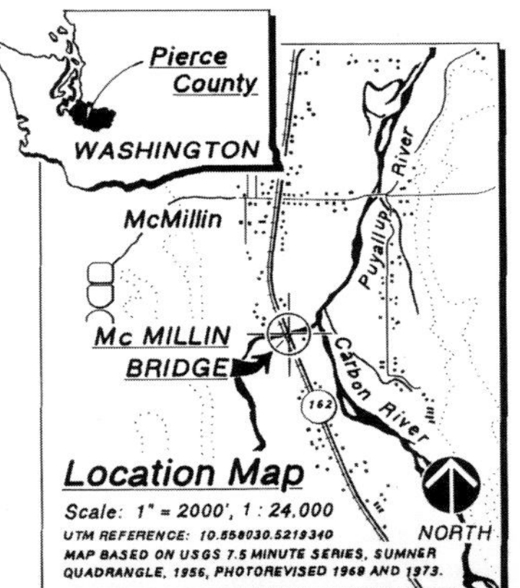

This recording project is part of the Historic American Engineering Record (HAER), a long-range program to document historically significant engineering and industrial works in the United States. The HAER program is administered by the Historic American Buildings Survey/Historic American Engineering Record Division (HABS/HAER) of the National Park Service, U.S. Department of the Interior. The Washington State Bridges Recording Project was cosponsored during the summer of 1993 by HABS/HAER under the general direction of Dr. Robert J. Kapsch, Chief, and by the Washington State Department of Transportation (WSDOT), Bernie L. Chaplin, Environmental Program Manager.

The field work, measured drawings, historical reports, and photographs were prepared under the direction of Project Leader Eric N. DeLony, Chief of HAER and HAER Historian Dean A. Herrin, Ph.D. The recording team consisted of Supervisory Architect Karl W. Stumpf (University of Illinois at Urbana-Champaign); Supervisory Historian Robert W. Hadlow, Ph.D. (Washington State University); Architects Vivian Chi (University of Maryland), Erin M. Doherty (Miami University), Catherine I. Kudlik (The Catholic University of America) and Wolfgang G. Mayr (US/ICOMOS, Technical University of Vienna, Austria); Historians Jonathan C. Clarke (US/ICOMOS, Ironbridge Institute, England) and Wm. Michael Lawrence (University of Illinois at Urbana-Champaign). Formal photography was done by HAER Photographer Jet Lowe. WSDOT Cultural Resources Specialist Elizabeth A. Robbins served as department liaison.

DELINEATED BY: Erin M. Doherty, Summer 1993

WASHINGTON STATE BRIDGES RECORDING PROJECT
NATIONAL PARK SERVICE
UNITED STATES DEPARTMENT OF THE INTERIOR

MC MILLIN

MC MILLIN BRIDGE - 1934
STATE ROUTE 162 SPANNING THE PUYALLUP RIVER, 0.8 MILES SOUTH OF MC MILLIN
PIERCE COUNTY WASHINGTON

SHEET 1 of 3

HISTORIC AMERICAN ENGINEERING RECORD
WA-73

McMillin Bridge, a concrete Pratt truss built over the Puyallup River, Pierce County, in 1934. Drawing 1993.

Erin M. Doherty, HAER

Pratt Truss

Also known as the "N-truss," this arrangement of bridge members places the vertical pieces in compression and the diagonals in tension. The Pratt truss, thus, has parallel chords and a web system composed of vertical posts with diagonal ties inclined outward and upward from the bottom-chord panel points toward the ends of the truss. Many bridge builders in the nineteenth and early twentieth centuries used timber for vertical members and steel for the diagonals. Patented in 1844, the Pratt truss design was used mainly for spans under 250 feet in length.

Notable examples in the state include the Peshastin Creek Bridge (a.k.a. "L. C. King" or "Old #604," ca. 1897) and the West Monitor Bridge (1907), both in Chelan County.

The Peshastin Creek (L. C. King) Bridge, Chelan County, perhaps the oldest Pratt truss in Washington, built ca. 1897. Photo 2005.

Pam Trautman, WSDOT

Parker Truss

This form is a Pratt truss with a polygonal top chord, which reaches its greatest height at the center panels. The arched top chord increased the structure's rigidity and strength, enabling the construction of longer spans.

Notable examples in Washington include the Mount Si Bridge (1904; moved 1955), King County; Curlew Bridge (1908), Ferry County; and the Orient Bridge (1909; demolished 1992), Ferry and Stevens counties.

The Mount Si Bridge over the Snoqualmie River, King County—one of the oldest Parker truss bridges in the state, built in 1904 on the White River near Buckley, Washington, and moved to its present location near North Bend in 1955. Photo 2002.

Craig Holstine, WSDOT

Petit Truss

The Petit truss was developed in the 1870s to achieve greater span lengths. It represented a refinement of the Pratt truss

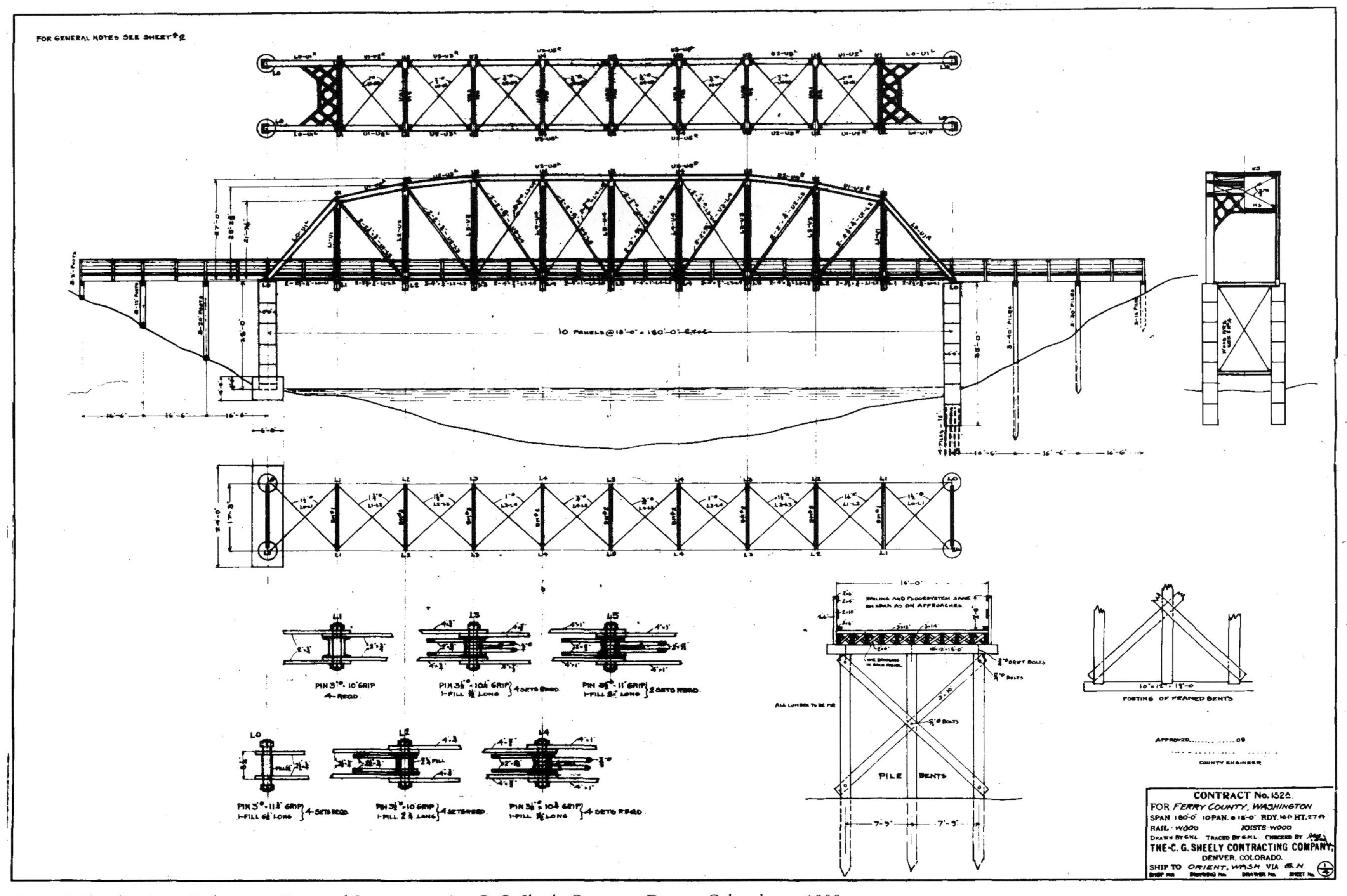

Orient Bridge drawing, a Parker truss, Ferry and Stevens counties. C. G. Sheely Company, Denver, Colorado, ca. 1909.

Office of Archaeology and Historic Preservation

design by adding struts and ties to strengthen each vertical diagonal panel. The Pennsylvania Petit truss, distinguished by its use of a polygonal or curved top chord, is also a Parker truss with sub-struts and/or sub-ties.

The Pennsylvania Petit truss was patented in 1875 and derived the name from its extensive adaptation by the Pennsylvania Railroad. The design was used most often for spans between 250 and 600 feet. Few bridges of this type were built after the 1920s.

Notable examples in Washington include the Middle Fork Nooksack River (Guide Meridian) Bridge (1915), Whatcom County, and the Vancouver-Portland/Interstate 5 Bridges (1917 and 1958), Clark County.

Stossel Bridge, King County, a Warren truss built on the Snoqualmie River in 1951. Piers of the previous bridge remain adjacent to the present bridge. Photo 2002.

Craig Holstine, WSDOT

Warren Truss

The Warren truss disperses the load stresses by alternating diagonals in compression and tension, with the truss members forming the letter W. Typically, Warren trusses include vertical members, which stiffen the structure. Patented in 1848, the design became the most commonly used truss style and remains so today.

Notable examples of Warren deck trusses include the Elwha River Bridge (1913), Clallam County, and the Lake Washington Ship Canal Bridge (1961), Seattle. The Stossel Bridge (1951) in King County is a Warren through-truss with vertical members.

Steel Cantilever Truss

This truss form is represented in some of the state's longest bridges. Each bridge is a combination of anchor spans, cantilevers, and suspended spans. The cantilever span itself projects beyond and overhangs its supporting pier and is balanced by an anchor span connecting the

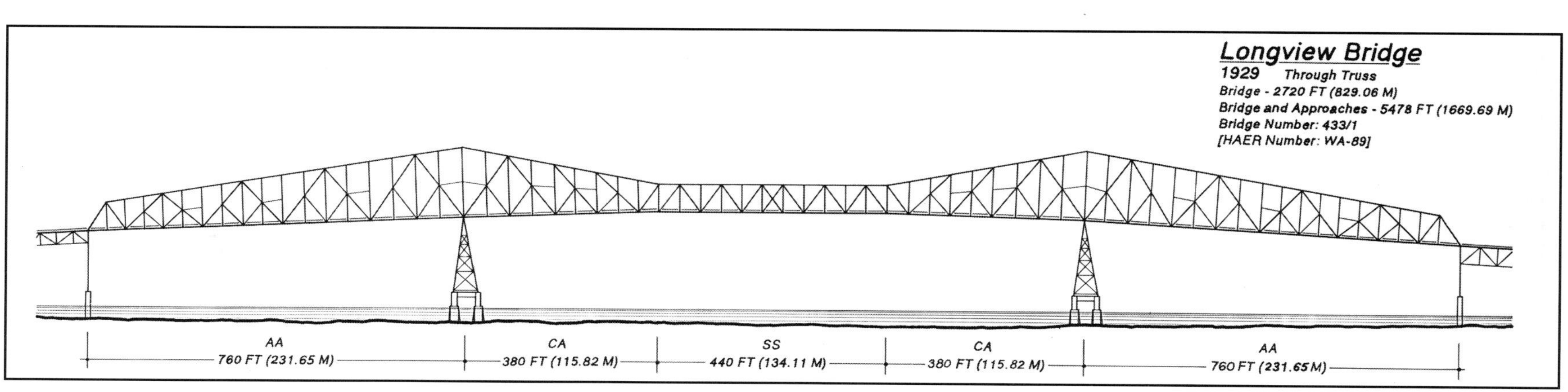

The Columbia River Bridge at Longview, a steel through-truss cantilever built in 1929. Drawing 1993.

Karl W. Stumpf, HAER

The Pasco-Kennewick Bridge over the Columbia River, ca. 1922. Cantilever spans under construction.
Washington State Archives, WSDOT Records

Deception Pass Bridge, a steel cantilever deck-truss built in 1935. Photo 1993.

Jet Lowe, HAER

pier with the land or shoreline. If the cantilever arms are not connected at mid-span, a simple truss is "suspended" between the two cantilevers. Bridges of this type are advantageous when the erection of falsework or scaffolding is not feasible, such as over deep gorges or across broad expanses of water.

Notable examples include the First Columbia River Bridge at Wenatchee (1907), Chelan and Douglas counties; Bridge of the Gods (1926), Skamania County, Washington, and Hood River County, Oregon; Vantage Bridge (1927; dismantled 1963; rebuilt at Lyons Ferry 1968), Columbia and Whitman counties; Longview Bridge (1930), Cowlitz County, Washington, and Columbia County, Oregon; George Washington Memorial (Aurora Avenue) Bridge (1932), Seattle; Deception Pass Bridge (1934), Island and Skagit counties; Grand Coulee Bridge (1935), Douglas and Okanogan counties; Kettle Falls Bridge (1941), Ferry and Stevens counties; Fort Spokane Bridge (1941), Lincoln and Stevens counties; Northport Bridge (1949), Stevens County; Bridgeport Bridge (1950), Douglas and Okanogan counties; and the Agate Pass Bridge (1950), Kitsap County.

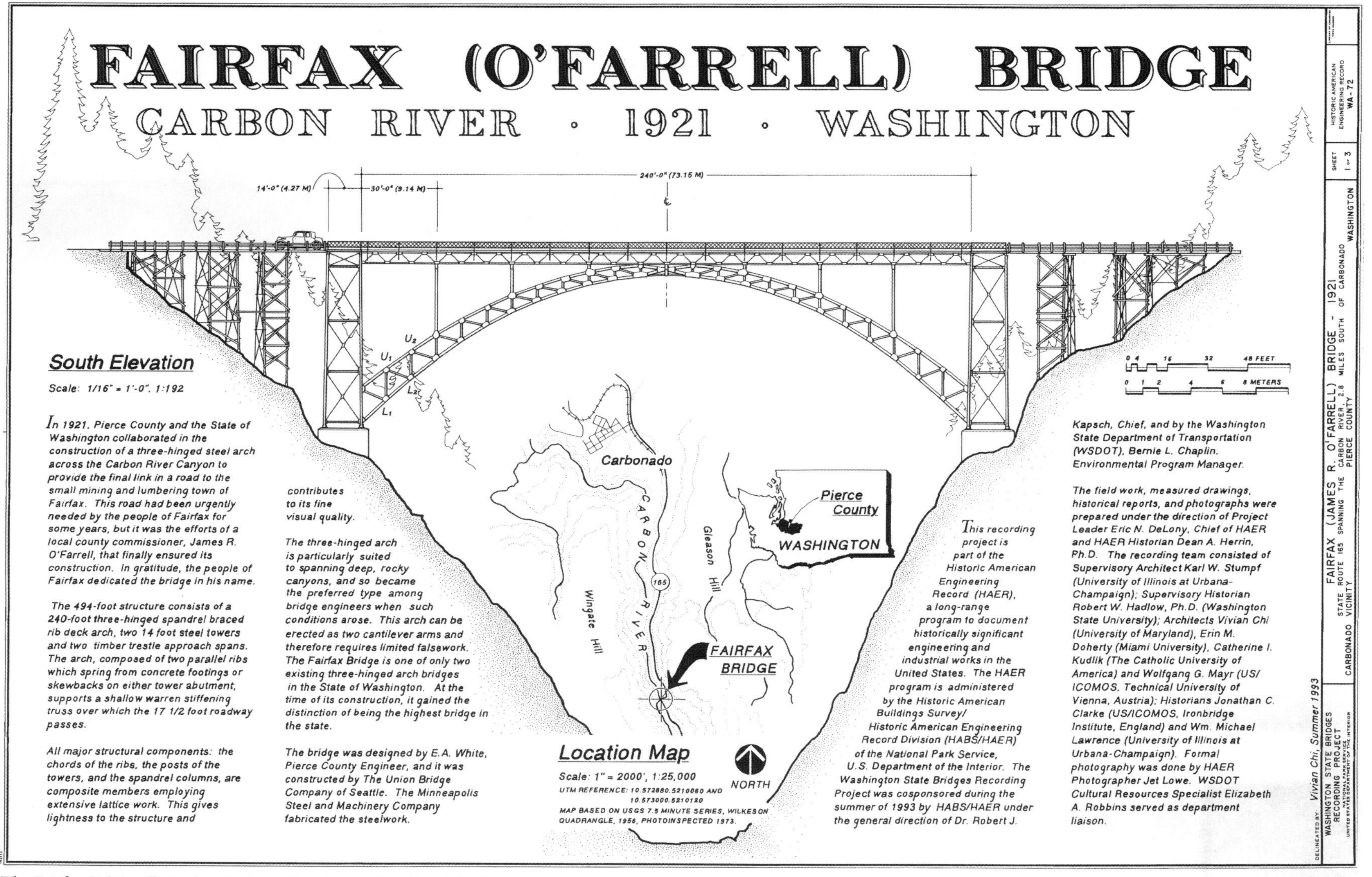

The Fairfax (O'Farrell) Bridge, Carbon River, is a steel deck-arch built in 1921. Drawing 1993.

Vivian Chi, HAER

Arch Bridges

The arch bridge, developed on a large scale by ancient Roman engineers using stone and cement, acts in compression. Early forms required massive foundations and abutments. Steel, because of its inherent strength, became a popular material for relatively short span arch bridges in the late nineteenth and early twentieth centuries. The steel arch's basic form changed little, always including ribs, spandrel posts, and stiffening trusses.

"Steel deck-arch" examples include the Ravenna Park Bridge (1913) and Queen Anne Drive Bridge (1936), Seattle; Fairfax Bridge (1921), Pierce County; Wenatchee Avenue Bridge (1950), Chelan County; Modrow Bridge (1957), Cowlitz County; and the Gorge Creek Bridge (1955), Whatcom County.

"Steel through-arch" examples, on the other hand, include the Pioneer Memorial ("Blue") Bridge (1954), Benton and Franklin counties, and the Rainbow Bridge (1957), Skagit County.

After 1890, the use of steel-reinforced concrete for arch bridges opened a new innovative era in bridge design. Daniel Luten was the most prominent American bridge designer in that genre. The Luten system improved on earlier designs by placing steel bars in the tension zones.

Monitor Bridge, an open-spandrel concrete arch built on the Wenatchee River in 1931. Photo 2005.

Zachary Dee Holstine

Previously known as the "Boston Street Bridge," the Lower Custer Way Bridge over the Deschutes River, Tumwater, was built by Charles G. Huber in 1915. The city replaced the deck, sidewalks, railings, and lights on this multi-span Luten arch in 2003–4. Image # 498.

State Capitol Museum, Washington State Historical Society

Concrete Luten arch examples include the Lower Custer Way Crossing (1915), Thurston County, and the Pickett Bridge (1920), Whatcom County.

As bridge designers learned to better combine the rigidity of concrete with the pliability of steel into an integrated unit, lighter open-spandrel concrete arch forms appeared.

Open-spandrel concrete arch examples include the Monroe Street Bridge (1911), Latah Creek/Sunset Bridge (1913), Post Street Bridge (1917), and Greene Street Bridge (1956), Spokane; Monitor Bridge (1931), Chelan County; and the Jim Creek Bridge (1945), Cowlitz County.

During the 1920s and 1930s, Washington highway bridge engineers built four reinforced-concrete tied arches and a concrete rainbow arch. These designs use a pair of arch ribs above the roadway to suspend the deck or floor system. The deck slab and other horizontal ties resist the strong horizontal tension on the arch, eliminating the need for massive abutments.

Concrete tied-arch examples include the Goldsborough Creek Bridge (1923) and North and South Hamma Hamma Bridges (1924), Mason County, and the Duckabush River Bridge (1934), Jefferson County. The Indian Timothy Bridge (1923), Asotin County, is the lone rainbow-arch example in the state.

Concrete Beam, Girder, and Slab Bridges

The concrete girder bridge uses a large beam member as the structure's main support and usually receives loads from the floor beams and stringers. Thus, the superstructure has two or more girders supporting a separate floor system, as differentiated from a multi-beam bridge or a slab bridge. A slab bridge has a superstructure of reinforced concrete, constructed either as a single unit or as a series of narrow slabs placed parallel to the roadway and spanning the space between the supporting abutments.

Although reinforced concrete was used in beam, girder, and slab bridges from the beginning of the twentieth century, only after World War II did it gain widespread application in response to changing styles and rising steel costs. Bridge designers sought economy and simplicity in structural features, clean lines, and a lack of ornamentation.

The Alaskan Way Viaduct (1953) in Seattle is an excellent example of concrete girder, beam, and slab construction. This complex, and in places double-decked, structure includes steel girder spans as well.

Europeans were the first to develop concrete hollow-box (cellular) construction, which by the mid-1930s began to be used in Washington. The design was economical, for concrete was poured around hollow forms and steel reinforcing was placed only at the structure's points of greatest live load.

Examples of concrete box-girder bridges include the Purdy Bridge (1936), Pierce County; Capitol Boulevard Bridge (1937), Thurston County; Donald-Wapato Bridge (1947; removed 2004) and Toppenish-Zillah Bridge (1947), Yakima County; and the Lake Washington Ship Canal/Interstate 5 Bridge (1961), Seattle.

Prestressed concrete construction was introduced in Washington in the 1950s and has revolutionized bridge building. Prestressed concrete beams and girders are those in which cracking and tensile forces have been greatly reduced by compression with tensioned cables or bars imbedded within the concrete. Today, the prestressed concrete girder is the most predominant feature on America's bridges. In Washington, 37 percent of all highway bridges are constructed of prestressed concrete.

A notable example of a historic-era, prestressed concrete bridge in Washington is the Klickitat River Bridge No. 142/9 (1954), Klickitat County.

Steel Beam and Girder Bridges

Although railroads used wrought-iron girder bridges as early as the 1840s, difficulties in transporting metal girders limited their use on roadways. However, improvements in steel manufacturing in the 1890s led to their increased use in highway bridges after the turn of the century. Rolled steel was widely used for multi-girder bridges of short (sixty feet and less) spans. In the 1940s, steel-plate girders first appeared in the state, providing for longer spans that, unlike truss bridges, could be widened fairly easily. As producers supplied tougher structural steel, welding and high-strength steel bolts replaced riveting and enhanced the popularity of steel beam and girder bridges.

Among the more prevalent in Washington were plate girder structures. A steel-plate girder consists of a large, I-shaped beam composed of a solid vertical section with horizontal flange plates welded to either end.

Examples of steel-plate girder bridges include the Grande Ronde Bridge (1941), Asotin County; Wenatchee Avenue Southbound Bridge (1955), Chelan County; Maple Street Bridge (1958), Spokane; and the Port Washington Narrows (Warren Avenue) Bridge (1958), Kitsap County.

Some of the state's more innovative bridges were constructed using steel-box girders. A steel-box girder consists of a hollow structural member with a square, rectangular, or trapezoidal cross section.

Significant examples of steel-box girder bridges include the Patton Bridge (1950), King County, and the Portage Canal Bridge (1951), Jefferson County.

Two types of movable bridges: vertical lift (upper left) and swing (lower right), Snohomish River, Everett.

Washington State Archives, WSDOT Records

Movable Bridges

The movable bridge is typically found in urban environments where there is a need for overhead clearance for vessels on a navigable waterway, and where it is not practical to construct a high, static structure with long approaches. Old versions typically are represented by the medieval drawbridge in Europe. Modern versions were developed mainly after 1890, when steel was readily available and engineers designed better electrical motors and counterbalancing methods. Movable bridges are of three basic types: swing, bascule, and lift. (Other kinds of movable spans exist on "floating" bridges.)

Movable bridges in Washington were once more numerous than they are today, reflecting the passing need for maritime transportation to accommodate both commercial and passenger traffic on our waterways. For instance, steamboats once plied rivers that today carry only recreational boat traffic. The bridges that raised and swung open for passenger steamboats are mostly long gone, but some movable bridges continue to operate on waterways still serving marine commerce.

Swing Bridges

A swing bridge rotates horizontally on a pivot point. This was the main type of movable bridge in use until the end of the nineteenth century. Some were jackknife style, with the pivot point on one shore, but more commonly, a swing bridge pivoted from a pier in the waterway's center.

Examples of historic highway swing bridges in the state include the Steamboat Slough Bridges (1926 and 1953) in Snohomish County.

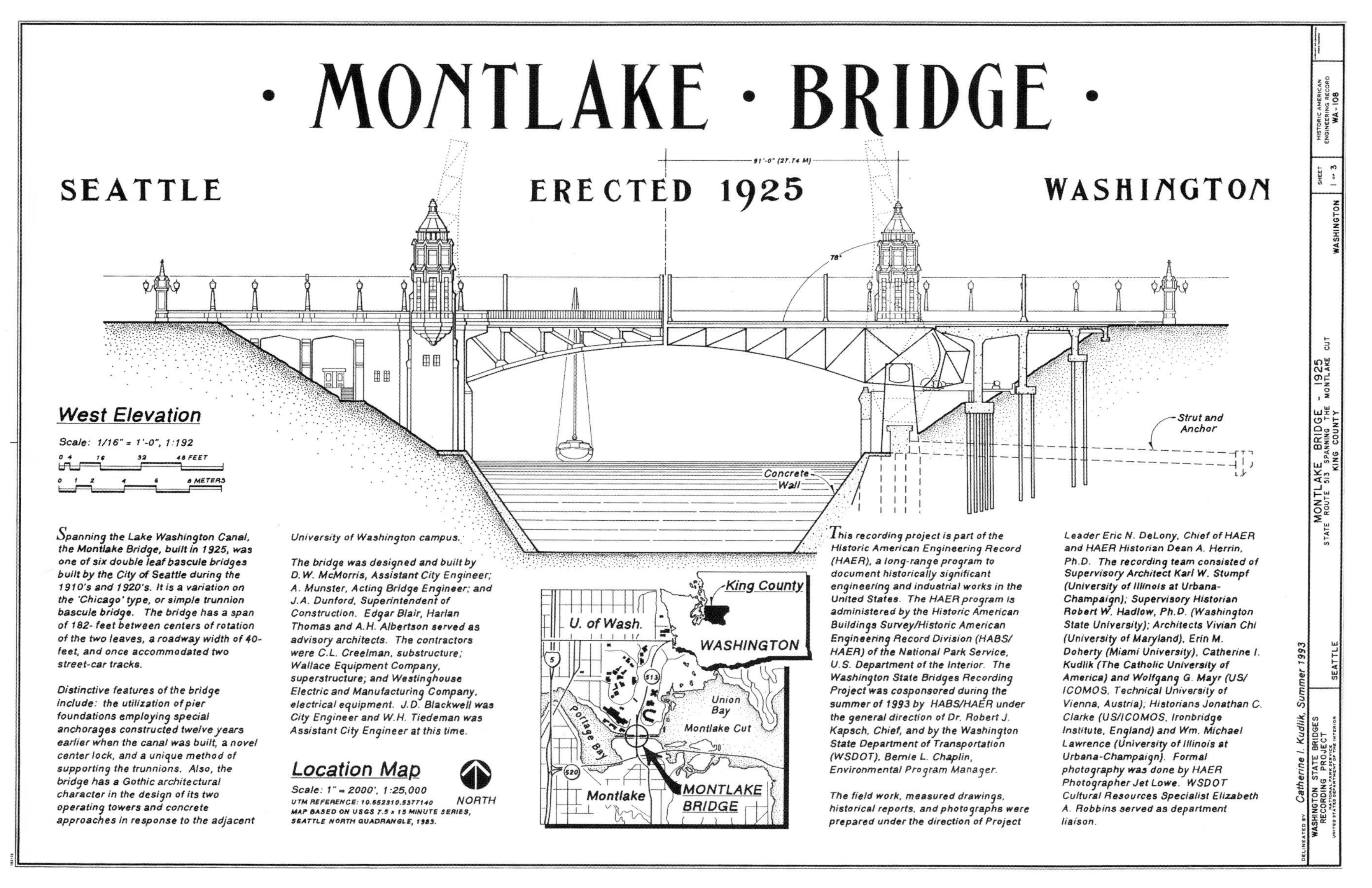

The Montlake Bridge, a bascule built on the Lake Washington Ship Canal in Seattle in 1925. Drawing 1993.

Catherine Kudlik, HAER

Bascule Bridges

The bascule bridge derives its name from the French word for "seesaw." Old forms in centuries past were called a "drawbridge." A bascule bridge employs a counterbalance and electric motor to raise the span, which may be a one (single) or two (double) "leaf" style. The bridge's one or two leaves rotate vertically on a heavy pin, or "trunnion."

A bascule bridge operates more quickly than a swing span. Leaves may be raised partially or fully to allow the passage of boats below. The most common type was

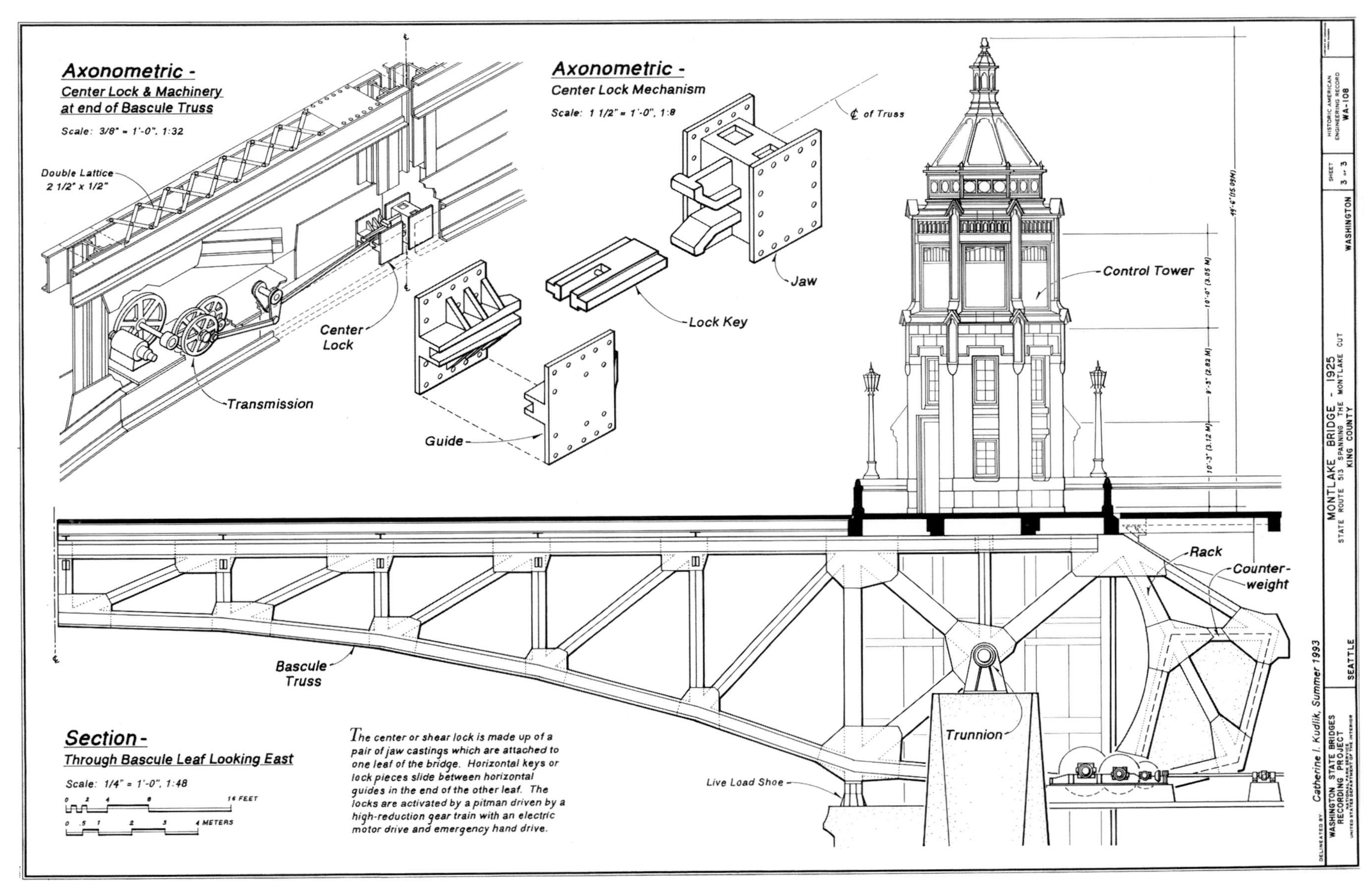

developed initially for railroads in Chicago in the 1890s by the J. B. Strauss Company. The double-leaf bascule's advantage over a single leaf is that its shorter leaves require smaller counterweights. The double-leaf bascule is used typically for longer distances and wider spans.

Notable examples in Washington include the Fremont Bridge (1917), Ballard Bridge (1917), University Bridge (1919), Montlake Bridge (1925), South Park Bridge (1930), and First Avenue South Bridge (1955), all in Seattle; and the Hoquiam River Bridge (1928) and Chehalis River Bridge (1954), Grays Harbor County.

Details of one of the Montlake Bridge's control towers and bascule leaves. Drawing 1993.

Catherine Kudlik, HAER

City Waterway (Murray Morgan/Thea Foss Waterway) Bridge, Tacoma, a vertical lift built in 1913. Photo 1993.
Jet Lowe, HAER

Lift Bridges

A vertical-lift bridge maintains its movable span in a horizontal position, while moving the span vertically using counterweights and pulleys in towers at each end of the movable deck.

Notable examples include the City Waterway Bridge (1917), Tacoma; the Snohomish River Bridges (1927 and 1954), Snohomish County; and the Vancouver-Portland Interstate 5 Bridges (1917 and 1958), Clark County, Washington, and Multnomah County, Oregon.

Floating Bridges

Floating pontoons have been adopted to support roadways, and the earliest forms, developed in ancient times, were made of wood. Permanent hollow or cellular concrete pontoons, however, were first used on the Lake Washington Floating Bridge in 1940. All modern floating bridges are held in place by cables attached to anchor blocks secured to the lake or channel bottom. Floating bridges in Washington are also movable bridges, because of their "slip spans" that open for ship traffic.

Notable examples in Washington include the Evergreen Point Floating Bridge (1963), Homer M. Hadley Floating Bridge (1989), and the Lake Washington Floating (Lacey V. Murrow) Bridge (1940/1993), all connecting Seattle with eastern King County; and the Hood Canal Floating Bridge (1961), connecting the Kitsap and Olympic peninsulas in Jefferson and Kitsap counties.

The first Lake Washington Floating Bridge nearing completion, Seattle, March 28, 1940.
Washington State Archives, WSDOT Records

OK
OK
OK
Griffiths Dairy
Simmer
711-
Unit 3. General View across Lake Washington, looking N.E'ly. 3-28-'40.

The Wind River Bridge, a steel suspension structure rebuilt in 1925, and replaced by the Conrad Lundy Jr. Bridge in 1960, Skamania County.

Washington State Archives, WSDOT Records

Suspension Bridges

In a suspension bridge, cables anchored at either end of the span are supported on towers to suspend and support the roadway. In the eighteenth century, iron chains were used to suspend roadways, but wire cables replaced these after 1801. America's first great suspension span, the Brooklyn Bridge, took more than fifteen years to complete, and opened in 1883. A suspension bridge's flexibility made it impractical for most heavy-load railroad use. However, as the automobile became the main means of transportation, suspension-bridge technology advanced rapidly in the early twentieth century, and soon came to be used for the longest highway spans.

Notable examples in Washington include the Tacoma Narrows Bridges (1940, 1950, and a third bridge now under construction), Pierce County; Wind River Canyon (timber, 1913; rebuilt in steel, 1925), Skamania County; and the Yale Bridge (1932), Clark and Cowlitz counties.

Cable-Stayed Bridges

This type of bridge utilizes cables or stays to support the superstructure. The cables pass over or attach to one or more towers at the main piers. Cable-stayed bridges look similar to suspension bridges. Both have roadways that hang from cables, and both have towers, but the two bridges support the roadway's load in very different ways. In suspension bridges, the cables ride freely across the towers, transmitting the load to the anchorages at either end. In cable-stayed bridges, the cables are attached to the towers, which alone bear the load.

European engineers developed the cable-stayed design from earlier models that used wrought iron or other material to "stay" bridge decks to towers. Advances in cable stays came in the decades after Professor F. Dischinger began investigating the technique in Germany in the late 1930s. Because it is statically indeterminate, the system

The Chow Chow Bridge, a cable-stayed structure built over the Quinault River in 1952, collapsed in 1988.
Office of Archaeology and Historic Preservation

The SR 509 (21st Street) Bridge, a cable-stayed structure built in 1996 over the Thea Foss Waterway in Tacoma. Photo 2005.

Craig Holstine, WSDOT

was difficult to analyze with a reasonable degree of accuracy until relatively recently.

In post-World War II Europe, with steel yet scarce, the design was ideal for rebuilding bombed-out bridges that still had standing foundations. In 1955 one of Dischinger's collaborators persuaded Swedish engineers to build the world's first modern cable-stayed bridge, the Stroemsund Bridge.

American engineers adopted the design in the last half of the twentieth century. Several notable examples have appeared in Washington, such as the Chow Chow Bridge (1952; lost 1988), Grays Harbor County; Benton City-Kiona Bridge (1957), Benton County; Ed Hendler Bridge (1977), Benton and Franklin counties; and the State Route 509 Bridge (1996), Tacoma.

NOTES

1. Eric DeLony, *Landmark American Bridges* (New York: American Society of Civil Engineers, 1993); www.cr.nps.gov/habshaer/haer.
2. Robert H. Krier, Craig Holstine, Robin Bruce, and J. Byron Barber, "Inventory, Evaluation, and National Register of Historic Places (NRHP) Nomination of Bridges in Washington State, 1941–1950: A Project Summary," Short Report DOT92-9, Archaeological and Historical Services, Eastern Washington University, Cheney, 1992; Oscar R. "Bob" George and Craig Holstine, "Evaluation and Nomination to the NRHP of Washington State Bridges Built 1951–1960," WSDOT Environmental Affairs Office, Olympia, 2001. Lisa Soderberg used slightly different criteria for evaluating Washington bridges. See Lisa Soderberg, "Historic Bridges and Tunnels in Washington State," NRHP Thematic Nomination, 1980.
3. See T. Allan Comp and Donald Jackson, "Bridge Truss Types: A Guide to Dating and Identifying," Technical Leaflet 95, American Association for State and Local History, *History News* 5, no. 5 (May 1977); Carl W. Condit, *American Building Art: The Twentieth Century* (New York: Oxford University Press, 1961); Henry Petroski, *Engineers of Dreams: Great Bridge Builders and the Spanning of America* (New York: Alfred A. Knopf, 1995); and David Plowden, *Bridges: The Spans of North America* (New York: W. W. Norton, 1974, 1984).

Part One: Historical Perspectives

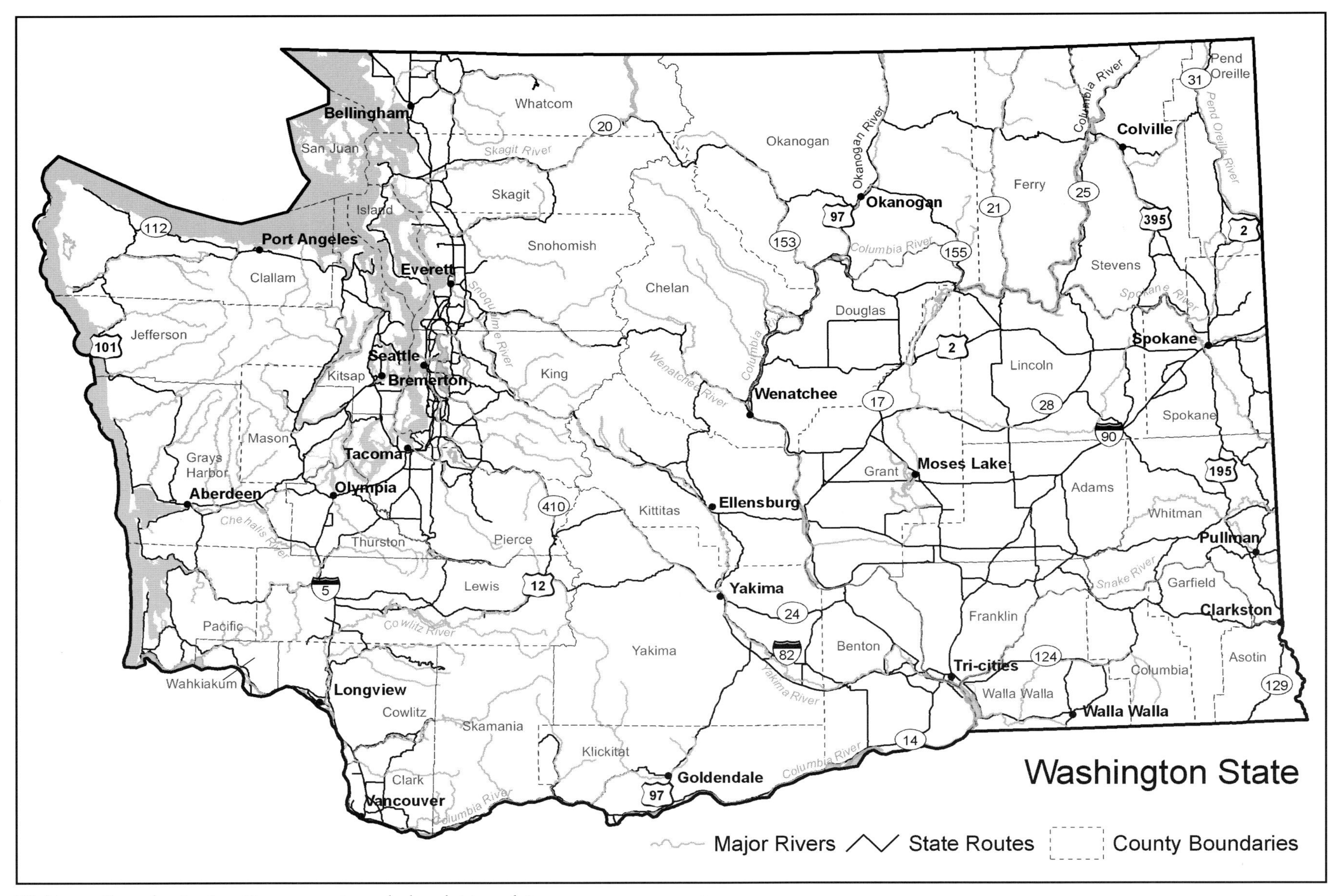

Washington State, showing State Routes, counties, and selected cities and rivers.

Tanya Johnson, WSDOT Environmental Services Office, GIS Program, Olympia

150 Years of Spanning Washington

C. H. Clemon's skid road bridge, measuring 300 feet long and 50 feet high, somewhere in the southern Puget Sound timberlands, 1897. E. A. Smith, photographer, image # 1214.

State Capitol Museum, Washington State Historical Society

The Early Years

Today we see few remnants of our earliest highway bridges. But in the last half of the nineteenth century, settlers moving to what is now Washington built roads and bridges that transformed the territory and knit its separate communities together by the time Washington became a state in 1889.

It was hard work. Washington's terrain varied widely from region to region and presented daunting challenges. Around the Puget Sound basin were thousands of inlets, coves, tide flats, streams, and lakes. Vast, spidery networks of rivers drained the snowcapped Olympic Mountains and the rugged Cascade Range, which effectively divided the state into eastern and western sections. The mighty Columbia River, with its canyoned tributaries, wound in great curves from Canada southward to the Oregon border, then westward until flowing into the blue waters of the Pacific Ocean. Dense evergreen forests blanketed much of the land, except for the semi-arid south-central part of the state.

From the time people first arrived in this area, waterways both

A sinuous, timber trestle on the Colville River, Stevens County, ca. 1905. Frank Palmer, photographer, image # L93-17.71.

Northwest Museum of Arts and Culture/Eastern Washington State Historical Society, Spokane

facilitated and hindered travel. In canoes usually carved from cedar trees, native peoples paddled off the Pacific Coast and in the Strait of Juan de Fuca, Puget Sound and its estuaries, and in tributary rivers. Trails led eastward over the Cascade Range to an inland region with a radically different climate. Major rivers draining the Columbia Plateau all led to the interior's first "major highway," the Columbia River, where natives, early explorers, and fur traders paddled its treacherous waters and followed trails along its banks. Bridges were unheard of and roads suitable for wagon passage were nonexistent until settlers arrived.

Captain George E. Pickett, U.S. Army, ca. 1857. Image # 1994.46.1.

Whatcom Museum of History and Art, Bellingham

Traveling difficulties actually contributed to Washington Territory's formation. Settlers north of the Columbia River, weary of the time-consuming journey to the Willamette Valley and the seat of Oregon territorial government, petitioned for a separate territory, and Congress obliged by creating Washington Territory in 1853. The next year, the first territorial assembly passed an act allowing individual landowners along roads to erect bridges and charge tolls, and counties soon hired contractors to build bridges and maintain them for at least two years. The law allowed private parties to build bridges, which then could be purchased by the territory or county governments. In a rare early commitment to bridge development, the assembly authorized construction of a toll drawbridge across the foot of Budd Inlet in Olympia in 1857 and again in 1859.[1]

The U.S. Army built some of the first roads and bridges in Washington. In the 1850s, George E. Pickett, who later became famous for the failed Confederate charge at Gettysburg, served as a young Army captain at Fort Bellingham.

Bridge built in 1857 by Captain George Pickett in what is now Bellingham. Hegg, photographer, image # 9143.

Whatcom Museum of History and Art, Bellingham

In 1857 he erected a wooden bridge over Whatcom Creek on a road between the fort and the town of Bellingham. The ramshackle remains of the first "Pickett Bridge" were removed in 1903. A concrete Luten arch erected in its place in 1920 still bears his name. Today the bridge, the

The Pickett Bridge, a Luten arch built in 1920 on the site of the 1857 structure, Bellingham. J. W. Sandison Collection, image # 5370.

Whatcom Museum of History and Art, Bellingham

A typical timber bridge built in the "King Post" style, near Crowell's Mill, Plumb Station, Deschutes River, Thurston County, 1903. Image # C1949.1230.16.5.

State Capitol Museum, Washington State Historical Society

house he shared with his Indian wife and child in Bellingham, and American Camp on San Juan Island are tangible reminders of Pickett's brief years on the northwest coast.[2]

In time, rutted wagon roads and rough-hewn log bridges gave way to maintained roadways and timber-beam spans. These early wooden structures played a critical role in Washington's early years. The region's wealth of natural resources drew lumbermen, miners, farmers, merchants, and others who relied on bridges for commerce and travel. Compared to expensive ferries, the primitive spans offered the best means of crossing waterways for most travelers. No highway bridges, however, have survived from the territorial era.

Most of Washington's earliest bridges were made of whole and sawed logs, especially on roads through forested country. High-quality timber in ample quantities was readily available, and bridge design and construction methods required minimal skill and engineering. Timber configurations were easily adapted to various relatively short crossings, and they were quick and easy to build.

Short, simple crossings usually required only parallel log stringers with a log or plank decking. Heavier traffic loads and longer crossings required more substantial structures, such as pile trestles and timber trusses, usually "King Post" or "Queen Post" configurations for shorter spans and Howe trusses for longer spans. Unfortunately,

A "queen post" style timber bridge.
Washington State Archives, WSDOT Records

these structures frequently fell victim to floods, fire, and other hazards. Because they were built low over the water, they often impeded navigation, too.

The Howe truss became the earliest and most popular truss type in Washington. The earliest Howe timber truss remaining in the state is the Grays River Bridge in Wahkiakum County, first built in 1905 and rebuilt by WSDOT in 1989. Wood walls and a shingled gable roof cover the truss. The bridge is the state's oldest covered bridge and the only covered bridge remaining in service on a public highway.

Other timber Howe trusses in use include the Harpole (Manning-Rye) Bridge, built as a railroad bridge

Timber trestle with king-post swing span on the Deschutes River, Tumwater, ca. pre-1889. Clark, photographer, image # 498.
State Capitol Museum, Washington State Historical Society, Olympia

in Whitman County in 1922. Now a private vehicular bridge, it is referred to as a "covered" bridge because its trusses are encased in frame coverings. Similarly, the Pe Ell Bridge, crossing the Chehalis River south of the town of Pe Ell in Lewis County, is covered. Built in 1934, the entire structure, including its Howe through-trusses and plank deck, is covered by corrugated tin. In King County, the Norman Bridge, demolished in 2004, was a wooden Howe through-truss first built over the Snoqualmie River in 1924 and reconstructed in 1950.

Other innovative timber bridges were used to span Washington's waterways. Timber suspension bridges crossed some of the state's most imposing chasms, such as the Wind River Canyon near Carson in Skamania County. The Wind River Bridge, built in 1913, consisted of a single-lane timber deck suspended by steel cables nearly 600 feet across the canyon. Timber towers with two pyramid-frame legs, one below each cable saddle, supported the cables, which in turn held up timber trusses on each side of the roadway and deck. The bridge was replaced by a steel suspension structure in 1925.[3]

On the Skagit River in northwestern Washington, one of the state's largest waterways, steam-powered vessels of

The Harpole Bridge on the Palouse River in Whitman County was built in 1922 as a railroad bridge, and since converted for vehicular access to private property. Photo 1993.

Jet Lowe, HAER.

A covered bridge on the Deschutes River, Tumwater. Plummer, photographer.

State Capitol Museum, Washington State Historical Society

Steamboat *North Star* passing through the Okanogan River Drawbridge at Riverside. The drawbridge collapsed during its inaugural opening in 1905, and was repaired. Frank Matsura, photographer.

Okanogan County Historical Museum, Okanogan

Steamboat *North Star* passing under the Okanogan High Bridge, 1910. Frank Matsura, photographer.

Okanogan County Historical Museum, Okanogan

the "Mosquito Fleet" hauled logs, grain, hay, and passengers. Not surprisingly, the first bridge to cross the Skagit was a movable structure, allowing unobstructed boat passage on the county's economic lifeline. Built in 1892–93 in Mount Vernon, the wooden truss drawbridge required a tender to open the draw span by hand, a laborious task by any measure. After floodwaters carrying a sizeable tree took out the structure's western span in 1906, city officials made repairs, but then decided to replace the entire bridge, complete with a swing span operated by a gasoline engine. Between 1914 and 1920, four other bridges were built across the Skagit, including a swing bridge on the north fork.[4]

In Eastern Washington, the Okanogan River provided access to mines, forests, and rich agricultural lands. Steamboats plied the relatively shallow waters, only one hundred feet wide in places, during springtime when water levels were highest. Reaching upstream to Riverside, nearly thirty miles from the Columbia, steamboats delivered annual spring freight shipments to settlers, miners, and merchants of the vast Okanogan Highlands. Travelers arriving on the river's east bank were compelled to use a cable ferry to reach the town and boat landing on the west bank, an inconvenience that involved summoning the bartender in the J. W. Jones Saloon, who also served as the ferry operator. Riverside citizens collected $1,000 for a bridge fund and petitioned the Board of Okanogan County Commissioners. The wooden drawbridge built as a result of the request collapsed at the height of its initial opening when welcoming the first steamboat of the 1905 season. Fortunately, the damage could be repaired, and the bridge was soon back in service.[5]

In 1910, Okanogan County chose an unlikely alternative to a movable bridge at Okanogan. The aptly named High Bridge was a steel structure built high enough to allow steamers to pass beneath its lofty deck. Arrival of a railroad just two years later ended steamboating in the Okanogan Valley, leaving the High Bridge as an artifact of a bygone era. Unsuited to the demands of automobile traffic, the bridge was demolished in 1920. Its concrete footings are still visible just north of the mouth of Salmon Creek.[6]

Through the last three-quarters of the nineteenth century, the art of bridge building had advanced with railroad development. Railroads offered the most convenient and predictable transportation for people and goods. And, by the 1890s, the rise of the American steel industry was making metal truss bridge technology a more feasible option in Washington.

This wood through-truss with metal tie rods is either the first or second bridge built on Wenatchee Avenue over the Wenatchee River in Wenatchee. Although tracks in the wood deck appear to be from wagons and bicycles, this structure probably carried early automobiles. Image # 85-0-87.

Wenatchee Valley Museum and Cultural Center, Wenatchee

Steel truss bridges designed by professional engineers—prefabricated for easy on-site assembly and transported by railroads to construction sites—revolutionized bridge building in Washington by the turn of the twentieth century. A bridge built in rural northeastern Washington is a good example. In 1909 the C. G. Sheely Construction Company of Denver, Colorado, shipped a pin-connected Parker through-truss via the Great Northern Railway's branchline and erected it over the Kettle River at the mining town of Orient.[7]

The Automobile Age

Washington's citizens increasingly became fed up with mud-clogged or rutted roads, and short-lived and unreliable bridges. They demanded better. At the turn of the century, the "Good Roads" movement, followed by creation of the State Roads Commission, heralded the dawn of a new era in Washington's bridge history. In 1905 the legislature passed a bill establishing a commissioner's office and a state highway fund of $100,000. Two years later, passage of the state aid road law relieved counties of financial obligations, and the state began furnishing all engineering and construction costs.[8]

As automobiles replaced horses and wagons, public pressure mounted for funding a more aggressive highway program. In 1910 some 9,300 Washingtonians owned cars. A mere ten years later, more than 186,000 automobiles cruised the roadways. They were part of everyday life and were here to stay. The proliferation of cars and trucks confronted bridge engineers with exciting new demands. Railroad traffic required heavy load capabilities and resulted in the nearly exclusive use of steel in truss spans. But automobiles were much lighter. Now, bridge designers could use a wide range of materials and forms. Steel remained preferable for long spans over waterways, but concrete arches, beams, and slabs predominated on shorter

In the early twentieth century, bridges built for horses and wagons were carrying automobiles. The Oyster Creek Bridge, shown here ca. 1915, on Chuckanut Drive in Whatcom County was replaced in 1924 by a reinforced-concrete T-beam bridge built on a seventy-two degree curve. J. W. Sandison Collection, image # WHG133.

Whatcom Museum of History and Art, Bellingham

crossings. The advance of pneumatic technology brought rivet guns into common use in bridge construction after 1910, making for stronger and longer spans.

The federal Good Roads Act of 1916 provided states with federal matching funds for road and bridge projects. This prompted Washington to launch an ambitious road and bridge-building program. Oregon became the first state to enact a gasoline tax in 1919, and, the same year, Washington created a motor vehicle fund for engineering, construction, improvement, and paving of the state's main highways. Other states followed suit, and soon gas taxes became the primary source of revenue for highway projects nationwide.

Bridge funding could be a creative art, particularly in the days before widespread federal support. Often, various funding sources were needed. In 1918, Cowlitz Valley citizens collected $4,000 in subscriptions for a bridge across the Cowlitz River at Riffe. The state appropriated $11,100, and the Cowlitz County Commissioners added $7,000. The combination procured a used 270-foot, pin-connected, steel Petit truss from the city of Salem, Oregon. Charles G. Huber of Seattle erected the bridge on concrete piers at its new site for $21,980 in 1918.

Constructed in 1915 on Guide Meridian Road in Whatcom County, this Pennsylvania Petit steel through-truss is an example of an early automobile-era bridge. It was moved in 1951 to the Middle Fork Nooksack River. J. W. Sandison Collection, image # 41444.

Whatcom Museum of History and Art, Bellingham

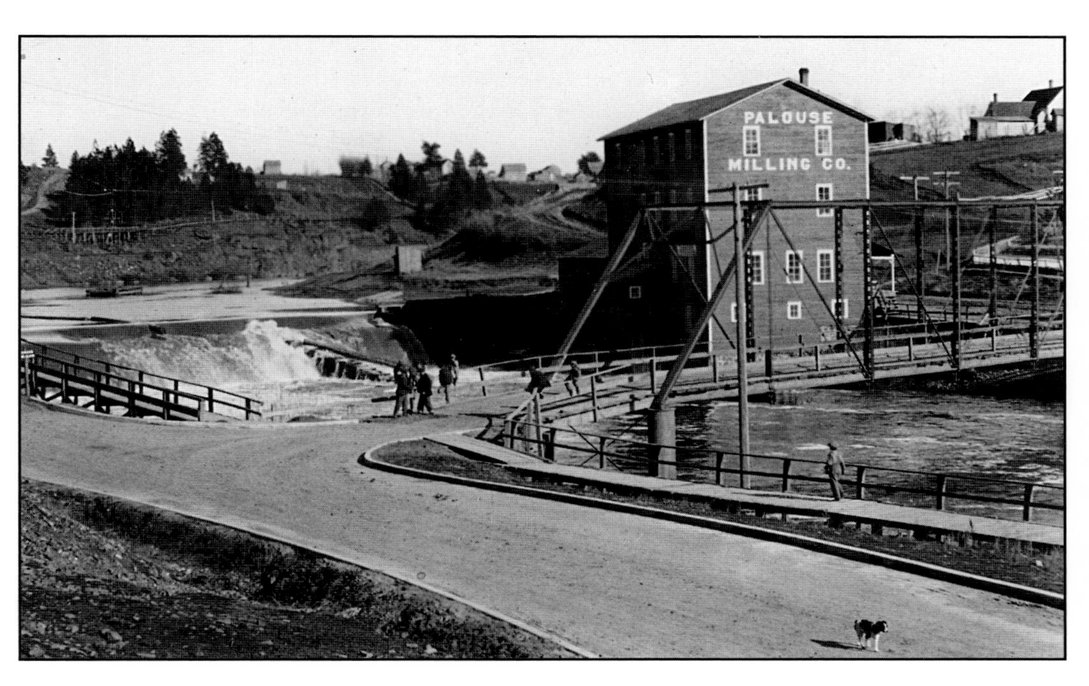

This steel Pratt through-truss was a very early automobile-age bridge in Washington. The "F" Street Bridge was built in 1901 in Palouse, Whitman County. Manuscripts, Archives, and Special Collections, photo ca. 1920, negative # 70-063.

Washington State University Libraries, Pullman

Once a highway entered a national forest, the project became a "federal aid forest road" eligible for federal funding. Such was the case with the Quinault River Bridge, a two-span steel structure. The Bureau of Public Roads awarded the construction contract to R. E. Meath of Portland in 1919.[9]

Completion of a highway across the northern slope of the Olympic Peninsula required the bridging of many challenging obstacles. In 1913 the Portland Bridge Company erected a two-span, riveted, carbon-steel Warren deck truss over the Elwha River in Clallam County. Today, the 576-foot Elwha River Bridge is the oldest deck truss on a highway in Washington, and the state's longest-surviving Warren truss in use as a highway bridge.

Building the Olympic Peninsula Highway was one thing, maintaining it was another, as this report by the state highway engineer in 1924 attests:

> Willful and Malicious damage is sometimes met. For instance, a logging company moving a logging donkey [steam winch] over the Olympic Highway found that a

timber bridge, by reason of overhead struts, did not provide sufficient clearance. The foreman in charge destroyed the struts with an axe and proceeded with his load.[10]

Apparently, the steam donkey and crew made the crossing safely, and presumably the bridge became a candidate for replacement.

In the early years of the twentieth century, the need to move people, trains, and then automobiles over navigable waterways brought movable steel bridges to Washington. In previous years, movable timber structures spanned waterways and allowed the passage of ships, boats, and barges. Navigation considerations required congressional, and sometimes state, approval for bridge construction. Some tributary rivers to saltwater ports served maritime shipping as well as local steamboat traffic. Falls and rapids might seasonally interrupt or restrict steamboat travel on the shallow waters of inland rivers, but movable bridges did not.

Where water depth and stream bottom conditions allowed, movable wooden bridges had been among the first structures built to serve roadway traffic in many riverside communities. Then, eventually, the use of steel for movable spans and improved designs led to the construction of several notable historic movable bridges. These include the "Jack Knife" Bridge, a single-leaf trunnion bascule constructed over Everett's Ebey Slough in 1914, and three double-leaf trunnion bascule bridges built between 1915 and 1919 across the Lake Washington Ship Canal in Seattle.

World War I brought reduced funding levels and labor shortages that delayed many highway and attendant bridge projects. When the war ended in 1918, demands for more and better roads rose dramatically. "Every city, town, hamlet, and outlying district in the nation voiced its appeal," proclaimed James Allen, the state highway commissioner, "and in response to this outcry there came the greatest era of road construction in the history of the United States."[11]

Congress directed the secretary of war to transfer war surplus materials to the Department of Agriculture for distribution to state highway departments for roads and bridges. Washington received mostly vehicles, which, along with explosives, pumps, and engineering instruments, contributed to a mild bridge-building boom beginning around 1920. Construction of the Richland Bridge, consisting of three 116-foot steel truss spans across the Yakima River, inaugurated the post-World War I era of notable bridge projects. Near the mouth of the Snake River, Franklin and Walla Walla counties jointly funded construction of a steel deck and through-truss on the highway connecting Pasco with Walla Walla and Oregon.

In 1920 the Washington State Department of Highways established a bridge division to "design all bridges according to the best modern practice." Charles E. Andrew became the first official bridge engineer in the department, heading up a staff of perhaps a half dozen professional engineers. Until his departure in 1927 (to become principal engineer for the San Francisco-Oakland Bay Bridge in California), Andrew oversaw the division's expansion and the design of many of the state's earliest significant bridges.

Between 1920 and 1922, the engineers designed two movable structures, a lift span at Kelso (using J. A. L. Waddell's patent), and a swing bridge at Raymond. They also directed construction of a 1,738-foot steel truss across the Snake River at Central Ferry that became the state's longest highway bridge. Highway Department engineers designed the elegant, hinged, steel arch Fairfax Bridge over the Carbon River in Pierce County, and the 240-foot, riveted, Petit truss bridge built in 1922–23 over the Dosewallips River on Hood Canal's west shore. On the state's eastern border, Washington and Idaho shared costs for the 830-foot Newport Bridge, built in 1926–27 over the Pend Oreille River. The state also built two steel swing spans and a vertical-lift bridge on State Road No. 1 between Everett and Marysville in Snohomish County, and two bridges over the Puyallup River, including one in Tacoma, an impressive, 2,821-foot-long structure consisting of five steel through-trusses spanning three rail lines and the river. From 1924 to 1928, some eighty-six bridges were built by or with the aid of the State Department of Highways.[12]

By the mid-1920s, in response to the continually growing number of automobiles and rising demands for better roads, the department again initiated a more aggressive highway and bridge-building program. During this time, it standardized designs for small bridges, allowing engineers to spend more time designing structures for more challenging crossings. Unlike in other states, which relied primarily on private consultants, the Washington State Department of Highways kept most of the bridge design work in-house, and the department's engineers designed the overwhelming majority of bridges built during succeeding decades. But, because many new bridges were of unprecedented dimensions, some requiring materials and construction techniques new to the state, private engineers and bridge builders were submitting increasingly sophisticated designs to win state contracts.[13]

The standardization of designs for small- and medium-length spans encouraged the increasing use of concrete in Washington earlier than in other parts of the country. While steel truss bridges were preferred for longer spans, shorter spans allowed the use of either steel or concrete. Washington lacked a "steel tradition," with few successful steel plants. In contrast, at least five Portland cement (the principal cementing agent in structural concrete) plants were operating in the state by 1925. In addition, unlike many other states, Washington possesses nearly inexhaustible deposits of incredibly hard gravel, with a compressive strength (exceeding 20,000 pounds per square inch)

An early automobile-era concrete bridge, the Baker River Bridge was built in the town of Concrete, Skagit County, in 1917. The donation of concrete by two local producers probably discouraged construction of a steel bridge. Photo 1993.

Jet Lowe, HAER

particularly suited as aggregate. Specifying concrete in bridge projects made sound economic sense and ensured that construction funds would support local industries rather than distant steel corporations. Other savings included labor costs: relatively unskilled workers could perform concrete work, whereas the erection of prefabricated steel structures often required skilled labor.[14]

An early example in Washington of decision makers opting for concrete over steel occurred in Skagit County. When the issue of replacing an unsafe wooden bridge across the Baker River in the small town of Concrete came to a vote, county commissioners chose a reinforced-concrete design over a steel truss, probably on a construction-cost basis. Two local manufacturers, the Superior Portland Cement Company and the Washington Portland Cement Company, donated their products to ensure that a concrete structure would be built. Completed in January 1917, the Baker River Bridge is an elegant, open-spandrel arch, consisting of two parabolic ribs springing from a pair of massive concrete piers.[15]

Spokane County Commissioners weighed the relative merits of steel and concrete when they awarded a construction contract in 1924 for a bridge over Latah Creek in the town of Waverly. J. C. Broad submitted two bids, one for a steel bridge, a second for a concrete structure. Surprisingly, the commissioners chose the slightly more expensive bid for the concrete bridge, perhaps for aesthetic reasons. (The open-spandrel concrete arch was considerably more attractive than the steel girder design submitted by the bidder.) It seems reasonable to assume the commissioners also considered concrete to be relatively maintenance free compared to steel, which required occasional painting. They may have been influenced by local industries, including the now-unknown supplier of the aggregate used in the bridge's concrete. (Spalling on the bridge's approaches indicated that local basalt was used in construction, a poor aggregate in concrete.) Thus, for a number of reasons, at the town of Waverly, like in many Washington communities, a concrete bridge was built at a stream crossing where steel would likely have been used elsewhere in the country.[16]

The technique of reinforcing a structure by imbedding steel bars in concrete arches was introduced in America about 1900. After 1914 the reinforced-concrete Luten arch was the style of choice for spans shorter than one hundred feet, and the three-span Luten arch built in 1915 on Lower Custer Way in Tumwater is a good example. Reinforced-concrete girders were in use by the 1920s. In 1926 the Colonial Building Company of Spokane built the longest concrete girder span in the state: a sixty-three-foot span on the 231-foot Tokul Creek Bridge in King County. (Much longer concrete girder spans had already been built in Minnesota and California, however.)[17]

Construction of the first roadway bridge across the Columbia River began with pouring concrete from a trestle for the bridge footings, ca. 1906. The bridge was built to carry a water pipe and road from Wenatchee to what later became East Wenatchee.

Washington State Archives, WSDOT Records

Advancements in Bridge Building

As the twentieth century progressed, engineers adopted advanced methods for building steel cantilever spans, the largest type of truss structure. Many of Washington's largest and most beautiful historic cantilever spans were built during the 1920s and 1930s. After the Wenatchee Bridge (1907), a decade passed before construction of the Vancouver-Portland Bridge, the second highway bridge to span the Columbia River in Washington. Several extraordinary bridges across the Columbia followed in the 1920s, including the Pasco-Kennewick Bridge (1922), Hood River Bridge (1924), Bridge of the Gods (1926), and the Vantage Bridge (1927).

When completed at Wenatchee in 1907, the first highway bridge over the Columbia River carried horse-drawn wagons and sleighs, and the first automobiles in the community.

Washington State Archives, WSDOT Records

Completed in 1917, the Columbia River Bridge (foreground) linked Vancouver, Washington, and Portland, Oregon, on the Pacific Coast Highway. A parallel bridge (background), also a vertical lift structure, was built to carry northbound traffic in 1958.

Washington State Archives, WSDOT Records

In the 1930s, designers progressively refined and simplified the cantilever form, merging functional and aesthetic elements. Most prominent was the 1930 Longview Bridge, which for years stood as the longest and highest cantilever bridge in North America. Other important representative spans included the George Washington Memorial (Aurora Avenue) Bridge (1932), Deception Pass Bridge (1934), and the Grand Coulee Bridge (1935). The following decade brought further advances, as featured in the Kettle Falls Bridge (1941), Fort Spokane Bridge (1941), Northport Bridge (1949), Bridgeport Bridge (1950), and the Agate Pass Bridge (1950).

Bridge-building trends in the 1930s included the increasing use of silicon steel, with greater strength compared to common steel and thus providing for lighter, less costly structures. Silicon steel of 24,000 pounds per square inch (psi) tensile strength (versus the 18,000 psi of carbon steel) determined the design of the Nisqually River Bridge, which in 1934 was the state's longest simple truss span at 322 feet.[18]

Engineers designed more continuous type structures, using what was termed "relatively new" rigid-frame construction, eliminating expansion joints, materials, and

The First Beebe Bridge over the Columbia near Wenatchee was built to carry a water pipe for the Beebe Orchards, but it also served automobile traffic. The wooden suspension structure completed in 1920 was replaced by a steel bridge in the 1960s.

Washington State Archives, WSDOT Records

DEVOTED TO CIVIL ENGINEERING AND CONTRACTING — McGRAW-HILL COMPANY, INC

April 30, 1925

Engineering News-Record

New Jersey Plans Arterial Road from Holland Tunnel—Building Wilson Dam at Muscle Shoals-Part II—Port of Manila Opens Large Shipping Pier

Safeguarding Long-Highway Bridge from Fire

Completed in 1924, the Columbia River Bridge linked White Salmon, Washington, and Hood River, Oregon. Originally a fixed-span bridge (shown here), it was later converted to a movable "lift" structure. From the *Engineering News-Record*, April 30, 1925.

McGraw-Hill Companies, New York, NY

some expense. Design costs for continuous beams and girders were higher, however, reflecting the complexity of the involved computations. Structural steel came into use for bridge piles where wood or concrete piles were impractical, thus saving costly excavation efforts. New "vibrating machines" were used in placing concrete, enhancing strength (up to 20 percent in some cases) and reducing the volume of required concrete. Plywood replaced lumber in concrete form work, cutting costs and improving appearances.[19]

Completion of the Longview Bridge in 1929 marked the last time a privately financed bridge was built across the Columbia River. It also signaled a turning point in bridge history. Bridge construction in Washington accelerated dramatically as federal funding became a staple in bridge financing. Before the New Deal programs began, between 1930 and 1932, the State Department of Highways spent just over $1 million on thirty-eight highway bridges. Between 1932 and 1934, however, federal funds provided by the Works Progress (later, Projects) Administration totaling more than $2 million arrived to help build eighty-nine bridges. Then, from 1934 to 1936, WPA again contributed more than $2 million toward the construction of forty-four of the ninety-eight bridges built.

Some federal funds borrowed by the state for major bridge projects were repaid by charging user tolls. Creation of the Washington Toll Bridge Authority (WTBA) in 1937 provided the financial vehicle for these projects on a grand scale. The WTBA used bonds and tolls to finance the state's two largest bridge projects of the 1930s (both completed in 1940), the first Tacoma Narrows Bridge and the first Lake Washington Floating Bridge (Lacey V. Murrow Bridge). Ironically, both bridges were doomed to failure, one almost

The first bridge at Vantage on the Columbia River was constructed in 1927. It was removed when the new Interstate 90 Bridge was built in 1963. In 1968 WSDOT reassembled the structure at Lyons Ferry on the Snake River, where it now carries traffic on State Route 261. Alfred Simmer, photographer, 1927.

Washington State Archives, WSDOT Records

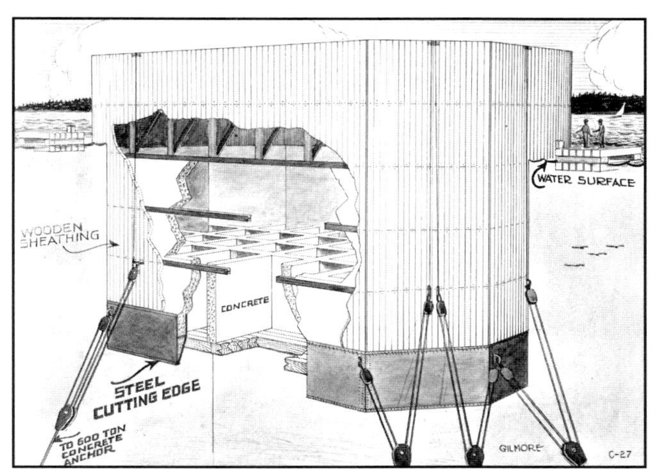

Advancements in construction techniques and materials contributed to the design of the first Tacoma Narrows Bridge. Shown in this drawing is a floating caisson used to construct the bridge piers.

Washington State Archives, WSDOT Records

immediately (the so-called "Galloping Gertie" at the Narrows) and the other a half century later.[20]

In 1936 the state's director of highways, Lacey V. Murrow, wrote in the department's sixteenth biennial report, "A considerable effort has been made toward obtaining structures which will be pleasing to the eye as well as structurally adequate, it being recognized that it is nearly always possible to combine strength with beauty by proper attention to the design." The report cited several noteworthy examples—the Cowen Park Bridge in Seattle, a concrete deck arch with fluted columns on ribbed arches supporting the deck and iron lamps on fluted concrete posts lighting the roadway; the similar parabolic arch ribs with concrete spandrel columns in the Wenatchee River Bridge near Leavenworth; and the East 34th Street Bridge in Tacoma.

Art Deco motifs flourished in the Capitol Boulevard Bridge crossing the Deschutes River in Tumwater. The

Although the Mount Baker Ridge Tunnel connecting the first Lake Washington Floating Bridge to downtown Seattle represented cutting-edge engineering, horse teams were still required in construction, as seen in this 1939 photo. Alfred Simmer, photographer.

Washington State Archives, WSDOT Records

Cement pouring for the Lake Washington Floating Bridge, the first floating concrete bridge in the world. Shown here is a "Koehring Mixer," used for pouring railings. Alfred Simmer, photographer, December 27, 1939.

Washington State Archives, WSDOT Records

bridge's chevrons, zigzags, totem poles, and polychromed concrete lampposts adorned a major highway entering the state capital.[21]

Major federal dam projects on the Columbia River resulted in the construction or reconfiguration of some of the state's most impressive steel bridges. As Bonneville Dam neared completion in the Columbia Gorge in 1938, the Hood River-White Salmon Bridge, built in 1924, was raised and a lift span added. A few miles downstream, the Bridge of the Gods, completed in 1926, was raised forty-four feet above the Bonneville pool. To help facilitate construction of Grand Coulee Dam on the upper Columbia, the Bureau of Reclamation erected a steel through-cantilever truss just below the dam site in 1935.

Totem pole and wall with bas relief on the Capitol Boulevard Bridge, commemorating Tumwater as the "Site of the First American Pioneer Settlement in Washington–1845."

Office of Archaeology and Historic Preservation

As the waters of Lake Franklin D. Roosevelt rose behind Grand Coulee Dam in 1941, the Highway Department installed the two longest bridges built in Washington in the 1940s at the mouth of the Spokane River and at Kettle Falls. The federal government reimbursed the state for construction costs as part of the Grand Coulee Dam-Columbia Basin Reclamation Project. These bridges provided vital links in Eastern Washington's transportation network while serving as vivid reminders of the region's benefits from the New Deal.

On the eve of World War II, Congress passed the Defense Highway Act of 1941. The law mandated construction of access roads to defense-related plants. The "Priorities Critical List" determined which materials could be used to build bridges on routes serving vital facilities. Projects deemed essential to national defense were funded, but other plans were postponed. More concrete was used to compensate for a lack of reinforcing steel. Limited funding, manpower, and materials forced the Department of Highways to rely heavily on timber to meet bridge construction needs.[22]

Since the early 1930s, the Highway Department had built bridges using combinations of steel, concrete, and timber. Improvements in methods of connecting structural members enabled engineers to design structures using timber that formerly required steel. Although the percentage for Washington is unknown, timber bridge construction accounted for more than 50 percent of Oregon's wartime bridges. Because the life span of a timber trestle bridge is generally less than forty years, few of those bridges survive on our state's highways today.[23]

Some bridge projects directly benefited from war surplus materials during World War II. In September 1943, the State Department of Highways acquired surplus steel reinforcing bars that made it possible to build several bridges that otherwise would have been postponed. The Jim Creek Bridge, a graceful, reinforced-concrete arch completed in Cowlitz County in 1945, is one prominent example.[24]

Despite postwar economic instability and continued material shortages, state and local agencies relied on ingenuity and thrift in constructing some notable bridges. A prefabricated Pratt pony truss span originally intended for use in Europe during the war became a single-lane

The elegant lines of the Jim Creek Bridge mask its significance. One of the few bridges built by the State Department of Highways during World War II using surplus steel reinforcing bars, the structure was a priority project due to its location on a highway (State Route 503) accessing vital timber around Mount St. Helens for the war effort.

WSDOT Bridge Preservation Office, Olympia

highway bridge over the Kettle River in northeast Washington. For nominal cost, Stevens County purchased the bridge from the War Assets Administration and erected it as the Barstow Bridge in 1947.[25]

Innovations in Bridge Design, Materials, and Construction

The Federal Aid Highway Act of 1938 introduced the concept of a national highway system. In a report to President Roosevelt, the Public Roads Administration (later renamed the Bureau of Public Roads) had recommended a 75,000-mile system of connected highways identified by the military as strategically important for national defense. As a result, in 1944 Congress passed a Federal Aid Highway Act that required the designation of a "National System of Interstate Highways," consisting of 40,000 miles connecting principal urban and industrial centers.

The federal mandate resulted in monumental changes in bridge building nationwide, including Washington. New bridges were needed across the Columbia River and in urban settings. To minimize traffic disruptions, engineers devised longer spans and faster construction methods. New design standards emphasized safety, including greater horizontal and vertical clearances. Bridges were designed to adapt to future traffic demands and heavier traffic loads. Continuous steel-plate girder bridges were built with significantly longer span lengths, providing a viable option to steel trusses for long spans. Also, the introduction of composite design enabled the concrete deck slab to work with steel girders in more slender spans.

The most significant development in bridge engineering technology in the 1950s, however, was the advent of prestressed concrete. Thirty-seven percent of all Washington bridges currently in service on state, county, and city highways are constructed of prestressed concrete, and one hundred and eighteen prestressed concrete highway bridges built in the 1950s remain in service today. The Klickitat River Bridge (No. 142/9), built for Klickitat County in 1954, is believed to be the oldest prestressed concrete bridge in the Washington State Bridge Inventory.[26]

Advances in steel quality and weldability increased interest in the design of steel girder bridges in the 1950s. Until the end of World War II, welding was not permitted on the major structural members of bridges in the federal aid system. When improved steel became widely available in the 1960s, welded structures began to replace riveted plate girders. High-strength steel bolts eventually replaced rivets for connecting structural members.

Built in 1954, the Klickitat River Bridge (No. 142/9) is believed to be the first prestressed concrete bridge on the state highway system. It was also the first use of segmentally precast, posttensioned concrete girders by the Highway Department.

Craig Holstine, WSDOT

Many of our state's most impressive structures are reminders of the revolution in bridge building that started in the 1950s. The "Blue Bridge" (Pioneer Memorial Bridge) crossing the Columbia River between Kennewick and Pasco was the first steel tied-arch design in the state. It resulted in a greatly improved appearance over the through-truss structures previously used for long river crossings. Homer M. Hadley, one of the state's great bridge innovators, designed the Portage Canal Bridge with a drop-in, steel box-girder segment as part of the main span, making it the second bridge in North America to use this precedent-setting design. Hadley's 1955 design of a steel-and-concrete "tied-cantilever" bridge crossing the Yakima River between Benton City and Kiona was truly innovative. The Benton City-Kiona Bridge, with its towers and stays, eventually became an American prototype for what came to be called "cable-stayed" bridges.

Seattle engineer Harry Powell designed bridges of aesthetic beauty that won prestigious national awards. One was the Rainbow Bridge over the Swinomish Channel at La Conner in Skagit County, the first fixed through-arch span in Washington, which also pioneered the use of high-strength structural steel. Other elegant, award-winning Powell designs included the Modrow Bridge in Cowlitz County and the B-Z Corner Bridge in Klickitat County. Both are three-hinged, open-spandrel, rib deck-arch spans incorporating the extensive use of welded steel.

The 1950s also brought improvements to the most important north-south highway on the west side of the Cascades, Primary State Highway 1 (PSH1), later known as Interstate 5. Of the numerous bridge and tunnel projects along the route, the most notable included the 3,538-

foot-long toll bridge across the Columbia River between Washington and Oregon at Vancouver (1958).

The other main PSH1 improvement was a bypass route along Seattle's industrial area and waterfront. Here, collaborative efforts between the State Highway Department's Bridge Division and the City of Seattle's Engineering Department produced two innovative bridges and a tunnel of unusual design and construction. The First Avenue South Bridge, a bascule span of state record-setting dimensions over the Duwamish River, rests on a cellular concrete floating foundation. Further north, State Route 99 rises above Seattle's historic waterfront on a 1½ mile long, double-deck concrete structure known as the Alaskan Way Viaduct. At the viaduct's north end, the Battery Street Tunnel was constructed to connect with Aurora Avenue to the north. Although damaged by the Nisqually Earthquake of 2001, the viaduct remains an integral link in Seattle's congested transportation network.

Another innovative bridge of the period was a double-deck, concrete-and-steel span structure with a carrying capacity of twelve lanes of traffic on Interstate 5 across the historic Lake Washington Ship Canal in north Seattle. Engineers used a computer to design the substructure for the Lake Washington Ship Canal Bridge in 1958–59, but many years would elapse before that technology completely replaced slide rules in bridge engineering. With the eventual advent of computers and handheld calculators, bridge designers entered vast new realms of possibilities unimaginable to previous generations.

The use of movable bridges has declined significantly since the 1940s, as a result of changes in the patterns of marine commerce and the costs associated with operating and maintaining these types of bridges. In many cases, movable bridges have been replaced by fixed spans. In 1944 the State Department of Highways managed the operation of thirty-two movable highway bridges. Of these, sixteen (50 percent) were swing bridges, nine (28 percent) were lift bridges, and seven (22 percent) were bascule bridges. In 2001, twenty-four operating movable bridges remained: eleven (46 percent) were bascules, eight (33 percent) were lifts, and only five (21 percent) were swing bridges. Since the end of the 1950s, only one lift bridge has been built on Washington highways (excluding those associated with the state ferry system)—the Hoquiam River Bridge at Riverside, constructed in 1970. It is unlikely that we will see many lift bridges in the future.[27]

Washington State has led the world in the design and construction of floating bridges since completion of the Lake Washington Floating Bridge in 1940. The Evergreen Point (or Albert D. Rosellini) Bridge (1963) connecting north Seattle with Bellevue on State Route 520 has an overall length of 12,355 feet, of which 7,578 feet are floating, making it the longest floating bridge in the world. The Lacey V. Murrow Interstate 90 Bridge (1993) across Lake Washington between central Seattle and Mercer Island measures 7,700 feet, with 6,620 feet of floating sections. Its parallel structure at the same location, the Homer M. Hadley Bridge (1989), is also 7,700 feet long, with a floating length of 5,811 feet. The Hood Canal Bridge (1961) measures 7,967 feet long with a floating length of 6,521 feet, and was the first structure of its type built on salt water. It remains the longest floating bridge on tidal water in the world, and of all floating bridges, it is the third longest (behind the Evergreen Point and Lacey V. Murrow bridges on Lake Washington).

Other floating bridges around the world are of lesser dimensions than Washington's four floating bridges. The Derwent River Bridge in Hobart, Tasmania, Australia, measures 3,807 feet in total length, of which 3,168 feet are floating components. The Lake Okanagan Bridge in Kelowna, British Columbia, Canada, has a total length of 2,885 feet, with a 2,100-foot floating length. The Salhus Fjord Bridge in Norway has a total length of 4,592 feet, of which 4,087 feet are floating. Also in Norway is the Bergsoy Sound Bridge at 3,060 feet, with a floating length of 2,771 feet. These latter two bridges are built over salt water, but both are shorter than the Hood Canal Bridge.

Conclusion

As automobile needs transformed our dirt roads into a modern highway system, bridge building in Washington evolved from reliance on an abundance of timber to more durable materials capable of carrying heavier loads over longer spans. Steel and concrete have replaced timber almost completely in bridge building. The Warren and Petit trusses survived, while other metal trusses popular in the eastern United States never appeared here. The State Department of Highways did adopt the Pennsylvania Petit truss for longer span designs. Steel continued to play an important role where concrete was unsuitable, such as in long spans, technically demanding crossings, and for movable structures. Riveting and the availability of silicon steel, too, helped maintain steel's popularity, especially in cantilever bridges. By the 1960s, welding and friction bolts resulted in the greater use of rigid-frame steel structures for relatively short spans. Lightweight, easily designed, and rapidly erected steel-plate and box girders facilitated the construction of the state's more innovative drop-in spans and cable-stayed bridges.

From roughly 1910 through the 1930s, engineers used steel trusses and reinforced-concrete arches for short- and medium-length spans. The appearance of concrete girders, T-beams, and flat slabs increased the use of concrete in Washington and other far western states much earlier than in other sections of the country, which continued to build mostly steel bridges. Because Washington lacked a "steel tradition" prevalent in the East and Midwest, and the Portland cement industry thrived in the Pacific Northwest, the Northwest anticipated what became a nationwide reliance on reinforced concrete for standard highway bridge designs. By the 1950s, when precasting and prestressing became available, reinforced concrete had almost completely replaced steel in bridge construction. Given Washington's pioneering use of concrete, some would view its floating bridges as the state's ultimate expression in concrete.[28]

Workmen pose on a steel through-truss bridge, ca. early twentieth century, probably in Whatcom County. J. W. Sandison Collection, image # 3571WHA601.

Whatcom Museum of History and Art, Bellingham

Throughout history, bridges have transported people across seemingly insurmountable barriers. They define our level of civilization, and illuminate our sense of beauty. Responding to changing demands and using state-of-the-art technologies and materials, bridge engineers have helped promote our economic well-being, while elevating Washington's profile in the world of bridge building. Our historic bridges are a testament to our journey through the past century and a half. Although fewer and fewer historic bridges remain in use, they help remind us of who we are and where we came from.

NOTES

1. Under the territorial system, the legislature appropriated funds for specific roadway construction, supervised by territorial highway commissioners from counties through which the roads passed. Commissioners were appointed by the legislature and served until a highway's completion. The legislature provided limited funding for state roads, and some construction had occurred by the turn of the century. After determining the system unsatisfactory, the legislature authorized establishment of the office of Washington State Highway Commissioner in 1903, but Governor Henry McBride vetoed the bill. Two years later, the legislature passed a similar measure, this time winning the governor's approval. The 1905 act establishing the highway commissioner authorized thirty-six state aid roads in twenty-five counties, with expenditures apportioned to each county at the rate of one-half of what it (the county) paid into the Public Highway Fund. Washington State Department of Transportation, Second Biennial Report, 1908, 19; WSDOT, Third Biennial Report, 1910, 11.
2. The Pickett House stands at 910 Bancroft Street in Bellingham. Lelah Jackson Edson, *The Fourth Corner: Highlights from the Early Northwest* (Bellingham, WA: Cox Brothers Inc., 1951), 60–61, 115–25; Lisa Soderberg, "Pickett Bridge," Historic American Engineering Record Inventory, Office of Archaeology and Historic Preservation, Olympia, ca. 1979 (hereafter cited as "HAER Inventory").
3. "Wind River's Bridge Crews Off for Winter," *Skamania County Pioneer*, December 19, 1958.
4. Margaret Willis, ed., *Skagit Settlers* (Mount Vernon, WA: Skagit County Historical Society, 1975), 4–6, 170.
5. Louise McKay, "Drawbridges across the Okanogan," *Okanogan County Heritage* 17, no. 3 (Summer 1979), 3–12.
6. Paul Moses, "The Bridges of Okanogan, 1910–2003," *Okanogan County Heritage*, June 2003, 2–3.
7. Lisa Soderberg, "Orient Bridge," HAER Inventory, 1979; Robin Bruce, "Orient Bridge," HAER Report, Archaeological and Historical Services, Eastern Washington University, Cheney, 1992.
8. WSDOT, Third Biennial Report, 1910, 11.
9. WSDOT, Seventh Biennial Report, 1918, 37, and Eighth Biennial Report, 1921, 7, 73; James Norman, et al., *Historic Highway Bridges of Oregon* (Portland, OR: Oregon Historical Society Press, 1985), 33.
10. WSDOT, Tenth Biennial Report, 1924, 39.
11. WSDOT, Eighth Biennial Report, 1921, 9, 73.
12. Mark S. Woodin, "Bridges, Now and Then," *Highway News* 3, no. 3 (September 1953), 61–64; WSDOT, Ninth Biennial Report, 1922, 31–33; WSDOT, Eleventh Biennial Report, 1926, 17; Charles Miles and O. B. Sperlin, eds., *Building a State: Washington, 1889–1939* (Tacoma, WA: Washington State Historical Society Publications, Vol. III, Pioneer Inc., 1940), 257–58.
13. Jonathan Clarke, "Material Concerns in the Pacific Northwest: Steel Versus Reinforced Concrete in Highway Bridge Design in Washington State, 1910–1930," *Construction History* 16 (2000), 33–61; WSDOT, Ninth Biennial Report, 1922, 37.
14. Ibid., 35–36.
15. Ibid., 40–41; Wm. Michael Lawrence, "Baker River Bridge," HAER No. WA-105, 1993.
16. Robert H. Krier and Craig Holstine, "Latah Creek Bridge No. 4102," HAER No. WA-163, Archaeological and Historical Services, Eastern Washington University, Cheney, 1998.
17. Clarke, "Material Concerns in the Pacific Northwest," 41, 43; WSDOT, Eleventh Biennial Report, 1926, 47.
18. WSDOT, Sixteenth Biennial Report, 1936, 27.
19. WSDOT, Fifteenth Biennial Report, 1934, 39; WSDOT, Sixteenth Biennial Report, 1936, 30.
20. WSDOT Fifteenth Biennial Report, 1934, 44; WSDOT, Sixteenth Biennial Report, 1936, 25, 27; WSDOT, Seventeenth Biennial Report, 9.
21. WSDOT, Sixteenth Biennial Report, 1936, 25.
22. Robert H. Krier, et al., "Bridges Built in Washington in the 1940s," National Register of Historic Places, Multiple Property Documentation (MPD), 1991.
23. James Norman, *Oregon Covered Bridges: A Study for the 1989–90 Legislature.* (Salem, OR: Oregon State Highway Division, Environmental Section, 1988), 34.
24. Robert H. Krier, et al., "Jim Creek Bridge," NRHP Nomination, 1991.
25. Robert H. Krier, et al., "Barstow Bridge," NRHP Nomination, 1991.
26. Oscar R. "Bob" George and Craig Holstine, "Bridges Built in Washington in the 1950s," NRHP MPD, 2001; George, "Klickitat River Bridge," NRHP Nomination, 2001. Text following endnote 26 is from George and Holtine's NRHP MPD.
27. Ibid.
28. Clarke, "Material Concerns in the Pacific Northwest," 54–56.

Failure and Progress

"The bridge seemed to be among the things that last forever; it was unthinkable that it should break."
—Thornton Wilder, *The Bridge of San Louis Rey*

The fictional failure of an eighteenth century Peruvian rope suspension bridge provided Thornton Wilder with the backdrop for his timeless discourse on the role of fate in human experience. He hastened to note that the bridge stood for more than a century as "the finest bridge in all Peru," and was believed to be "protected" by St. Louis of France and "by the little mud church on the further side." Its failure was inconceivable to those traveling the road between Lima and Cuzco. Stability may sometimes be in the eye of the beholder. The laws of physics, and structural conditions and composition, determine a bridge's longevity. Unless fate intervenes.

Flooding in the spring of 1948 wreaked havoc with bridges in north central Washington. Shown here, the Chewack River Bridge near Winthrop in Okanogan County. Photo by Alfred Simmer, August 26, 1948.

Washington State Archives, WSDOT Records

Throughout time, bridges have represented achievement, success, and progress. Proud designers and builders have proclaimed that their engineering feats represent obstacles overcome, and isolation and separation ended forever; that bridges are eternal monuments to man's skill and determination. Yet inevitably, bridges, like mortals, meet their end. Nowadays, those ends usually come on schedules designed to satisfy safety and traffic needs.

Historically, bridges have rarely outlasted their design or structural lifetimes, and have met untimely, unexpected ends. Floods, fires, winds, overloading, vehicular crashes, and other phenomena of man and nature have destroyed bridges, sometimes with tragic results. Although fate's role could not be discounted, witnesses or engineers usually could identify plausible causes for structural failures.

Some bridge failures can be traced to design shortcomings. The first Tacoma Narrows Bridge was the most infamous disaster of this kind, but not the state's first. Early bridges failed with disturbing regularity. Few were designed by professional engineers, and many were built of inferior local materials. For the most part, the causes of their failures have been lost to history. At least two resulted in multiple fatalities—the Division Street Bridge collapse in Spokane in 1915 (see Chapter 6), and the Allen Street Bridge failure in Kelso in 1923. A consequence of the latter, or so it is now believed, resulted in more careful and regular bridge inspections statewide.

A flock of sheep brought down this wooden bridge over the Spokane River below Long Lake Dam in September 1942.

Washington State Archives, WSDOT Records

This chapter will not address all bridge failures in Washington. For one reason, not all failures were documented. Some bridge failures are mentioned in other chapters.

Unexpected Losses

WSDOT has compiled an extensive, but incomplete, list of bridge failures in Washington. It begins with the 1923 collapse of the Allen Street/Kelso-Longview Bridge; notably, none of the many earlier bridge losses are included. Nor are failed structures owned by federal or other state agencies, or by counties and cities (other than the Allen Street Bridge). Nevertheless, between 1923 and 1998, at least seventy bridges were lost to various causes, not including demolition, removal, or replacement. Floods destroyed forty-two bridges; fires claimed eight; mud and debris flows caused by the Mount St. Helens eruption in 1980 took out five bridges; four were lost to collisions with either vehicles or boats; storms and overloads claimed three each; tsunamis destroyed two structures; wind brought down our most famous bridge (see below); and it is not known why two other bridges failed.

Man's behavior has caused the demise of otherwise sound structures. In 1850 the bridge over the Maine River in France collapsed, killing as many as half of the 500 soldiers crossing the structure. Failure was due, it was believed, to the undulating motion caused by the troops marching in step. (Since then, military formations customarily break step when crossing bridges to avoid repeating the Maine River disaster.) Rhythmic dancing by party revelers may have caused skywalks not engineered to withstand undulations to collapse at the Hyatt Regency Hotel in Kansas City, Missouri, in 1981; more than one hundred people died in the worst structural failure in American history. As engineer Henry Petroski observed, "dancing a bridge to destruction is not so farfetched an idea as at first it might seem."[1]

Unbelievable as it may seem, sheep trotting home from grasslands on the Spokane Indian Reservation brought down a bridge in 1942 (see Chapter 6). Undulation caused by the woolly travelers, not their combined weight, is thought to have sent the wooden structure crashing into the Spokane River. Although many sheep were killed, no human life is known to have been lost in what was perhaps the most bizarre bridge failure in the state's history.

Located only one hundred yards downstream from a steamboat landing at Riverside on the Okanogan River, a log-and-sawn-lumber drawbridge was completed just in time to welcome spring steamboat arrivals. On March 15, 1905, as the first steamer, the *Enterprise,* whistled its approach, the drawspan lifted as planned for the first time. As it reached full height, the drawspan twisted and crashed through one of the approach spans, damaging the *Enterprise,* not to mention the community's pride.

An investigation revealed the failure's cause—an improper weight ratio of the drawspan to its counterbalance, which was weighted with rock. Apparently the ratio was calculated when the timbers were green and wet, but when the wood dried, it lightened the drawspan's weight, creating the imbalance. Steamboat company officials complained of the narrow passage through the drawspan, where strong currents pushed boats toward pilings. Furthermore, they said, the bridge's construction was "so faulty as to warrant no thrills of pride in the Okanogan taxpayer." Even after its repair, the bridge failed to meet standards set by the U.S. government. Responding to federal demands to fix the structure, county commissioners stated emphatically that they "had already spent $1,519 for repairs and they would spend no more money on the hoodoo bridge." Although imperfect, the bridge remained in service until 1916.[2]

Omak, eight miles downstream from Riverside, had similar problems with its movable bridge. One week after its completion in June 1911, the steel structure opened to allow passage of the steamboat *North Star.* The swing spans promptly collapsed under the weight of two wagons loaded with ore. When the county refused the last payment due the builder, charging faulty construction, the Oliver Bridge Company sued. Courts eventually awarded Oliver's widow a modest settlement, and by June 1912, the swing span returned to service.[3]

Early in the automobile era, steel pony trusses typically spanned short crossings. Not uncommonly, concrete or macadam decking added atop plate girders exceeded dead-load design weights. This caused buckling of the top

The Omak Bridge's swing span collapsed under the weight of wagons loaded with ore in June 1911, only two weeks after the bridge opened.

Okanogan County Historical Museum, Okanogan

The Omak Bridge's swing span (far right) is shown in repaired condition in 1912 after its collapse the previous year. Frank Matsura, photographer, image # 3174.

Okanogan County Historical Museum, Okanogan

Forty tons of live load on a pony truss built to carry five tons had predictable results when a loaded logging truck collapsed the Hoko River Bridge on the Olympic Penisula, April 22, 1947.

Washington State Archives, WSDOT Records

chords, occasionally even before a bridge opened to traffic. As *Engineering News* noted in 1913, "It is a fact that a great many low-truss highway bridges built for erection in remote country districts are almost devoid of bracketing or other stiffening of the top chord." The publication recommended "provision must be made for probable future increase in thickness of the macadam surfacing, for road repair on country highways is not carried on by bridge engineers, and is often conducted on the plan of simply piling on more stone without regard to a set of long forgotten blueprints."[4]

Disregard of a posted load limit on a pony truss bridge built in rural Clallam County in 1931 resulted in its destruction. On April 22, 1947, F. H. Jarnagin drove a loaded log truck onto the wood-deck steel pony truss four miles west of Sekiu. The bridge collapsed, sending logs into the Hoko River as the truck burned. Jarnagin escaped unharmed, but faced legal action for carelessness. "The bridge was posted warning users that five tons was the limit and Mr. Jarnagin destroyed the bridge by undertaking to go over it with approximately forty tons," said Clallam County Prosecuting Attorney D. E. Harper.[5]

Some river crossings seem to have been cursed. The Allen Street (or Kelso-Longview) Bridge across the Cowlitz River collapsed on January 3, 1923. Fourteen years earlier, a flood destroyed the previous structure that had stood at

The wooden Allen Street Drawbridge in Kelso is shown here on September 15, 1922, four months before its failure. Concrete piers for a new steel bridge are under construction in front of the drawbridge.

Washington State Archives, WSDOT Records

the site in Kelso for only two years. The 1923 failure of the wooden, movable drawbridge spans was attributed to overloading and deterioration of its wooden deck. As the four-inch-thick planks on the deck began to wear down, a three-inch layer of new planks were added. Rain soaking into the worn planks added weight to the deck, straining the supporting cables, which snapped, dropping the span into the icy river.

The structure collapsed in late afternoon that fateful day. Mill workers and downtown shoppers clogged the one-lane bridge on their way home before the dinner hour. At least thirteen cars spilled off the bridge. Passengers scrambled for safety on floating logs, debris, and bridge timbers. Boats navigated through the wreckage rescuing survivors, but at least seventeen people were lost. Some bodies were never recovered. The state's worst bridge disaster in terms of fatalities reportedly prompted the state to begin regular bridge inspections.[6]

Another timber bridge failure in Cowlitz County led to the construction of one of the state's award-winning bridges. When county commissioners rejected a bid to replace an aging timber structure over the Kalama River, the Hart Company lowered its bid and built a new Modrow Bridge in 1950, presumably of inferior materials or to lesser specifications than the costlier design. On the afternoon of July 9, 1958, an automobile and a heavily loaded truck were crossing the bridge when the sound of splitting timber warned of impending disaster. The bridge collapsed, sending the vehicles into the river, with broken timbers crashing down atop the occupants, who miraculously survived. Later, it was reported that the bridge was "insured against everything but collapsing." Harry Powell and Associates, among the region's premier bridge engineers, designed the elegant, steel-arch structure that replaced the ill-fated, "low-bid" Modrow Bridge in 1959 (see Chapter 11).[7]

Collapse of the wooden drawbridge spans and towers in Kelso on January 3, 1923, killed at least seventeen people in the state's worst bridge disaster. One fallen span is shown on the left; timbers from a tower lie atop the steel trusses of the new bridge.

Washington State Archives, WSDOT Records

Record-setting floodwaters in 1894 washed away the first timber bridge to span the Wenatchee River at the north end of town. Fire then destroyed the replacement bridge, a combination timber-and-steel structure. On July 4, 1917, fireworks ignited a blaze that consumed the wooden deck of the third bridge to cross the river at the site. Repaired in short order, that bridge lasted until 1931, when it was replaced by one of the two steel bridges that today carry the combined four lanes of Wenatchee Avenue.[8]

Fireworks ignited a blaze that burned the deck on the Wenatchee Avenue Bridge in Wenatchee, July 4, 1917.

Wenatchee Valley Museum and Cultural Center, Wenatchee

Fire has claimed several of the state's major bridges. On February 6, 1947, a hay truck collided with a fully loaded fuel truck on the Washougal River Bridge (built in 1926) between Camas and Washougal. The crash ignited an inferno that collapsed the steel truss into the river. Miraculously, the drivers of the trucks escaped. Another driver was not so lucky when his vehicle knocked a utility pole onto the wooden deck of the Fairfax Bridge over the Carbon River in 1998. The collision started a fire that consumed the vehicle, its driver, and part of the bridge. (The timber approach span has since been replaced, leaving the 1921 steel arch in place.) In August 1968, a worker's cutting torch ignited newly creosoted deck timbers of the bridge across the Columbia River at Brewster. About 1,400 feet of steel, weakened by the flames, collapsed into the river, leaving grotesquely twisted trusses hanging in place. The 1928 steel cantilever structure had only recently been raised on its piers above waters inundated by the Wells Dam downstream.[9]

The Columbia River Bridge at Brewster, a steel cantilever built in 1928, and destroyed by fire, 1968.

WSDOT

The Brewster Bridge conflagration began when a welding torch ignited the wood deck, August 1968.

WSDOT

Rivers draining the North Cascades have always been prone to flooding. Heavy rains or snowfall combined with rapidly melting snow and no dams to restrain the raging torrents have damaged and destroyed numerous bridges. "Low-water" bridges, that is, those that could be constructed only during times of low water, were especially susceptible to loss on the Methow River. The flood of 1894 took out most timber bridges. Their replacements fared little better. Upon collapse of a bridge near the river's mouth in 1906, the *Methow Valley News* observed, "The collapse of the Pateros bridge is a calamity that will be noticed over the entire county.... Such peculiar antics on the part of the Pateros bridge at this critical time is enough to cause the average citizen to look with suspicion on every bridge in the county and rightly wonder at what moment the rest of them will turn topsy-turvy." By December of that year, another bridge built to replace the lamented structure had washed out.[10]

The 1948 flood brought down most steel bridges on the Methow River (seven in all), although the bridge at Winthrop remains in place today. Originally built in 1929, the steel truss structure was rebuilt in 1950 to fix the temporary repairs made after the 1948 flood. That flood also seriously damaged the Methow River Bridge at Twisp, a concrete T-beam built in 1931 and rebuilt in 1949, which has since been replaced with a new bridge.

Floods have shaped the physiographic characteristics of much of the Pacific Northwest. Melting snows and turbulent runoffs carry not only eroded soils, but also uprooted large trees. When floating trees hit bridge piers, damage

The Methow River Bridge in Twisp was severely damaged by flooding in the spring of 1948. Shown here is a Bailey bridge crossing over the collapsed span, a temporary solution until the bridge could be repaired. Alfred Simmer, photographer, August 26, 1948.

Washington State Archives, WSDOT Records

Salmon Creek Bridge, collapsed by scouring action that undercut a pier, January 4, 1956.

Washington State Archives, WSDOT Records

can occur from the collisions alone, but more often debris collects and can change the direction of stronger-than-normal currents. The resulting erosion around piers and abutments, called "scour," has caused the great majority of Washington's bridge failures.

A classic example occurred on the night of January 4, 1956, when a concrete T-beam bridge collapsed into Salmon Creek north of Vancouver. A large fir tree lodged at the base of the piers, redirecting the current so that it undercut the downstream end of the south pier. A State Highway Department crew attempted unsuccessfully to clear the debris, then closed the bridge to traffic. Minutes later, the pier settled, the deck slab broke free from the curb, and the spans separated at the expansion joint, crashing into the streambed. This occurred only thirteen days after the bridge and the newly completed, four-lane U.S. Highway 99 had been officially opened between Vancouver and Tumwater.[11]

On November 12, 1943, a heavy load of logs shifted and destroyed the Cora Bridge on the Cowlitz River.

Washington State Archives, WSDOT Records

Flooding also claimed one of the state's truly unique bridges on the Quinault River. The Chow Chow Bridge was the first use of cable-stayed technology in the state. Because of its remote location, the bridge's clever design went unnoticed by the engineering world. Yet the Chow Chow Bridge anticipated more elegant cable-stayed structures that later came to grace the state's public roadways.

It was built by Frank Milward, a logging company superintendent determined to move logs from deep within the Olympic Peninsula's rain forests. Although not a trained engineer, Milward designed the structure for the Aloha Logging Company, which in the 1950s was cutting timber on the Quinault Indian Reservation. Milward faced the daunting challenge of transporting heavily loaded log trucks across the Quinault River. He abandoned the traditional use of pile trestles, which repeatedly washed away in floods.

Instead, he accomplished the task with an innovative design that did not require supporting piers standing in the formidable river. Without blueprints, Milward constructed a twelve-foot wooden model of a cable-stayed bridge before the formal theory was developed and proven. In 1950 he supervised construction of the first of two cable-stayed bridges on the Quinault River. For reasons now unknown, the first bridge was replaced approximately two miles downstream by the second cable-stayed bridge, the Chow Chow Bridge, built for $44,000 in 1952. Both were constructed of western red cedar and were among the earliest of the type anywhere in the United States.

The underside view of the Chow Chow Bridge deck, revealing some of the innovative engineering in this precursor of the cable-stayed bridge type.

Office of Archaeology and Historic Preservation

The Chow Chow Bridge's deck consisted of log sills supported by timber "A" frames hung from the cables. Six 2½-inch galvanized steel cables stretched between two 56-foot-high cedar timber towers. Attached to the sills at six different points, the cables created a series of independently suspended cross sections, allowing for repair or replacement of individual cables or timber supports. Four concrete "deadmen" buried twelve feet deep anchored the cables at each end of the span.

At 250 feet in length, the bridge endured heavy log truck traffic and the ill effects of the rain forest climate until its collapse in 1964. It was rebuilt, but collapsed again in 1973. The Chow Chow Bridge failed for the third and final time in 1988. What could be recovered of its large cedar components were sawed into blocks and split into shingles at the Quinault Tribal Shake Mill. The shingles were then used as siding for the Quinault Indian Nation Community Center in Taholah. Perhaps this form of historic preservation could only happen in the Pacific Northwest.[12]

Until 1919, wooden drawbridges crossed the Deschutes River estuary at the extreme southern end of Puget Sound. That year the City of Olympia built the 4th Street Bridge, a 1,000-foot-long, concrete triple Luten arch, to straddle the wide, muddy tide flats at the river's mouth. The bridge stood for decades on timber piles in the soft mud, receiving little modernization. In 1993 the Olympia City Council passed a resolution to implement transportation management plans, which would determine the need to increase the 4th Street Bridge's vehicle capacity. A 1995 study showed that the bridge was suffering structural deterioration, and replacement efforts began. Six years later, the Nisqually earthquake struck the area. The concrete columns under the west approach failed, and several sections of the concrete baluster-type railings toppled from the deck into the sound. Severely damaged, the bridge was permanently closed.

The city council replaced the bridge with a haunched-girder design reflecting the deck arches of the 1919 structure. The new bridge opened in December 2003. It sports eleven-foot-wide vehicle lanes, five- and six-foot-wide bike lanes, and a public park overlooking the new bridge from the elevated western approach. A section of the 1919 bridge railing and three original light poles and fixtures were salvaged and erected, with interpretive information, in the Park of the Seven Oars. Some may call it "virtual preservation," but at least history is not forgotten at this historic crossing in the state capital.[13]

In February 2001 the Nisqually Earthquake severely damaged the 4th Street Bridge in Olympia.

Office of Archaeology and Historic Preservation

In December 2003 a new concrete bridge opened on 4th Street in Olympia. Its haunched-girder design reflects the deck arches of the 1919 structure lost to the Nisqually earthquake.

Craig Holstine, WSDOT

Light standard and railing salvaged from the damaged 4th Street Bridge and erected in the Park of the Seven Oars, Olympia.

Craig Holstine, WSDOT

THE PRICE OF PROGRESS

Progress in Washington's highway bridge building history has not come easily at times. Bridge failures have brought progress as well as heartache. Several noteworthy examples tell the story.

First Tacoma Narrows Bridge, November 7, 1940

Elegant and graceful, the slender Tacoma Narrows Bridge stretched like a steel ribbon for nearly a mile across Puget Sound. The third-longest suspension span in the world opened July 1, 1940, to noisy fanfare and a boisterous crowd of politicians, engineers, reporters, and citizens. A mere four months later, exuberance and pride gave way to shock and astonishment as the great span's short life ended in disaster.

When the bridge was first completed, officials and engineers had good reason to feel triumphant. Just the year before, New York's Bronx-Whitestone Bridge was hailed as "the ultimate in suspension-bridge design." The Tacoma Narrows Bridge, however, brought suspension bridge development to its next logical step. Increasingly, designers put a premium on economy, lightness, slimness, and flexibility. The great span represented a culmination of the trend to extend length while narrowing deck width and minimizing stiffening members. Its width-to-length ratio was 1 to 72, a dramatic doubling over the previous record holder, the Golden Gate Bridge (1 to 35 ratio), which was completed just three years earlier in 1937.

The Tacoma Narrows Bridge's initial designer was the project's lead engineer, Clark Eldridge. However, when the state applied for federal support from the Public Works Administration, that changed. During PWA reviews, the cost-cutting recommendations of renowned bridge architect Leon Moisseiff were adopted. Eldridge's twenty-five-foot-deep truss deck was scrapped and an eight-foot solid plate girder substituted by Moisseiff.

The bridge's 425-foot-high flexible towers and slender suspended roadway were susceptible to wind action. The solid deck and plate girders proved highly sensitive to aerodynamic forces, behaving like an airplane wing, lifting and falling in even a light breeze. Thus, the bridge's slender design drew the acclaim of designers, but also made it vulnerable to the gusty winds of The Narrows.

The first travelers across the Narrows Bridge immediately noticed the undulating roadway. Cars one hundred yards in front disappeared from view as the deck rhythmically rose and dropped. Even before the bridge's completion, workmen affectionately dubbed it as "Galloping Gertie." Soon, Gertie was attracting thrill seekers as well as normal travelers. A carnival atmosphere seemed to numb drivers to the potential danger.

Intensely interested, though not amused, was F. B. Farquharson, professor of engineering at the University of Washington in Seattle. At the request of the Washington Toll Bridge Authority, Farquharson launched studies to understand the bridge's oscillations and to recommend steps for reducing them to less dangerous levels. Before changes could be made, fate intervened.

In the early morning of November 7, 1940, winds of up to forty-two miles per hour raced through the narrow channel. The bridge's undulations increased alarmingly. Farquharson drove to the site as quickly as he could, and began taking motion picture footage. A reporter for the *Tacoma News Tribune*, Leonard Coatsworth, loaded his daughter's black spaniel, Tubby, into his car and headed across the bridge toward the family's summer cottage. He never reached the center of the bridge as the undulations grew more severe. The bridge deck began to twist as it rose, and it fell more violently with each wave. The roadway's extreme twisting, increassingly magnified by the aerodynamic effect of steady winds on the bridge's sides, began to rip the span apart.

Massive steel girders twisted like rubber, wire cables more than a foot in diameter began to snap, and bolts sheered and flew into the wind. The catastrophe unfolded slowly enough so that much of it was photographed, and no human lives were lost.

The first Tacoma Narrows Bridge was hailed as "the ultimate in suspension-bridge design." It stood for only four months after its opening.

Washington State Archives, WDOT Records

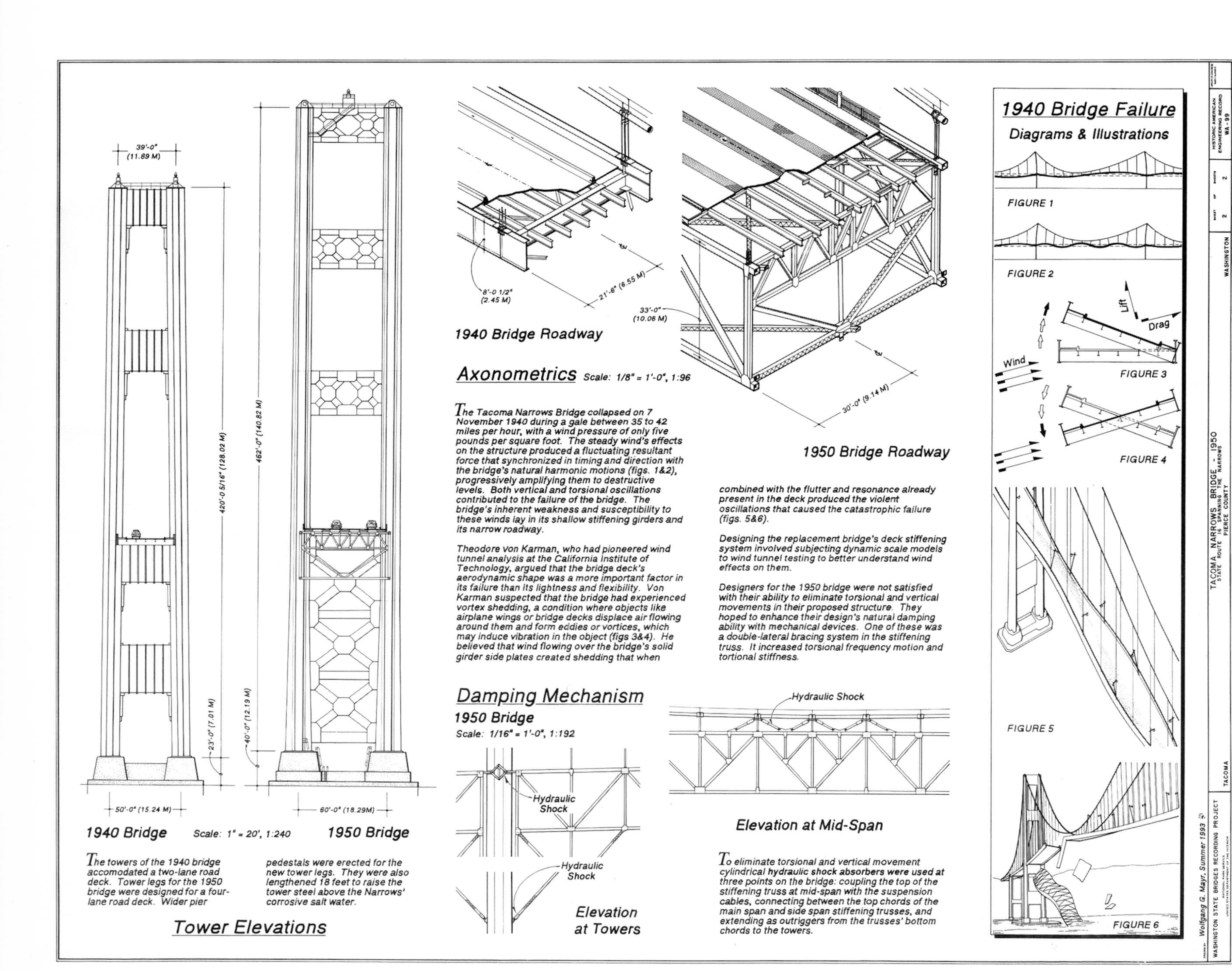

The towers of the 1940 bridge accomodated a two-lane road deck. Tower legs for the 1950 bridge were designed for a four-lane road deck. Wider pier pedestals were erected for the new tower legs. They were also lengthened 18 feet to raise the tower steel above the Narrows' corrosive salt water.

The Tacoma Narrows Bridge collapsed on 7 November 1940 during a gale between 35 to 42 miles per hour, with a wind pressure of only five pounds per square foot. The steady wind's effects on the structure produced a fluctuating resultant force that synchronized in timing and direction with the bridge's natural harmonic motions (figs. 1&2), progressively amplifying them to destructive levels. Both vertical and torsional oscillations contributed to the failure of the bridge. The bridge's inherent weakness and susceptibility to these winds lay in its shallow stiffening girders and its narrow roadway.

Theodore von Karman, who had pioneered wind tunnel analysis at the California Institute of Technology, argued that the bridge deck's aerodynamic shape was a more important factor in its failure than its lightness and flexibility. Von Karman suspected that the bridge had experienced vortex shedding, a condition where objects like airplane wings or bridge decks displace air flowing around them and form eddies or vortices, which may induce vibration in the object (figs 3&4). He believed that wind flowing over the bridge's solid girder side plates created shedding that when combined with the flutter and resonance already present in the deck produced the violent oscillations that caused the catastrophic failure (figs. 5&6).

Designing the replacement bridge's deck stiffening system involved subjecting dynamic scale models to wind tunnel testing to better understand wind effects on them.

Designers for the 1950 bridge were not satisfied with their ability to eliminate torsional and vertical movements in their proposed structure. They hoped to enhance their design's natural damping ability with mechanical devices. One of these was a double-lateral bracing system in the stiffening truss. It increased torsional frequency motion and tortional stiffness.

To eliminate torsional and vertical movement cylindrical hydraulic shock absorbers were used at three points on the bridge: coupling the top of the stiffening truss at mid-span with the suspension cables, connecting between the top chords of the main span and side span stiffening trusses, and extending as outriggers from the trusses' bottom chords to the towers.

Farquharson and Coatsworth made their way safely to the bridge's eastern end just in time. When the heaving bridge deck broke free from the towers, the reporter's abandoned car and its lone passenger, the unfortunate Tubby, followed the plummeting center span into Puget Sound, 190 feet below.

In a mere seventy minutes, the suspension bridge suffered "the most dramatic failure in bridge engineering history." Today, the ruins of "Galloping Gertie" resting on The Narrows bottom are listed in the National Register of Historic Places.

In the wake of this calamity, engineers searched for ways to understand aerodynamic instability and to design safer and better suspension bridges. A new Tacoma Narrows Bridge, completed in 1950, reflected the first important discoveries and innovations in this effort. The lessons revealed by research benefited the entire structural engineering profession, contributing a wealth of data and knowledge about the role and importance of aerodynamics in suspension bridge design. Since the disaster, aerodynamic testing in wind tunnels has become a standard procedure in the structural analysis of suspension bridges.[14]

The undulating Tacoma Narrows Bridge deck was ripped apart by 42-mph winds on November 7, 1940, in the most spectacular bridge failure in United States history. The disaster led to a revolution in suspension-bridge engineering.
WSDOT

Hood Canal Floating Bridge, February 13, 1979

Not long after its opening in August 1961, stability problems already had appeared in the Hood Canal Floating Bridge. Then on February 13, 1979, a "storm within a storm" whipped winds between 80 and 120 miles per hour over the bridge. The storm raged for several hours, when around 7 a.m., pontoons on the bridge's 3,775-foot west half ripped free and sank or floated away. No lives were lost, but truck driver "Red" Taylor barely escaped.

Afterwards, consulting engineers concluded that the bridge failed mainly because of the "high magnitude of the dynamic loading from the wind generated waves [causing] stresses from wave action [that] alone exceeded the functional structural capacity" of the bridge. Also contributing to the span's demise were slippage of the anchors on the channel bottom, the ponding of water on the pontoon decks, and water entering the pontoons.

Within a month after the sinking, Governor Dixie Lee Ray signed legislation authorizing WSDOT to rebuild the bridge. Senator Warren G. Magnuson was instrumental in obtaining the needed federal funds for the replacement. The new west half was engineered to be three to four times stronger than the original. Because no design criteria existed for what was characterized as the bridge's "unique design and construction," WSDOT applied architectural and technological principles developed for the offshore oil industry to create a computer program for predicting responses of floating structures in heavy seas.

HAER drawing showing causes for the failure of the first Tacoma Narrows Bridge. Drawing 1993.
Wolfgang G. Mayr, HAER

A storm of unforeseen magnitude hit the Hood Canal Floating Bridge on February 13, 1979. Torn from their anchor cables, concrete pontoons sank or floated away.

Washington State Archives, WSDOT Records

Engineers then conducted tests on an "elastic model" of a new Hood Canal Bridge design at the Netherlands Ship Model Basin to assess the program's accuracy. The data gave bridge designers "a high degree of confidence" in their computer predictions. Those calculations indicated "wave loadings were much higher than ever predicted by previous designers." The new bridge consisted of several distinctive features: stronger, watertight pontoons divided into compartments accessible only through a single hatch; stronger mooring lines attached to anchors two to three times heavier and less prone to slippage than the originals; and open railings on the pontoon decks to avoid ponding during storms.

On October 25, 1982, the restored Hood Canal Floating Bridge opened to the public. First over the new bridge

The steel and concrete girder approach and a fallen truss were all that remained visible of the Hood Canal Bridge's west half after the storm of February 13, 1979. Cutting-edge marine engineering was used in designing the replacement spans, and the bridge reopened in 1982.

Washington State Archives, WSDOT Records

was "Red" Taylor, driving the same truck in which he had narrowly escaped death. His truck sported a banner fitting the spirit of the occasion: "I'm gonna make it this time."[15]

First Lake Washington Floating Bridge, November 25, 1990

The world's first floating concrete bridge had served the region faithfully for half a century when disaster struck. Completion in 1989 of a second floating bridge over Lake Washington, the Homer M. Hadley Floating Bridge, meant the fifty-year-old span could be closed for making major repairs. With the bridge closed to traffic, pontoon renovation began in late spring 1990.

In the midst of the $36.5 million renovation, the bridge that made history became history. On November 22 and 23, after workers went home for Thanksgiving break, a weekend storm with high winds and heavy rain sent waves pounding into the bridge's sides. Water poured over the roadway and flooded one pontoon. Efforts to pump out the water failed. On Sunday, November 25, a chain reaction took down one pontoon after another. The bridge sagged; sections broke off and sank. By the end of the day, seven pontoons had sunk, and the entire floating section had to be scrapped.

The sinking pontoons severed thirteen anchor cables on the nearby, newly finished Interstate 90 Homer M. Hadley Bridge. Despite initial concern that it too might sink, it remained intact. Fortunately, no fatalities or injuries resulted from the bridge's collapse. Extensive study of the mishap by a blue-ribbon governor's commission and the state attorney general's office concluded that fault for the bridge failure lay with both the WSDOT and the contractor.[16]

Today, all that remains of the original bridge are the approaches to the transition spans on the new structure that replaced the original bridge in 1993. These pier-supported, 215-foot, steel tied-arch spans provide a vertical clearance of thirty-five feet for the passage of small craft, and they are the bridge's most visually distinctive features.

Pontoons sinking as emergency personnel watch from the adjacent Homer M. Hadley Floating Bridge, November 25, 1990.

WSDOT

NOTES

1. Henry Petroski, *To Engineer Is Human: The Role of Failure in Successful Design* (New York: St. Martin's Press, 1985), 161. See also Petroski, *Design Paradigms: Case Histories of Error and Judgment in Engineering* (New York: Cambridge University Press, 1994), and David Plowden, *Bridges: The Spans of North America* (New York: W. W. Norton, 1974).
2. Louise McKay, "Drawbridges across the Okanogan," *Okanogan County Heritage* 17, no. 3 (Summer 1979), 3–7.
3. Ibid., 7–12.
4. "Point of Weakness in Low-truss Highway Bridges," *Engineering News* 69, no. 11 (March 13, 1913), 523–24.
5. Correspondence to R. W. Finke, Washington State Department of Highways, June 25, 1947, Department of Highways Records, Washington State Archives, Olympia.
6. Camilla G. Summers, *About Kelso: An Historical Gem* (Kelso: Self-published, 1982), 41–42; Greg Kolle, former WSDOT Bridge Engineer, personal communication, 2003.
7. Margot Vaughn, ed., "The Modrow Bridge," *Cowlitz County Quarterly* 26 (Spring 1984), 35–36.
8. "Wenatchee Bridge Burned," *Wenatchee Daily World*, July 4, 1917; "Temporary Crossings First to be Arranged," *Wenatchee Daily World*, July 6, 1917; Robert H. Krier and Craig Holstine, "Wenatchee Bridge No. 285/20E National Register of Historic Places Eligibility Evaluation," Archaeological and Historical Services Short Report DOT 98-20, Eastern Washington University, Cheney, May 1998.
9. Portland *Oregonian*, February 7, 1947; *Daily Okanogan*, August 6, 1968; *The Olympian*, March 5, 1998.
10. Issue of June 29, 1906, quoted in Jessie Schmidt, "Methow Valley Fords and Bridges," *Okanogan County Heritage*, December 1968, 7–10.
11. Darrell Smith and Wesley Karney, "Salmon Creek Bridge," *Highway News* 5 (March 1956), 8–9, 28.
12. Lisa Soderberg, "Chow Chow Bridge," HAER Inventory, 1979; Richard Wells, Director of Planning and Development, Quinault Indian Nation, response to WSDOT Historic Bridge Inventory correspondence from Craig Holstine, 1993; Robert H. Krier and Craig Holstine, "An Assessment of the Current Status and Condition of Bridges and Tunnels in Washington State Listed in the NRHP," Archaeological and Historical Services Short Report DOT 93-10, Eastern Washington University, Cheney, August 1993, 14.
13. Jay Burney, City of Olympia, personal communication, November 2004.
14. Sources for the failure of the 1940 Tacoma Narrows Bridge are abundant. Most notable are WSDOT, "Tacoma Narrows Bridge History," www.wsdot.wa.gov/TNBhistory (narrative by Richard Hobbs); Richard Scott, *In the Wake of Tacoma: Suspension Bridge Design and the Search for Aerodynamic Stability* (Reston, Virginia: ASCE Press, 2001); Valerie Sivinski, Penny Chatfield Sodhi, and John M. Simpson, "Tacoma Narrows Bridge Ruins," NRHP Nomination, 1991; Advisory Board on the Investigation of Suspension Bridges, *The Failure of the Tacoma Narrows Bridge*, a reprint of original reports; "A Contribution to the Work of the Advisory Board on the Investigation of Suspension Bridges by the United States Public Roads Administration and the Agricultural and Mechanical College of Texas," School of Engineering, Texas Engineering Experiment Station, College Station, Texas, 1944; Walter A. Averill, "Collapse of the Tacoma Narrows Bridge," *Pacific Builder and Engineer* 46 (December 1940), 20–27.
15. Principal sources for the Hood Canal Floating Bridge failure include: Craig Holstine, "Hood Canal Floating Bridge," NRHP Determination of Eligibility, WSDOT, Environmental Affairs Office, 2002; Richard Hobbs, "Hood Canal Floating Bridge," HAER report, June 2004; A. D. Andreas, "SR 104 Hood Canal Bridge: Task Force Preliminary Report, July 1979, and Summary Report, October 1979," WSDOT, 1979; C. S. Gloyd, "Concrete Floating Bridges," *Concrete International*, May 1988, 17–24; Hood Canal Bridge Facts, WSDOT Web site.
16. Principal sources for the first Lake Washington Floating Bridge failure include: "The Fall of the Bridge," Special Report section, *Seattle Post-Intelligencer*, December 20, 1990; *Seattle Times*, November 26 and 27, 1990; "Panel Finds Holes Helped Sink I-90 Bridge," *Seattle Times*, May 2, 1991, A1; M. Myint Lwin and Donald O. Dusenberry, "Responding to a Floating Bridge Failure," *Public Works*, January 1994, 39–43; Donald Dusenberry, "What Sank the Lacey Murrow?" *Civil Engineering* 63, no. 11 (November 1993), 54–59.

Loss and Preservation

Many losses of bridges officially deemed "historic" have been unavoidable, although ardent bridge preservationists might argue otherwise. Age, deterioration, and obsolescence have caused the removal of most historic bridges lost in the last half century. Bridge replacement is a fact of life. Not all structures can be saved. Rehabilitation costs can be a deterrent, and engineering challenges posed by safety considerations and increased capacity needs oftentimes drive decisions to remove bridges. Thus, numerous historic bridges in Washington have met unavoidable fates determined by factors ranging from demographic dynamics to physical deterioration.

A complete listing of all historic bridge demolitions in the state is not available. However, at least eighteen National Register of Historic Places bridges (listed or eligible), and many more so-called "Category 2" bridges, have been removed in the past twenty-five years. Some of the more notable examples of recent losses are worth mentioning. These lost bridges—most of which were listed in or determined eligible for the NRHP—appear below in order of construction dates.

So as not to imply that no historic bridges have been saved or rebuilt, we also offer some recent examples of historic bridge success stories. Award-winning rehabilitations are part of our bridge history. Typically, these resulted from compliance with the National Historic Preservation Act and with the ethic it has inspired. Public agencies and their private consultants have earned prestigious awards for using modern materials in ways designed to preserve an historic structure's original appearance. Preserving a sense of the past is not always easy, however, especially where safety and modern traffic loads are concerned.

Losses of Historic Bridges

"F" Street Bridge, City of Palouse, Whitman County

When constructed in 1901, this structure was known as the "Palouse Flour Mill Bridge," due to its proximity to a facility that dominated the eastern skyline of the small farming community of Palouse. After the mill burned in 1923, the bridge took on the name of the nearest north-south street, although F Street did not actually cross the bridge. The short, steep approach from Fourth Street presented a minor traffic hazard when the bridge was built, but became a greater problem as transportation evolved into the automobile age.

A seven-panel, pin-connected, steel Pratt through-truss, the "F" Street Bridge spanned 140 feet across the Palouse River. Its fifteen-foot, nine-inch-wide wood-plank deck had no sidewalks or curbs. Resting on cylindrical steel piers filled with concrete, the floor system consisted of timber stringers attached to steel I beams. By 1990, the truss showed its age in metal fatigue, with the west truss being out of alignment and slightly twisted. A split stringer, plus a narrow roadway and load capacity that prevented grain trucks and farm equipment from crossing, led the City of Palouse to replace the bridge. At that time, it was one of the two oldest Pratt truss bridges in the state's highway system.[1]

The "F" Street Bridge (originally "Palouse Flour Mill Bridge"), constructed in 1901, was a steel Pratt through-truss.

Office of Archaeology and Historic Preservation

The 120-foot long McClure Bridge was a Pratt through-truss built in 1908 or early 1909. Extensive renovations over the years replaced most of the original components before its removal in 1989.

Office of Archaeology and Historic Preservation

McClure Bridge, Whitman County

Situated only four miles northwest of Palouse, the McClure Bridge also crossed the Palouse River. Constructed of timber probably cut at the Potlatch Mill in Palouse, the bridge was a Pratt through-truss, a popular design of the times, although normally built using iron and steel. The single-span structure was 120 feet long, 18 feet wide, and its truss was 20 feet high. O. H. Horton of Colfax constructed the bridge in late 1908 or early 1909. Extensive renovation occurred over the years, involving replacement of most of its original materials.

A 1988–89 study noted that "the steel loop welded eyebars and the adjustable cylindrical rods which serve as diagonals and counters in the Pratt configuration are the only remaining original members." With the passage of time, the bridge could not support modern loads, and its width restrictions reportedly encouraged some farmers to drive their combines through the river. Whitman County replaced the bridge in 1989.[2]

Orient Bridge, Ferry and Stevens Counties

When miners filed the first claims in the area in 1896, the future town site of Orient was still within the Colville Indian Reservation. A few years later, the area's substantial mineral wealth prompted the U.S. president to cut the reservation in half, opening the "north half" to settlement by non-Indians. The Townsend brothers renamed the "Never Tell" claim as the "Orient Mine," reportedly for the numerous Chinese miners who had panned gold along the Kettle River several decades before. The Orient and several other mines were soon producing ore, and a 320-foot-long wooden bridge met the need for a connection to the bustling new town of Orient.

High water in May 1902 swept the wooden structure "majestically downstream," as the local newspaper reported. A Howe truss replaced it the following year. Ore-laden wagons and other traffic rapidly weakened that bridge. In 1909 a third bridge was erected on the site, crossing between Ferry and Stevens counties. Ferry County bore the cost of the new bridge and awarded the construction contract to the Charles G. Sheely Company of Denver, Colorado, a large firm operating throughout the western United States. The company purchased completed trusses directly from steel foundries in the East and Midwest (something local firms apparently could not do) and shipped them by rail to construction sites. The Orient Bridge's Parker truss arrived via the Great Northern Railway.

The 180-foot, pin-connected Parker truss and its approaches were repaired and strengthened numerous times over the years. Both Ferry and Stevens counties contributed to its upkeep, but its loading capacity (H-9) was inadequate for the logging trucks and other commercial vehicles crossing the river to reach Spokane and points south, or Canadian destinations, via US 395. In 1992 the bridge was removed and replaced with a new structure.[3]

The Orient Bridge, a 180-foot long Parker truss span completed in 1909 over the Kettle River, was replaced in 1992, shortly after this photo was taken. Harvey S. Rice, photographer.

Office of Archaeology and Historic Preservation

Grant Avenue (Prosser Steel) Bridge, City of Prosser, Benton County

This early example of the Parker truss type in Washington was built over the Yakima River at Prosser in 1911. D. C. Maloney erected the three-span, 458-foot structure. It consisted of two pin-connected, 189-foot Parker through-trusses and one pin-connected Pratt steel pony truss.

The Grant Avenue Bridge in Prosser served the area for seven decades after its completion in 1911. This Parker truss span over the Yakima River was replaced in the mid-1980s.

Office of Archaeology and Historic Preservation

The bridge's narrow, timber deck and eight-ton load limit, and the substandard north approach, limited its usefulness and hastened replacement in the mid-1980s. Prior to its removal, the structure was the only existing highway bridge in Washington that demonstrated the once-common practice of using pony truss approaches in combination with longer main span trusses. It was also the oldest two-span, pin-connected Parker truss in the state.[4]

Jack Knife Bridge, Snohomish County

The Jack Knife Bridge was an unaltered example of a Strauss heel trunnion bascule bridge. A few movable bridges of this type were built in Washington in the early twentieth century, but this one was the first to be installed for highway use. (Its construction in late 1913 and early 1914 preceded that of the Wishkah River Bridge, also a Strauss heel trunnion bascule built in Aberdeen in 1924.) The structure was manufactured by the Milwaukee Bridge Company and shipped west via railroad. Erected on the west bank of Ebey Slough in Everett by the International Contract Company of Seattle, the bridge was operated manually "by one man on a hand crank." Its 112-foot-long, 18-foot-wide lift span provided 100 feet of clearance in the slough for boat passage.

The 1914 Jack Knife Bridge was the state's first bascule span built for highway use. It has been demolished.

Office of Archaeology and Historic Preservation, Olympia

In the 1950s, the government declared Ebey Slough to be non-navigable. Subsequently, the lift mechanism was neglected, although it was raised with the aid of "make-shift power" in 1969. That year a study concluded that a new bridge was needed because the Jack Knife could not be brought up to current requirements or modern traffic needs. The bridge was later removed.[5]

Wishkah River Bridge, Grays Harbor County

This bridge was not the movable bridge by the same name currently standing in Aberdeen, but rather a steel through-truss constructed in 1915 in a remote location in Grays Harbor County. Designed by F. D. Sheffield, it was reportedly a "variation of the Warren truss," with "members which act in both compression and tension." The 120-foot, riveted-steel, "double-intersection" Warren through-truss had two timber approach spans. Asphalt covered the fourteen-foot-wide, wood-plank deck. The bridge was believed to be the only one of its kind in the state. It was removed in the 1980s.[6]

Built in 1915, the Wishkah River Bridge was a double intersection Warren through truss. It served a remote section of Grays Harbor County until its removal in the 1980s.

Office of Archaeology and Historic Preservation

Mellen Street Bridge, City of Centralia, Lewis County

Built in 1911, this bridge earned notoriety in Washington history on November 11, 1919, when World War I veterans hanged Industrial Workers of the World member Wesley Everest from the structure. This occurred after a parade organized and led by the Centralia Post of the American Legion to celebrate the first anniversary of World War I's end turned into an assault on an IWW office. Armed "Wobblies," as the IWW men were known, opened fire, killing three veterans and wounding several more.

The Mellen Street Bridge in Centralia, built in 1911, from which Wesley Everest, Industrial Workers of the World member, was lynched on November 11, 1919. The bridge was replaced in the 1950s.

Lewis County Historical Museum, Chehalis

After police jailed several Wobblies, the mob dragged Everest from his cell, drove to the bridge west of town, and hanged him from the structure's narrow wooden deck over the Chehalis River. The original steel truss bridge stood just west of Interstate 5, Exit 81, in Centralia. The city replaced the structure in 1959 with a modern steel truss.[7]

Novelty Bridge, a Pennsylvania Petit truss over the Snoqualmie River, was completed in 1920 and removed in 1999. John Stamets, photographer, 1999 photo.
Office of Archaeology and Historic Preservation

Novelty Bridge, King County

C. C. Snyder and Company of Seattle built this Pennsylvania Petit bridge over the Snoqualmie River in 1920. Unlike the straight, horizontal top chord of the Pratt truss, the Pennsylvania truss had a polygonal top chord. The company used both pinned and riveted connections in constructing what renowned engineer J. A. L. Waddell referred to as a "hybrid" truss. In this bridge, the vertical posts and top and bottom chords were riveted, while the diagonal braces were pinned.

The bridge served King County Road 404, now 124th Street. Workers removed a wooden truss span to make way for the new bridge at the site. They then constructed a wooden trestle to support the new steel structure until concrete piers could be installed. Over the years, the bridge underwent numerous repairs and replacement of parts, including top lateral rods, timber deck stringers, and deck planking, which was eventually covered with asphalt. Despite the improvements, the structure's insufficient load-carrying capacity and inadequate deck width led to its removal in 1999.[8]

Pasco-Kennewick Bridge, Benton and Franklin Counties

In the late 1970s preservationists failed in their efforts to save the Pasco-Kennewick Bridge across the Columbia River. Constructed in 1922, it originally consisted of wooden trestle approaches connected to Warren and cantilever through-trusses. The trestles were later replaced with steel deck trusses. With the establishment and expansion of the nearby Hanford atomic facilities during World War II, traffic between the two cities mushroomed. Local commuters unaccustomed to delays demanded a modern bridge after a spilled hay load stopped traffic on the narrow structure. A new bridge, the so-called "Blue Bridge,"

The eventual removal of the 1922 Pasco-Kennewick Bridge (foreground) left an unobstructed view of the nearby Ed Hendler Bridge.
WSDOT

was then built a few miles west. The old bridge remained in place for more than twenty years until construction of the Ed Hendler Bridge immediately adjacent. Although monumental in length, the older bridge suffered elegance envy in the shadow of its cable-stayed neighbor. After prolonged efforts to save it, the 1922 bridge was removed.[9]

Dosewallips Bridge, Jefferson County

The Dosewallips Bridge, completed 1922–23, was a pin-connected Pennsylvania Petit truss. In 1946 its portals were raised about seven feet in hopes of reducing logging truck damage. Jet Lowe, photographer, 1993 photo.

HAER

Designed by the Washington Department of Highways for the U.S. Bureau of Public Roads, this bridge constituted a vital link in the Olympic Peninsula Highway, now State Route 101. The structure included a 240-foot, pin-connected Pennsylvania Petit truss. The truss arrangement on the Dosewallips Bridge seems to have reduced the amount of steel required for construction. Ward and Ward Inc. built the structure in 1922–23.

In 1946 the portals were raised approximately seven feet to provide additional vertical clearance. That proved insufficient, however, as serious collisions involving loaded logging trucks continued to damage the portals and sway braces for the next forty-five years. Portals and sway framing were raised repeatedly, but collisions continued. Hairline cracking of the concrete T-beam approaches and deck deterioration were factors in WSDOT's decision to replace the bridge.[10]

Latah Creek Bridge, Spokane County

Until the late 1990s, an attractive, open-spandrel, concrete arch bridge complemented the rolling Palouse hills scenery surrounding the farming community of Waverly in south Spokane County. Constructed in 1924, the bridge's parabolic arch spanned Latah Creek, which had washed away the first wooden bridge built there in 1912. A steel structure that replaced it sufficed for more than a decade before Spokane County decided to replace it. The Board of County Commissioners chose J. C. Broad's bid for building the arch bridge over his lower-cost bid to construct a "girder" bridge (probably of steel). Commissioners may have considered concrete to be a more cost-effective alternative, given the anticipated maintenance needed for a steel structure, or simply wanted to support a local gravel quarry that would supply aggregate.

The 1924 Latah Creek Bridge, an open-spandrel concrete arch, survived until the 1990s, when it was replaced. Pamela K. McKinney, photographer, 1998 photo.

Office of Archaeology and Historic Preservation

The bridge consisted of two steel-reinforced, concrete arch spans fifty-seven and sixty-three feet long and a concrete girder approach span. The spans met on a concrete pier near the creek's center. Spandrel columns rising from the arch ribs supported a concrete-slab roadway without sidewalks or curbs. An attractive, concrete, baluster-type handrail with arched openings adorned the deck. Although sturdy enough to carry mostly agricultural traffic on Prairie View Road, the bridge was too narrow for modern combines. One railing was broken, apparently due to a vehicular collision.

After seventy years of exposure to the elements, the structure suffered from deterioration of its concrete-and-steel reinforcing components. Perhaps improper ratios of water and cement contributed to the inadequately consolidated concrete, as evidenced by spalling and cracking. Spokane County replaced the structure.[11]

Mora Road Bridge, Clallam County

To ensure a secondary or detour route from State Route 101 to the small communities of Mora and Rialto Beach, Clallam County contracted with Oliver S. Reed and Creech Brothers Contracting Company of Aberdeen to build a bridge across the Soleduck River. The new route served the area's few residents well when another Soleduck River Bridge on the Quillayute Road was later replaced and the U.S. Navy closed that road during World War II

Mora Road Bridge, built in 1935 across the Soleduck River in Clallam County, was a 219-foot pin-connected steel Petit truss span. Craig Holstine, photographer, 1998 photo.

Office of Archaeology and Historic Preservation

while pilots destined for aircraft carrier duty trained on a nearby airfield.

The Mora Road Bridge was constructed in 1935, apparently using a steel company's standardized plans. This 219-foot, pin-connected, steel Petit truss structure was a less impressive example of the same type of bridge on the Dosewallips. The Mora Road Bridge portals were raised in 1946 to provide additional vertical clearance, but damage from logging trucks continued. The county replaced the bridge in the late 1990s.[12]

Outlet Creek Bridge, Pend Oreille County

Outlet Creek drains picturesque Sullivan Lake in the state's extreme northeastern corner. The Outlet Creek Bridge was one of many such structures built in our national forests by the Civilian Conservation Corps during the Great Depression. Shortly after establishing Camp Sullivan Lake at the lake's north end, CCC enrollees and Pend Oreille County laborers graded a gravel road to connect the camp with the nearest town, Metalline Falls, a few miles to the west. By 1935, CCC crews were at work on another road, this one along the lake's west side that led to the town of Ione several miles south. The new road required a bridge across Outlet Creek.

The Outlet Creek Bridge was built by CCC workers in 1935. Harvey S. Rice, photographer, 1993 photo.

Office of Archaeology and Historic Preservation

Engineers of the U.S. Forest Service Region One office in Missoula, Montana, prepared plans for the structure. The 190-foot timber trestle consisted of ten 19-foot spans, with a nail-laminated wood-plank deck 12 feet wide. Ten bents of five piles each (plus one bent of three piles) supported six-by-sixteen-inch timber stringers measuring twenty feet long. All wood, except that used on the railings, was Douglas fir, pressure treated with creosote oil. By May 1935, the CCC men and their Forest Service supervisors, with a contingent of county laborers, had completed the bridge.

Although the bridge's substructural elements were replaced in 1981, and new running planks installed from time to time on the deck, deterioration had become widespread by the early 1990s. An estimated seventy-five percent of the substructure suffered from advanced decay. That and the bridge's load-bearing capacity being substantially below modern safety requirements led the Forest Service to replace the bridge with a new concrete structure in 1994.[13]

First built in 1924, the Norman Bridge was a timber Howe truss reconstructed in 1950 at a new site some forty-five feet downstream on the Middle Fork of the Snoqualmie River. John Stamets, photographer, 2004 photo.

King County Department of Transportation, Seattle

Norman Bridge, King County

Henry and Bessie Norman paddled their canoe to North Bend for supplies until 1909, when a timber suspension bridge was built across the Middle Fork of the Snoqualmie River to provide access to logging camps up the valley. Replaced in 1924 by a timber Howe truss, that bridge partially collapsed in a 1944 flood. As a cost-saving measure,

King County rebuilt the structure in 1950 using timber from the old bridge at a new site forty-five feet downstream. By the 1980s, the bridge became too deteriorated to remain open to vehicular traffic and was transferred to the county Parks Division for use as a pedestrian bridge. Finally, in October 2004, the bridge was dismantled, a victim of Western Washington's climate.

The Norman Bridge measured 295 feet in length. Its 171-foot Howe through-truss consisted of vertical metal components and diagonal treated timbers. Four cedar pile trestles supported approach spans on both sides of the truss. An asphalt surface covered the three-inch cedar planks in the sixteen-foot, eight-inch wide deck, which accommodated two lanes of traffic. The Norman Bridge was one of the last wooden through-truss bridges to carry automobile traffic in Washington.[14]

Donald-Wapato Bridge, Yakima County

Designed by renowned bridge engineer Homer M. Hadley, the Donald-Wapato Bridge was one of the earliest multiple-cell, concrete, box-girder bridges in the state. It carried traffic from the village of Donald (a.k.a. Parker Bottom) across the Yakima River to the town of Wapato. At least four bridges preceded the Hadley structure at this difficult river crossing on an important fruit-hauling route and community connection. Floodwaters washed away the first three bridges and seriously undermined the fourth, a steel bridge consisting of two Pratt trusses. When federal support again became available after World War II, Yakima County received funds to replace the structure.

In 1947 the county hired Hadley, who had just opened his own consulting firm. Benefiting from his years of experimenting with concrete box construction while with the Portland Cement Association, Hadley borrowed the hollow-box idea from Europe, where concrete girders were made by pouring cement around hollow-box forms. For the Donald-Wapato Bridge, he produced a design that featured two single-cell boxes joined by a seven and one-half-inch-thick concrete slab serving as the bridge deck.

The Donald-Wapato Bridge, one of Washington's first concrete box girder spans, was designed by Homer M. Hadley in 1947. Lawrence M. Jacobson, photographer, 1994 photo.
WSDOT

Consisting of three main spans totaling 247 feet and a 41-foot approach span, the bridge had metal handrails with cross-hatched patterns along both sides of its 24-foot-wide roadway. A load-rating analysis revealed the bridge's capacity was inadequate for current needs, and consequently the county replaced the bridge in 2004.[15]

Preservation of Fabric and Design

Fortunately, not all bridge stories end unhappily for preservationists. Increasing public awareness of the values of historic preservation, both for enhancing our quality of life and for potentially saving taxpayer dollars, has helped create vocal constituencies for retaining historic bridges. Federal, state, and local funding sources have combined to save many of the state's oldest and most significant bridges. It could be said that all older bridges still in use are the result of historic preservation efforts. All have received maintenance, repairs, and modifications to varying degrees. Below are but a few examples of the more intensive efforts to retain historic bridges in Washington.

Saving Our Timber Bridges

Grays River Covered Bridge, Wahkiakum County

One of the state's last covered bridges remains in service immediately south of State Route 4, several miles inland from the mouth of the Columbia River. The structure was completely rebuilt by WSDOT in 1989. It is believed that the firm of Ferguson and Houston of Astoria, Oregon, originally designed the bridge, which was built by local residents donating their skills and labor.

The 1905 bridge was constructed as an open Howe pony truss at a cost of $2,700. Three years later, it was

Grays River Covered Bridge, Wakiakum County.

Washington State Archives, WSDOT Records

The Grays River Covered Bridge, one of Washington's last covered bridges, was completely rebuilt by WSDOT in 1989.

WSDOT

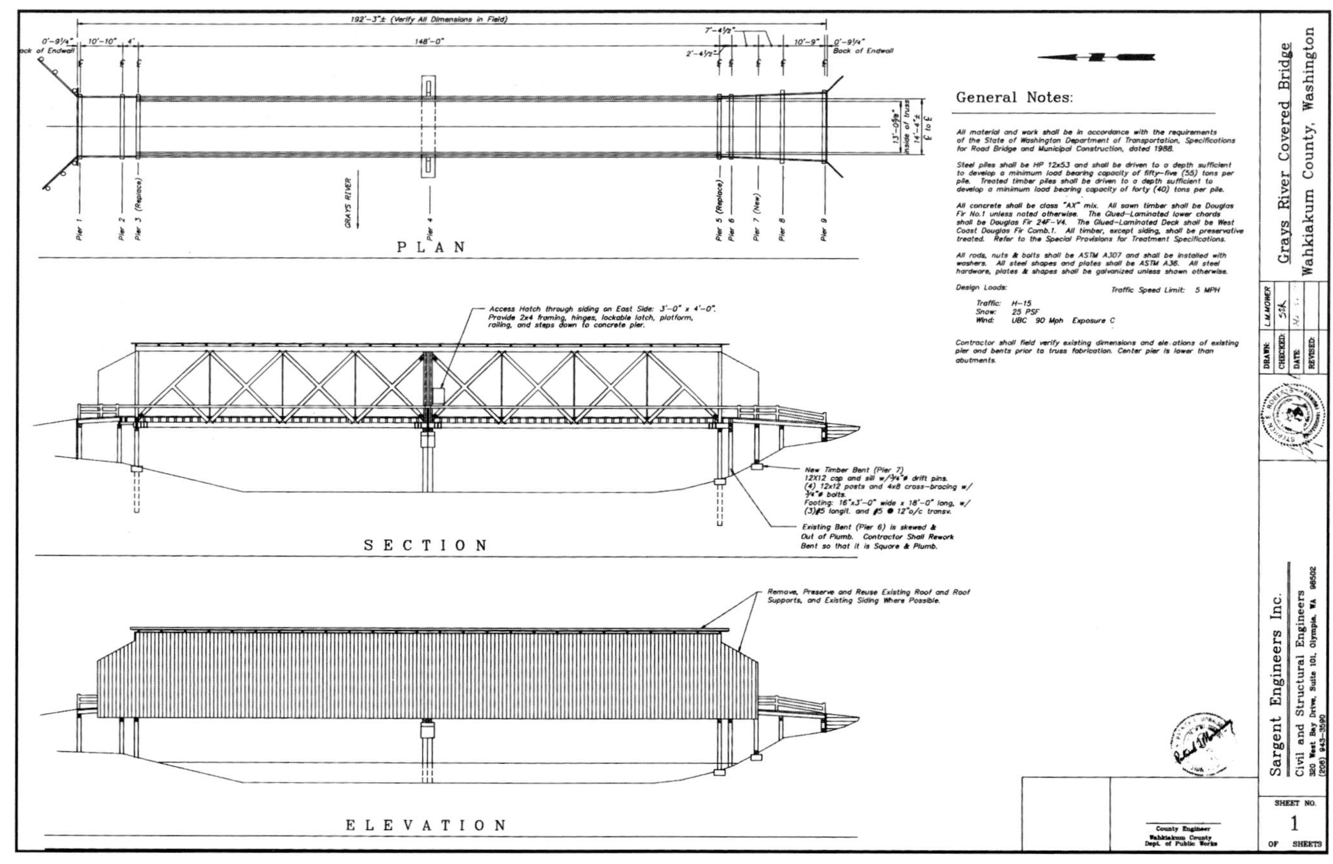

Plans for rebuilding the Grays River Covered Bridge. Sargent Engineers Inc., Olympia, 1989.

WSDOT

enclosed in cedar siding and covered with a tin roof to protect its untreated timbers from the heavy rains typical of southwest Washington. Farmers hauling dairy products to market constituted the bulk of traffic over the bridge in its early years.

In the late 1930s the single span began to sag. Consequently, a center pier was added for support. The pier was replaced in the 1940s and a new timber deck installed in 1952–53. A steel deck was added and a new approach constructed in 1964, and the center pier was again replaced in 1972. Wahkiakum County received a $10,000 grant in 1976 to repair the roof, sidewalls, abutments, and pilings, and the center pier was again replaced in 1984. By 1986, the pier had washed out, and the bridge was closed to traffic. Local citizen support and the 1989 State Centennial Celebration contributed to the decision to completely rebuild the structure.

To meet an H-15 loading threshold, Sargent Engineers of Olympia designed a two-span, timber through-truss that, when sided with board, batten, and cedar shingles and covered with a metal roof, looked virtually identical to the original bridge. Glue-laminated timbers were used in the upper and lower chords, and steel beams framed the portals, all to provide the necessary sway bracing for adequate wind loading. Tension rods, bearing plates, and the exterior boards that could be salvaged were used in the refabrication, which Dulin Construction of Centralia completed for $295,980 in 1989.

Its structural dimensions are as follows: 155.5 feet long; 22.5 feet high, with a 16.75-foot interior clearance; 14 feet wide, with a 12.5-foot lane width; and "porch" awnings with cedar shakes over the portals, providing 12.7 feet of clearance on the approaches. The reconstruction has given what the community once called "Sorenson's Covered Bridge" an estimated fifty years of longevity. For now, at least, it can be said that Washington still has three historic-era covered bridges—the Grays River, Pe Ell, and Harpole structures.[16]

Baring Bridge, King County

Yet another rare bridge type in Washington spans the South Fork Skykomish River in north King County. Known as the Baring Bridge for the small settlement about half a mile to the west, the timber suspension bridge stands in one of the state's most beautiful settings. In 1899 a mining company erected a suspension bridge here. It lasted (in rebuilt condition) until 1952, when the cables failed and its suspension span collapsed into the river.

The structure was rebuilt with stronger anchorages and stiffening trusses in 1953 and new towers in 1958. Rehabilitation in 1978 included two additional main suspension cables and anchorages. Nevertheless, the bridge faced replacement by the end of the decade. When King County

The Baring Bridge in King County is a 334-foot timber suspension structure with a 272-foot main span. The only remaining vehicular timber and cable suspension bridge in Washington, it has been rebuilt several times, most recently in 1994. Robert H. Krier, photographer, 1995.

King County Office of Cultural Resources, Seattle

announced plans to replace the timber suspension bridge with a concrete structure, local citizens objected.

In response to public pressure, King County rebuilt the single-lane, 334-foot timber bridge in 1994. The bridge now has an increased load capacity and widened deck, with its historic appearance largely preserved. The 272-foot timber truss main span is supported by twin cables suspended between two timber towers. Smaller suspender cables are attached to the "glulam" (glued-laminated) floor beams with steel eyebolt and ring connections. Wood-slat railings replaced chain links, enhancing the bridge's aesthetic appeal.

The Baring Bridge stands as a testament to the success of extensive reconstruction in maintaining basic form and function. Today, it is the only vehicular wood-and-cable suspension bridge left in Washington. The bridge is also a source of community pride.[17]

Chief Joseph Dam Bridge, Douglas County

One of the state's last remaining wood-deck truss bridges recently received a new lease on life. The U.S. Army Corps of Engineers originally constructed the Chief Joseph Dam Bridge (so named for its proximity to Chief Joseph Dam) in 1958 across Foster Creek, a Columbia River tributary. The 126-foot Howe deck truss was the only structure of its kind built in the 1950s in Washington for highway use. By the late 1990s, it needed numerous repairs and improvements.

Given the bridge's eligibility for NRHP listing, Douglas County embarked on an ambitious rehabilitation effort designed to increase load capacity, bring the structure up to current safety standards, and preserve the bridge's historic character. The timber railings failed to meet modern safety requirements, and the deck width of twenty-three feet was considered too narrow, causing the bridge to be posted for single-truck crossings.

Rehabilitation entailed the replacement of the deteriorated deck with a fiber-reinforced polymer deck, the first of its kind in Washington State. Use of this innovative material allowed for widening the deck without a significant reconfiguration of the bridge's deck truss. The structure's load-carrying capacity was not reduced and its historic character maintained.

"This is an excellent example of selectively using advanced materials to meet significant project challenges resulting in an efficient and successful project," said bridge engineer Barry Brecto, of the Federal Highway Administration, Washington Division.

Funding for the Chief Joseph Dam Bridge rehabilitation project included $436,000 in local money and approximately $2.4 million in federal funds.[18]

The 1958 Chief Joseph Dam Bridge is a 126-foot Howe deck truss.

Douglas County Public Works, Waterville

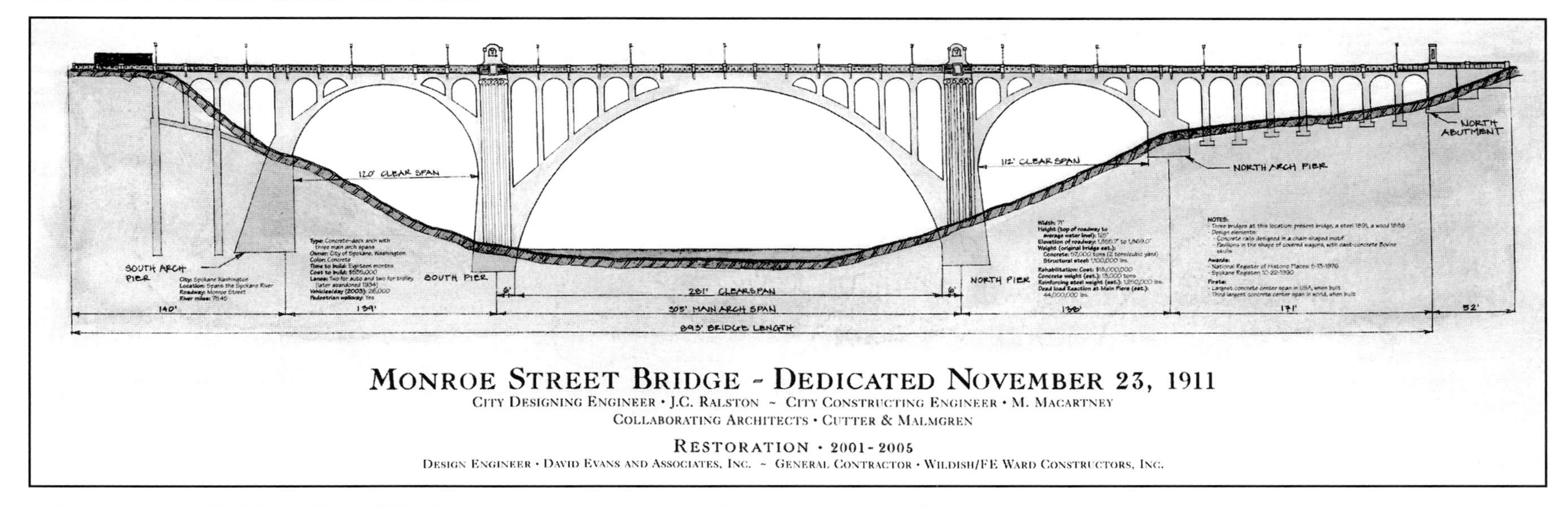

The Monroe Street Bridge in Spokane, completed in 1911. E. Michael Beard, drawing. © 1999 www.errolgraphics.com.
Errol Graphics, Portland, Oregon

Monroe Street Bridge, City of Spokane, Spokane County

Framing the lower Spokane River Falls with its elegant, open-spandrel, concrete arch, this bridge has been one of the defining features of the city of Spokane since the structure's completion in 1911. Its indelible image has become part of the community's logo, gracing city stationery with a blend of history, engineering, and natural beauty in an urban setting.

Ninety years after its construction—although still the gem in the city's crown—the Monroe Street Bridge neared the end of its service life. Faced with costly repairs that would preserve the original fabric, but leave structural problems unresolved, city engineers opted to implement what has been characterized as a "reconstruction in replica." The plan calls for the bridge to be rebuilt to allow widening from four to six lanes while essentially retaining the structure's original appearance.

The rebuild includes removal of all but the 305-foot, double-rib main arches, and replacing the 900-foot deck, two 120-foot arch spans, and the approaches with precast and cast-in-place concrete. Even the Conestoga wagon-shaped pavilions adorned with cast-concrete bison skulls will be reconstructed on the deck, thus preserving the contribution of the city's most notable architects, Cutter and Malmgren. According to projections by designers David Evans and Associates Inc., the rebuild will extend the structure's life another seventy-five years.[19]

Monroe Street Bridge undergoing renovation, March 2005, Spokane.
Craig Holstine, WSDOT

Henry Thompson/Baker River Bridge, Town of Concrete, Skagit County

The town of Concrete sits at the confluence of the Baker and Skagit rivers in the North Cascades. Its name is derived from the concrete manufacturing that employed the majority of its working population, which peaked at 1,500 in 1912. The Baker River Bridge was built in 1917 at the low cost of $21,946. At the time, this open-spandrel, reinforced-concrete arch was reportedly one of the longest single spans of its type. A steel truss was considered for the site, but two local concrete producers donated cement to ensure the town would have a bridge composed largely of its primary industrial product. The year after the structure's completion, a logging train killed the bridge's most ardent backer, Henry Thompson. On June 15, 1918, citizens dedicated the bridge in his honor.

Over the years, climate and age took a toll on the bridge. To correct extensive deterioration, Concrete officials obtained funding from the Federal Highway Administration and launched an extensive restoration project to return the bridge to its original structural integrity and beauty. In 2004, along with additional partners—WSDOT, the Washington State Transportation Improvement Board, and Puget Sound Energy—the city transformed the bridge, replacing the deck and concrete railings, inserting steel reinforcing into the forebeams and girders, and restoring the historic lighting with luminaries and pillars donated by PSE. Rehabilitation costs were expected to exceed $2 million.[20]

The 1917 Henry Thompson/Baker River Bridge at Concrete received extensive renovation, completed in 2004.
Skagit County Public Works, Mt. Vernon

Winnifred Street Bridge, Town of Ruston, Pierce County

The 1941 Winnifred Street Bridge in Ruston is a concrete box girder 215 feet long. After substantial rehabilitation by WSDOT, the bridge reopened in June 2003.
WSDOT, Tacoma Project Office

The small community of Ruston sits on the shores of Puget Sound within the northwest part of Tacoma. Before 1941, residents traveling to the nearby Point Defiance-Tahlequah Ferry to Vashon Island had to cross two bridges predating the present structure at this location. In 1941 contractor S. R. Gray built this 215-foot-long, concrete, box-girder bridge over the Northern Pacific Railway's spur line and tunnel located at the bottom of a deep ravine. At seventy-four feet above the railbed, the bridge is unusually high for a box-girder structure, and is believed to be the second of its type built in Washington. (The Purdy Bridge in northwestern Pierce County is the oldest.)

Recent inspections revealed an undermined abutment wall, and consequently the city closed the bridge. WSDOT then hired OTAK, a consulting engineering firm, and secured state and federal funding to correct the deficiencies. Concrete "Texas ornamentation" type guardrails replaced the original substandard rails, the hillside was stabilized behind the abutment, and period-style lighting was installed. The bridge also received a new deck surface after the old concrete was pulverized by an innovative method using high-pressure jets of water that left aggregate and reinforcing steel virtually unaltered. The Winnifred Street Bridge reopened in June 2003.[21]

Saving Our Steel Bridges

Howard Street (Steel) Bridge, City of Spokane, Spokane County

Howard Street crosses over three channels of the Spokane River above the lower falls. Spokane built its first bridge at this difficult crossing in 1881. The location was determined not by engineers, but by donors who raised the most money (as taxation had not yet come to the unincorporated settlement). S. G. Havermale's $500 meant that the bridge would cross to the island bearing his name. That structure was "rebuilt so many times exact count has been lost in musty City Hall records," or so the *Spokesman-Review* once observed. Bridges crossing the other channels of the Spokane River were equally troublesome.

As a first step in providing reliability at Howard Street, D. Boyington constructed a concrete arch bridge across the north channel in 1909. Designed by the Strauss-Bascule Bridge Company, that structure consisted of two spans totaling 202 feet long. In 1916 the Beers Building Company of Seattle built a 192-foot-long steel through-truss bridge on Howard Street across the river's middle channel. The company used some steel falsework trusses salvaged from the construction of the nearby Monroe Street Bridge five years earlier. Over the next several decades, the structure's wooden deck deteriorated, leading the city to install a precast concrete deck and sidewalks in 1963. When the city hosted Expo '74, Havermale Island and the surrounding channels and falls became centerpieces in a celebration of the natural environment. Both Howard Street bridges were closed to vehicles and dedicated to pedestrians visiting the exposition's venues astride the river's tumbling waters. After the

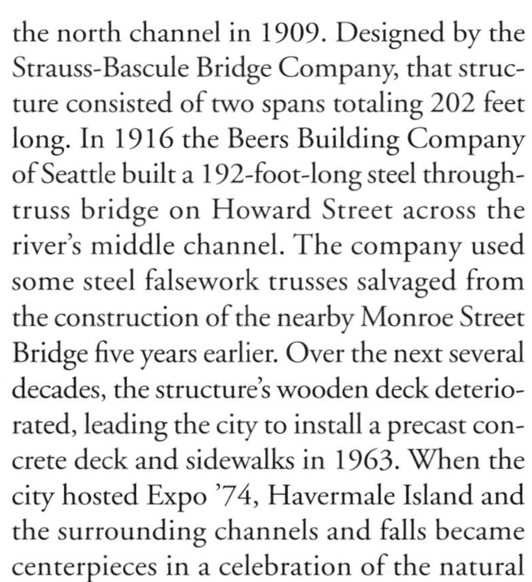

In 1916 the Beers Building Company of Seattle used steel falsework trusses, salvaged from construction of the nearby Monroe Street Bridge, to build this truss bridge over the middle channel of the Spokane River in what is now Riverfront Park in Spokane.

Marsha Reilly, Tumwater

On Howard Street in Spokane, D. Boyington constructed a concrete arch bridge across the north channel of the Spokane River in 1909. The bridge stands today in Riverfront Park in Spokane.

Marsha Reilly, Tumwater

fair, the bridges were permanently closed to automobile traffic and incorporated into Riverfront Park.[22]

Hood River-White Salmon Bridge, Klickitat County, Washington, and Hood River County, Oregon

Completed in 1924, this bridge could be called a poster child for perpetual preservation, although "historic preservation" is not entirely accurate, considering the multitude of alterations that have changed the structure's function and appearance over the years. Rising waters behind Bonneville Dam required replacing its fixed span with a movable lift span in 1938. The replacement of timber approach spans and numerous other changes throughout the 1940s, 1950s, and 1960s further altered this historic bridge. The Port of Hood River has recently undertaken extensive improvements designed to enhance the bridge's longevity, including deck replacement and guardrail updates.[23]

The Hood River-White Salmon Bridge over the Columbia River was completed in 1924 and has experienced numerous renovations since, including the 1938 replacement of one fixed span with a movable lift span.

Craig Holstine, WSDOT

Curlew Bridge, Town of Curlew, Ferry County

For nearly a century, this pin-connected Parker truss bridge has provided the citizens of the small town of Curlew with the most direct access across the Kettle River to the area's primary north-south transportation route (now State Route 21). Beginning in 1897, a cable ferry transported miners and settlers to a general store on the river's east side. The first bridge washed out at the site before William Oliver of Spokane erected the present bridge in 1908. The St. Paul Foundry Company supplied the steel for the 182-foot structure, whose polygonal top chord enabled construction of a span thirty feet longer than the standard Pratt truss using a top chord parallel with its bottom chord.

The Curlew Bridge, completed in 1908 over the Kettle River, is a pin-connected Parker truss scheduled to undergo major renovation.

Office of Archaeology and Historic Preservation

Its pivotal role in the community became apparent when the county closed the bridge in 2004. It suffered from cracked and rusted pins, corroded eyebars frozen in place, deteriorated wooden deck planks, and steel-encased cylindrical concrete piers out of plumb. When funding sources (WSDOT and federal) were identified, the county commissioners decided to have the structure disassembled and rebuilt with new pins, deck, approaches, guardrails (to federal safety standards), and higher-strength steel stringers to increase the load capacity to H-15. To avoid affecting the Kettle River, work on the bridge will be done off-site once the truss is swung away from the river crossing to dry land. Sandblasting and painting of all steel members will be completed as part of this exemplary preservation effort.[24]

NOTES

1. Robin Bruce, "F Street Bridge," HAER No. WA-31, 1990; Lisa Soderberg, "F Street Bridge," HAER Inventory, 1979.
2. Robin Bruce, "McClure Bridge," HAER No. WA-25, 1989; Lisa Soderberg, "McClure Bridge," HAER Inventory, 1979.
3. *Kettle River Journal* (Orient), May 16, 1902, 1; Robin Bruce, "Orient Bridge," HAER No. WA-32, 1992.
4. Jean P. Yearby, "Grant Avenue Bridge (Prosser Steel Bridge)," HAER No. WA-4, 1985; R. B. Davidson, "NRHP Determination of Eligibility for the Grant Avenue Bridge," n. d.; Lisa Soderberg, "Prosser Steel Bridge," HAER Inventory, 1979.
5. Lisa Soderberg, "Jack Knife Bridge," HAER Inventory, 1979; William T. Belshaw, "Jack Knife Bridge," NRHP Nomination, 1973.
6. Lisa Soderberg, "Wishkah River Bridge," HAER Inventory, 1979.
7. Tom Copeland, *The Centralia Tragedy of 1919: Elmer Smith and the Wobblies* (Seattle: University of Washington Press, 1993), 48–54; Margaret Shields, Historian, Lewis County Historical Museum, Chehalis, personal communication, November 2004.
8. Marcia Babcock Montgomery and Lisa Mighetto, "Novelty Bridge 404B," HAER No. OAHP 110598-02-KI, 1999.
9. Lisa Soderberg, "Pasco-Kennewick Bridge," HAER WA-8, 1980; Paul Dorpat and Genevieve McCoy, *Building Washington: A History of Washington State Public Works* (Seattle: Tartu Publications, 1998), 112–13.
10. Jonathan Clarke, "Dosewallips River Bridge," HAER No. WA-94, 1993.
11. Robert H. Krier and Craig Holstine, "Latah Creek Bridge No. 4102," HAER No. WA-163, 1998.
12. Robert H. Krier and Craig Holstine, "Mora Road Bridge No. 110/25: An Evaluation of Significance," Archaeological and Historical Services, Eastern Washington University, Short Report DOT98-03, 1998.
13. Craig Holstine and Darcy Fellin, "Outlet Creek Bridge," HAER No. WA-117, 1994.
14. Flo Lentz and Leonard Garfield, "Norman Bridge," NRHP Nomination, 1994.
15. Robert H. Krier, J. Byron Barber, Robin Bruce, and Craig Holstine, "Donald-Wapato Bridge," NRHP Nomination, 1991; Sharon A. Boswell, "Documentation of the Historic Donald-Wapato Bridge No. 396, Yakima County, Washington," Northwest Archaeological Associates Report No. WA 02-25, Seattle, November 8, 2002; WSDOT Bridge Preservation Office files, Olympia.
16. Wayne L. Rickert Jr., "Grays River Covered Bridge," HAER Report, August 1988; Robin Bruce, "Grays River Covered Bridge," HAER No. WA-28, 1991; Lisa Soderberg, "Grays River Covered Bridge," HAER Inventory, 1980; John Stites, Federal Highway Administration, BROS-2035 (004), State Contract SA-1146, November 1989; *Highway News*, August 1953, 45.
17. Robert H. Krier, "King County Historic Bridge Inventory Phase 3: Final Evaluation and Documentation," Short Report 485, Archaeological and Historical Services, Eastern Washington University, Cheney, 1995; Mimi Sheridan, "Baring Bridge," King County Landmark Registration Form, King County Cultural Resources Division, Seattle, 1999.
18. Oscar R. "Bob" George, "Chief Joseph Dam Bridge," NRHP Nomination, 2001; Carole Hardie, Engineering Coordinator, Douglas County, personal communication, October 2004.
19. Craig Holstine, "Monroe Street Bridge," Spokane Register of Historic Places Nomination, 1990; "Historic Spokane Bridge Rebuilt," *Pacific Builder and Engineer*, September 6, 2004, 10–11.
20. Wm. Michael Lawrence, "Baker River/Henry Thompson Bridge," HAER No. WA-105, 1993; Henry Thompson/Baker River Bridge Fact Sheet, Skagit County Public Works Department, August 21, 2004; Barbara Hathaway, Field Engineer/Geologist, Skagit County Public Works Department, personal communication, October 2004.
21. Leroy Slemmer, WSDOT Tacoma Project Engineer's Office, personal communication, November 2004; Robert H. Krier, J. Byron Barber, Robin Bruce, and Craig Holstine, "Winnifred Street Bridge," NRHP Nomination, 1991.
22. "$60,000 Repair Project Set for Historic Howard Street Bridge," *Spokesman-Review*, October 20, 1963; Lisa Soderberg, "Howard Street Bridges," HAER Inventories, 1979.
23. Bridge files, Hood River County Historical Museum, Hood River, Oregon; *Engineering News Record*, April 30, 1925; Judith A. Chapman and Elizabeth O'Brien, "Hood River-White Salmon NRHP Determination of Eligibility," Archaeological Investigations Northwest, Inc., Portland, Oregon, June 21, 2004; photo file, Port of Hood River, Hood River, Oregon.
24. Lisa Soderberg, "Curlew Bridge," HAER Inventory, 1979; Keith Muggoch, Ferry County Engineer and Director of Public Works, personal communication, November 2004.

Caro Reese
Jean Glebov
F.T. Evans
L.R. Durkee
Tom Smith
L.V. Murrow
R.S. Fluent
C.E. Andrew
Jack Taylor
R.H. Thomson
R.B. McMinn
R.M. Murray
P.H. Winston
C.C. Arnold
Simmer
-50-
"Shovel Crew" at start of Tunnel Excavation at East Portal. Lake Washington Bridge Project

CHAPTER 4

Designers of Dreams: Bridge Engineers

Many of Washington's first successful highway bridge engineers began their careers in railroad construction. Historic railway bridges are not discussed in this volume, although the state's early highway history owes much to railroad bridge-building technology. These men demonstrated not only talent and ingenuity, but also courage. Numerous examples here highlight those who stepped creatively beyond the boundaries of mainstream American bridge architecture.

Some were mavericks with little formal education, yet they built surprisingly innovative structures. Others used the most modern and sophisticated structural technology and science to create commanding bridge architecture. In seeking to harmonize aesthetics and structural design with a bridge's setting, they left enduring landmarks—symbols of evolving technology.

Charles E. Andrew (1884–1969)

An uncommon, visionary engineer, Charles E. Andrew left an unusual, if controversial, bridge legacy in Washington. As chief consulting engineer for the Washington State Toll Bridge Authority for some twenty years, Andrew guided construction of the first two Lake Washington floating bridges, the two Tacoma Narrows bridges, and the Hood Canal Floating Bridge. His admirers called him "a genius," but critics claimed he built "blow away bridges."

Born in Illinois in 1884, Andrew graduated from the University of Illinois in 1906 with a bachelor of science degree in civil engineering. That year he moved to Oregon, beginning his career building railroad bridges. In 1921 he moved to Washington and, until 1927, served as the first officially designated bridge engineer for the State Department of Highways. Better opportunities drew him to Sacramento and the San Francisco Bay area for the next decade. In 1931 Andrew became principal engineer for the San Francisco-Oakland Bay Bridge, leading the construction and design on what was at the time the world's largest bridge.

As the first official "Bridge Engineer" in the Washington State Department of Highways, Charles E. Andrew supervised the design of many of the state's earliest significant bridges.
Washington State Archives, WSDOT Records

A soft-spoken man with great energy, confidence, and determination, Andrew was a celebrity in engineering circles when he returned to Washington in March 1938. He immediately set to work on the Tacoma Narrows Bridge as chief consulting engineer. When "Galloping Gertie" failed in November 1940, many observers blamed Andrew. In fact, as an investigation soon revealed, he fought hard against the cost-cutting design compromises that led to the bridge's collapse. The second Narrows span, designed to Andrew's standards and specifications, has withstood Northwest storms for more than half a century.

The Evergreen Point Floating Bridge, also designed by Andrew, offers a different example of the controversies that tended to follow him. He accurately foresaw that the future growth of the greater Seattle area would put heavy demands on the bridge's capacity. Unfortunately, limited funding forced construction of a bridge narrower than Andrew knew would be needed.

When part of the Hood Canal Floating Bridge sank in a 1979 storm, Andrew again became the mark of critics, although he had died ten years earlier. As before, the blame was unjust. Andrew opposed the design

"Shovel Crew" awaiting excavation of the Mount Baker Ridge Tunnel's east portal, Lake Washington Floating Bridge Project, Seattle, April 24, 1939. Top, in the cab, Caro Reese (left) and Jean Glebov; Lacey V. Murrow (left) and Charles E. Andrew stand below the women; on ground, left to right: F. T. Evans, L. R. Durkee, Tom Smith, R. S. Fluent, Jack Taylor, R. H. Thomson, R. B. McMinn, R. M. Murray, P. H. Winston, and C. C. Arnold. Alfred Simmer, photographer.
Washington State Archives, WSDOT Records

compromises forced by budget limits, though whether these caused the bridge's failure is uncertain. When an investigation revealed that the bridge might not have failed if pontoon hatches were closed and not accidentally left open, Andrew's reputation began to regain its former stature.

Among Andrew's unrealized transportation visions were a tunnel crossing under Puget Sound from Alki Point to Bainbridge Island, a floating bridge to Vashon Island from Fauntleroy, and a large suspension bridge over Colvos Passage.[1]

Cecil C. Arnold (1899–1983)

In a career that spanned more than fifty years, Cecil C. Arnold compiled a distinguished record as a structural engineer in both the private and public sectors. Born in Walla Walla in 1899, Arnold graduated from the State College of Washington (now Washington State University) in civil engineering in 1925 and worked for eleven years with the Washington State Department of Highways. During World War II, Arnold served with the Army Corps of Engineers, contributing to bridge and airport construction projects in Seattle and Everett, then in Brazil.

He had moved to Seattle in 1927, where later, in 1946, he launched his own civil engineering consulting firm. In 1970, the firm became Arnold, Arnold, and Associates when son C. Adrian Arnold joined in partnership with his father.

While employed with the State Department of Highways, Arnold made contributions to such projects as the original Lake Washington Floating Bridge and the Aurora Avenue Bridge. The many bridges designed by his firm include the Gorge Creek Bridge just north of Newhalem, the Dalles Bridge near Concrete, and the Stossel Bridge in King County. Other notable designs by Arnold were portions of the Whidbey Island Naval Air Station, several ferry terminals around Puget Sound, Waterfront Park in Seattle, Sea-Tac Airport runways, and segments of the Interstate 5 and Interstate 90 freeways.

In 1961 Arnold was named "Engineer of the Year" by the Consulting Engineers' Council of Washington. In 1980 he received the Architect-Engineer Alumni Achievement Award from the Washington State University Alumni Association.[2]

William A. Bugge, Director of the Washington State Department of Highways, 1949–63. The Hood Canal Floating Bridge now bears his name.
Washington State Archives, WSDOT Records

William Adair Bugge (1900–1992)

Born in Hadlock, Washington, William Bugge graduated from the State College of Washington in 1923 in civil engineering. He worked as a surveyor for the Washington State Department of Highways off and on between 1922 and 1925, when he became a bridge inspector for the department. After leaving state employment in 1927, Bugge gained valuable experience as a construction engineer and supervisor in Seattle, Shelton, and Oregon City, Oregon. In February 1933, he became the Jefferson County engineer in Port Townsend, Washington.

Bugge served in that position until 1949 when he was named director of the Washington State Department of Highways, where he remained until retirement in 1963. During his fourteen-year tenure, many of the state's most notable bridges and its first interstate highways were constructed. After leaving WSDOT, he directed the design and construction of the Bay Area Rapid Transit (BART) system in San Francisco.

Bugge was named national highway "Man of the Year" in 1961 by the American Public Works Association. In 1977 the Washington State Highway Commission named the Hood Canal Floating Bridge in his honor. Washington State University gave Bugge the Alumni Achievement Award in 1973 for "distinguished service as Washington Director of Highways and for internationally recognized contributions." In 1980 WSU presented him with a Regents' Distinguished Alumnus Award.[3]

Arthur Herbert Dimock (1865–1929)

Arthur Dimock became one of the pioneer bridge builders in turn-of-the-century Seattle. Born in Nova Scotia in 1865, he completed college with a bachelor of engineering degree at age twenty. For several years, Dimock worked as a railroad engineer in Quebec and Cape Breton. Then in the late 1880s he set off for California, where he continued railroad work.

By 1897, Dimock had settled in Seattle and soon found employment with the City Engineering Department. His education and abilities won not only the admiration of his coworkers, but also promotion to principal assistant city engineer in charge of designing and constructing the city's main sewer trunk lines.

In 1911 Dimock was appointed to succeed R. H. Thomson as city engineer. During his term in office, the city completed the Lake Washington Ship Canal and the first three bascule spans—the Fremont, Ballard, and University bridges. In 1922 Dimock's career in Seattle abruptly ended. Articles in the *Seattle Post-Intelligencer* assailed Dimock for mismanaging the department and failing to help the city avoid a costly patent suit. His eleven years of service to the city ended in public embarrassment.

Dimock moved to Skagit County, where he opened a private engineering practice. In November 1929, just after completing a survey of Mount Vernon's city sewer system, Dimock suffered a stroke and died at age sixty-four.[4]

L. R. Durkee (1899–1988)

Structural engineer Linley Rathburn "L. R." Durkee was born in Delta, Colorado, on September 14, 1899. He graduated from Montana State College in Bozeman in 1921 and became a prominent Seattle builder from the 1930s to the 1960s. In the late 1920s and early 1930s, he supervised construction of Harborview Hospital, the Bon Marche department store, several schools and fire stations, and the Federal Office Building in Seattle. In 1933 Durkee began working as an engineer with the Public Works Administration, and he remained with the federal government until his retirement in the early 1960s. His appointment in 1937 as chief engineer for the Lake Washington Floating Bridge and Tacoma Narrows Bridge projects brought him to new levels of responsibility and professional accomplishment.

In 1941 Durkee took a job as regional director of the Defense Public Works Division of the Federal Works Agency, where he served throughout World War II. He supervised federal construction projects in Washington, Oregon, Idaho, Montana, and Alaska. Promotions in the federal bureaucracy led to his final post, regional director of the Federal Housing and Home Finance Agency.

Durkee received several distinguished awards. In 1948, his alma mater, Montana State College, granted him an honorary doctorate. The Washington Society of Professional Engineers honored Durkee in 1959 as "State Engineer of the Year," and in 1960 the Puget Sound Engineering Council selected him as "Engineer of the Year." He died in 1988 in Seattle.[5]

L. R. Durkee, while with the federal Public Works Administration, served as chief engineer for the Lake Washington Floating Bridge and Tacoma Narrows Bridge projects.

Washington State Archives, WSDOT Record

Clark Eldridge (1896–1990)

From modest beginnings in the small Western Washington town of Lake Stevens, Clark Eldridge became one the state's most accomplished and noted bridge engineers, leading a career punctuated by both acclaim and controversy.

Bright, energetic, and dedicated, Eldridge eagerly set about engineering studies in 1915 at Washington State College. After two years, however, World War I intervened. In 1917 he entered the U.S. Army, serving in both France and Germany. On returning home, Eldridge completed his engineering degree by correspondence.

Over the next fifteen years, from the 1920s until 1936, he distinguished himself in the Seattle City Bridge Engineer's office. Eldridge designed numerous notable spans for the city, including the Montlake, Garfield Street, and

As a prisoner of the Japanese in World War II, Clark Eldridge was reminded that he could not escape his association with "Galloping Gertie," the first Tacoma Narrows Bridge.

Washington State Archives, WSDOT Records

Spokane Street bridges, and the reconstruction of the University Bridge. By 1932, Eldridge was head of the department and oversaw construction of the Aurora Avenue (George Washington Memorial) Bridge.

In 1936 Eldridge joined the State Department of Highways. The forty-year-old engineer found himself designing two of the state's most colossal bridges, the Lake Washington Floating Bridge and the Tacoma Narrows Bridge.

He considered the latter "his bridge" from the time the Highway Department issued him the challenge of finding money to help build it. Eldridge found the money, but then things became complicated. His boss, State Highway Director Lacey V. Murrow, took Eldridge's design and cost estimates to the Public Works Administration in Washington, D.C. There, federal agency officials decided that Eldridge's $11 million plan was too expensive. They agreed to loan Washington the money for the Narrows Bridge, but only if they hired Leon Moisseiff, the noted New York engineer-consultant. Moisseiff redesigned the structure and said it could be built for $7 million. Moisseiff's recommendations were adopted, and the Tacoma Narrows Bridge moved forward, with Eldridge remaining in charge of supervising the mammoth project's construction.

Luther E. Gregory directed construction of the Naval Shipyards in Bremerton before serving as consulting engineer for the Tacoma Narrows Bridge and Lake Washington Floating Bridge projects.
Washington State Archives, WSDOT Records

Even before completion of the Tacoma Narrows Bridge, its deck showed significant movement, oscillating in the lightest breezes through The Narrows. Soon, workmen nicknamed the rippling span "Galloping Gertie." In November 1940, barely four months after the bridge opened, Eldridge received the telephone call that would change his life forever.

"I was in my office about a mile away when word came that the bridge was in trouble," he later wrote in an unpublished autobiography. He drove to the Tacoma anchorage, and had to be lead off the leaping roadway by a fellow engineer. "There, we watched the final collapse," Eldridge wrote with dismay.

Ripple effects from Galloping Gertie's fall were felt long after the catastrophe and far from the frigid waters of the Tacoma Narrows, and Eldridge learned firsthand just how far those ripples went. He accepted some blame for the failure and took work with San Francisco construction contractors for the U.S. Navy on Guam. At the outbreak of World War II, he was in the wrong place at the wrong time. Captured and imprisoned by the Japanese for the remainder of the war (three years and nine months), Eldridge assumed he had outdistanced his association with Galloping Gertie. One day, to his astonishment, a Japanese officer recognized Eldridge, walked up, and said simply, "Tacoma bridge!"

On returning to the Northwest after the war, Eldridge resumed work as a consulting engineer and contractor in the Tacoma and Portland areas, then as Skamania County engineer until his retirement in 1970. But retirement was only an official departure from public employment. Eldridge worked tirelessly until his death two decades later. Among the more notable projects was his engineering work for a retirement community near his home in Lacey, Washington.

Eldridge's memories of Galloping Gertie were a source of sadness for much of the second half of his life. As he wrote in his memoirs, "I go over the Tacoma Bridge frequently and always with an ache in my heart. It was my bridge."

The second Narrows Bridge that rose in its place, however, must have been some consolation. The bridge today stands on the piers he designed. And it bears a striking resemblance, with the exception of its four lanes, to the bridge he designed before fate crossed his path.[6]

Luther Elwood Gregory (1872–1960)

Few of Washington State's bridge builders can equal the record, stature, or wide variety of professional contributions of Luther Gregory. Born in 1872 in Newark, New Jersey, he graduated from the School of Mines at Columbia University in 1893. A year later, Gregory married Anna Roome, and their family eventually included five daughters and one son.

The 1898 Spanish-American War opened a new chapter for Gregory. He joined the U.S. Navy as a civil engineer, and began a career that took him from postings as the public works officer at various naval stations to the rank of rear admiral by the time he retired in 1929. In the 1920s, Gregory worked for five years as head of the Navy's Yards and Docks Department and spent a year as director of exhibits for the World's Fair in Chicago. Gregory's bridge work began in California as a member of the location commission appointed by President Herbert Hoover for the San Francisco-Oakland Bay Bridge.

By the early 1930s, he moved to the Puget Sound area. The Naval Shipyards were constructed in Bremerton under his direction. In 1937 he took on the task of consulting engineer for the State Toll Bridge Authority on the Tacoma Narrows Bridge and the Lake Washington Floating Bridge projects. The versatile Gregory also was active as first chairman of the State Liquor Control Board, serving 1934–42 and 1945–57. In 1938 the Rensselaer Polytechnic Institute awarded Gregory an honorary doctorate. Gregory remained active in local professional organizations and social clubs up until his death at age eighty-eight.[7]

Homer M. Hadley (1885–1967)

One of the state's most innovative and influential bridge designers was Homer M. Hadley. Born in Cincinnati, Ohio, in November 1885, Hadley attended the University of Washington between 1908 and 1916, intermittently taking classes in civil engineering and working for several railroad companies between Seattle and Alaska. He developed an interest in structural design work, and during World War I took a job designing concrete barges in Philadelphia before returning to Seattle.

It was in 1920 that he first proposed a concrete pontoon floating bridge to span Lake Washington between Seattle and Mercer Island. As a fledgling engineer in a local architect's office, Hadley's enthusiasm could not convince local bankers nor other potential investors. Hadley soon went to work as regional structural engineer for the Portland Cement Association in Seattle. But he persistently advocated the practicality and cost-effectiveness of a floating bridge.

By 1931 the Seattle City Council formed a committee to study applications for a franchise to build a toll bridge across Lake Washington, and by 1937, plans were set to construct a high bridge with a jackknife draw span from Seward Park eastward. In June 1937 came a fateful meeting with State Highway Director Lacey V. Murrow that at last moved Hadley's dream toward reality.

Murrow favored Hadley's route and a floating structure as the most practical, though a different design was selected. Hadley was disappointed that he never became a consultant on the bridge project because of his affiliation with Portland Cement. Nevertheless, he felt delighted to see his vision of a floating bridge across Lake Washington opened to the public in July 1940.

Hadley designed or suggested designs for several distinctive concrete bridges of historical significance. Both the McMillin Bridge south of Puyallup and the Purdy Bridge near Gig Harbor were Hadley's unique adaptation of the hollow concrete box design. In 1963 Hadley's design of the Parker Bridge near Yakima received honors from the American Institute of Steel Construction as the year's most beautiful short span.

In 1967, a few months after the death of Lacey V. Murrow, the legislature named the Lake Washington Floating Bridge after the former director. If Hadley felt disappointed, he kept it to himself. At age eighty-one, while vacationing with his wife in Eastern Washington in July 1967, Hadley drowned in Soap Lake. It was an ironic end for the man who convinced others that floating concrete would be the answer to some of the state's most challenging transportation needs.

In the years that followed, Hadley's family, friends, and the University of Washington's Mortar Board chapter took up the cause of gaining recognition for his key role in developing the floating bridge. In July 1993, some seventy

Homer M. Hadley is today revered as the state's most innovative bridge engineer. Credited with the idea for the world's first concrete floating bridge, the third bridge built across Lake Washington is named in his honor.

Courtesy of Eleanor M. Hadley

Engineering staff on the Lake Washington Floating Bridge Project, February 14, 1940. Left to right: R. H. Thomson, L. E. Gregory, R. M. Murray, C. E. Andrew, and R. B. McMinn. Alfred Simmer, photographer.

Washington State Archives, WSDOT Records

years after Hadley envisioned the world's first concrete floating bridge, he finally received public acknowledgment when the newly completed Interstate 90 floating span was named in his honor.[8]

Samuel J. Humes (1883–1941)

Seattle native Samuel J. Humes grew up in the two worlds of politics and engineering. The son of former city mayor Thomas J. Humes, he came to public administration by birth.

Humes graduated from Broadway High School, then briefly attended the University of Washington before taking a job in the city engineer's office in 1902. He worked there for fifteen years under the leadership of R. H. Thomson and Arthur Dimock. In 1917, at age thirty-four, the ambitious Humes, an active and staunch Republican Party member, was appointed King County engineer. Over the next five years, his farsighted policy of building bigger roads to accommodate the growth of automobile traffic earned him the nickname "Wider Roads Humes."

From 1921 to 1927, Humes focused on building his private engineering practice. In 1927 he took a job with the State Department of Highways in Olympia, and two years later, Governor Roland Hartley appointed Humes to head the department. He served in the post until March 1933. In those six years, Humes oversaw some of the state's most important transportation projects, including the four-lane Pacific Highway (State Route 99) between Tacoma and Everett (and the Aurora Avenue Bridge on that road), the Sunset Highway from North Bend to Yakima (part of which became Interstate 90), the Olympic Loop Highway (State Route 101), the Naches Highway (State Route 410), and thousands of miles of roads in Seattle and King County.

When Humes moved back to Seattle, he returned to private engineering practice. In 1938 he threw his hat into the political arena and won a seat on the city council. He was reelected in 1940; however, shortly after the start of his second term, Humes fell ill. A month's recuperation in California failed to return his health, and Humes died in June 1941.[9]

Daniel Webster McMorris (1864–1942)

Self-taught engineer Daniel W. McMorris never attended college, and earned a reputation as a builder of fortresses, not bridges. Yet he helped guide decades of public works in Seattle. He was particularly recognized for his work as one of the engineers of the Corregidor Fortress in the Philippines.

Born in Illinois in 1864, McMorris by age thirteen was residing in Dayton, Washington, where he graduated from high school. At twenty-three, he secured a job in the engineering office of the Oregon and Washington Territory Railroad Company. The ambitious McMorris rose from draftsman to resident engineer in charge of construction for bridges and buildings in the company's Grays Harbor division.

In 1891 McMorris arrived in Seattle and within two months started work as a draftsman and street inspector in the city engineer's office. He resigned as assistant city engineer in 1898 to enter federal government service as an Army civilian engineer. McMorris worked on the fortifications at Fort Ward, Fort Flagler, Fort Worden, and Fort Casey on Puget Sound, and harbor and river improvements in the Bellingham area. From 1904 to 1907, McMorris took advantage of an unusual opportunity in the Philippine Islands, as the U.S. Engineer's office began construction of the Corregidor Fortress. When his supervisor returned to the States, McMorris was promoted and placed in local charge of the fortification work.

McMorris's principal contributions to Seattle bridge construction followed his return to the city engineer's office. He earned a distinguished reputation as assistant city engineer, court engineer, and supervisor of bridge construction. Under Mayor Robert Harlin from 1931 to 1932, he headed the city engineer's office. The soft-spoken McMorris played a central role in virtually every major public engineering project in Seattle for more than twenty-five years. Most notable were his contributions to the Montlake Bridge, the first two West Seattle bascule bridges at Spokane Street, the West Garfield Street Bridge, and the 12th Avenue South Bridge. McMorris retired from the city engineer's office in 1934.

In 1942, though long retired, McMorris came into the local spotlight when General Jonathan Wainwright's troops withstood, for a time, overwhelmingly heavy Japanese assaults on the Filipino fortress, and "Corregidor" became a household word in America.[10]

Andreas W. Munster (1853–1929)

Admirers of several landmark bridges in Seattle might well give homage to the talented Norwegian-born engineer Andreas Munster. Born in Bergen in 1853 into a military family, Munster received an education at Charlmarske Institute in Sweden. Following his graduation in 1873, the aspiring engineer worked for eight years with the national railroad department in Norway.

Then, at age twenty-nine, Munster decided to seek his fortune in America. In 1882 he resigned as resident construction engineer and boarded a boat for the United States. The talented Munster landed a position as assistant engineer with the Northern Pacific Railroad's Rocky Mountain Division, then headquartered in Helena, Montana. He worked for the next year on completion of the transcontinental railroad. A couple of years later, an attractive post as bridge engineer with the city of St. Paul, Minnesota, moved Munster to the Midwest.

By 1906, Munster found himself working as a consulting engineer for the City of Seattle. From 1923 until his death six years later, Munster held the post of chief engineer for the Bridge Department. He designed and supervised the construction of some of the city's most recognizable historic structures. Most notable are the Montlake Bridge over the Lake Washington Ship Canal, Magnolia Street Bridge Viaduct (West Garfield Street), Fremont Bridge, University Bridge, and West Seattle's movable Spokane Street Bridge.[11]

Daniel W. McMorris specialized in designing coastal fortifications before joining the Seattle city engineer's office, where he contributed to numerous bridge projects, including the Montlake Bridge at the entrance to the University of Washington. UW # 23816

University of Washington Libraries, Special Collections

Ray M. Murray (1876–1955)

As a structural engineer in Seattle, Ray Murray contributed to some of the city's most outstanding achievements in bridge construction in the 1930s and 1940s. Born in Ada, Ohio, in 1876, Murray received his education there, graduating from Northern Ohio University in 1897 in civil and mechanical engineering. For a year, he designed roads and bridges in Hardin County, Ohio, then returned to the university in Ada to teach civil engineering. In his four years there, he served as dean of the College of Engineering and taught classes in advanced mathematics, geometry, mechanics, and structural design.

Ray M. Murray's career spanned construction of the Aurora Bridge in 1930 to the Alaskan Way Viaduct in the early 1950s.

Washington State Archives, WSDOT Records

Better pay lured Murray to jobs with railroads and other firms in Pennsylvania, Indiana, Montana, Spokane (1923–25), and then Portland, Oregon. In 1929 he moved to Seattle to take a position as a design engineer for the Aurora Bridge project with Jacobs and Ober, consulting engineers. From 1930 to 1946, Murray worked on various bridges in Seattle as resident engineer for the State Department of Highways. Most notable in those years was his involvement as engineer for the State Toll Bridge Authority on the Lake Washington Floating Bridge project from 1938 to 1941. From 1946 to 1950, Murray served as construction engineer for the City of Seattle on the Alaskan Way Viaduct.

In 1950 Murray retired from public employment and began private consulting. For five years, he contributed his expertise to structural engineering projects, until death claimed him in October 1955 at age seventy-nine. In the course of his professional career, Murray invented a reversible propeller, and wrote articles appearing in the *Engineering News-Record, Pacific Builder and Engineer*, and other professional journals. He was a member of the Seattle Engineers Club (and served a term as president), and a lifelong member of the Professional Engineers of Oregon Club.[12]

Lacey Van Buren Murrow (1904–1966)

He was the younger brother of Edward R. Murrow, the noted journalist, radio and television news commentator, and onetime head of the U.S. Information Agency. Lacey V. Murrow seemed to many to stand in his brother's shadow. Neither Lacey nor Ed ever felt that way. Each man made unique and distinctive contributions in his own right, and each received many accolades during his lifetime.

Born in Greensboro, North Carolina, in 1904, Lacey grew up with his brothers Dewey and Ed in Blanchard, Washington. The Murrow brothers attended Washington State College; Lacey graduated with a bachelor of science degree in civil engineering in 1923. His role with the State Department of Highways would be one of the most influential in Washington State bridge construction history.

Murrow had worked for the Highway Department intermittently beginning in 1919, then regularly after 1923. He moved steadily into positions of higher responsibility, from resident engineer, to location engineer, construction engineer, and then district engineer. In this period (1931), he met and married Olympia native Margaret Goodpasture. In 1933 Murrow began an eight-year term as director of the State Department of Highways. It proved a turning point in his career, and in the history of Washington bridge work.

The twenty-eight-year-old Murrow, concurrently appointed as chief engineer for the State Toll Bridge Authority, brought a dynamic and productive energy to transportation projects. In 1937, after a fateful meeting with engineer Homer Hadley, who first proposed a floating bridge across Lake Washington, Murrow became the idea's most ardent and visible advocate. His leadership brought Hadley's dream into reality in 1940 when the Lake Washington Floating Bridge was finished. The same year, Murrow oversaw the department's completion of the first, and ill-fated, Tacoma Narrows Bridge.

U.S. entry into World War II in 1941 turned Murrow's professional life in a new direction. Before the war, his favorite sport was flying. In the war, it became his profession. He entered the military and distinguished himself on many fronts in both WWII and the Korean conflict. From 1941 to 1946, he served as a command pilot, winning military honors that included a presidential citation with four cluster decorations, the Legion of Merit, the Order of the British Empire, and the Croix de Guerre. In 1951 Murrow became a brigadier general in the Air Force and served in Korea, Japan, and the United States before retirement. From 1954 until his death, Murrow lived in Washington, D.C., working for Transportation Consultants Inc. Most of those years he served as the firm's president.

Like his brother Ed, Lacey Murrow was a lifelong smoker. In the early 1960s, he began suffering from lung cancer. He underwent an operation shortly before his brother's death from the disease in 1965. Little more than a year later, in December 1966 at age sixty-two, Murrow was found shot to death in his room at the Lord Baltimore Hotel, a twelve-gauge shotgun propped against the bed.

Soon after Murrow's suicide, the Washington State Legislature passed a resolution requesting that the State Highway Commission rename the first Lake Washington Floating Bridge in his honor. The commission agreed and in March 1967 paid tribute to Murrow, declaring, "This notable engineering achievement received world-wide recognition for its pioneering of a new concept in over-water structures."[13]

Harry R. Powell (1901–1991)

During his lifetime, Seattle engineer Harry R. Powell became one of the most honored, respected, and award-winning bridge designers in the state's history. Born in 1901 and raised in Saskatchewan, Canada, Powell graduated from the University of Toronto in 1922 with an engineering degree in hydraulics and reinforced concrete. In 1926 he moved to Seattle, where he remained in private practice for more than forty years.

A stocky man with a quiet demeanor, Powell soon gained a reputation for his three loves—his work, duck hunting, and pipe smoking. By his fiftieth birthday, he had designed more than forty concrete bridges and as many timber spans. When Powell turned his attention to steel bridges in the late 1950s, he quickly began winning recognition from the prestigious American Institute for Steel Construction and other industry organizations. Between 1958 and 1963, Powell's bridge designs earned four national prizes and one international award.

Most notable in this roster were the Rainbow Bridge across the Swinomish Slough at La Conner (a first-place winner in 1958), the White Salmon River (B-Z Corner) steel arch bridge in Klickitat County (which won third place in a 1958 competition), and the steel arch bridge across the Kalama River in Cowlitz County (Modrow Bridge), which judges in the 1960 contest called "the most imaginative and sensitive design in the competition."

The Rainbow Bridge, Powell's personal favorite, has become one of the state's most photographed spans. Powell is also credited with designing Washington's first prestressed concrete bridge, located on the east fork of the Humptulips River in the Olympic National Forest. In 1954 he designed (with his associate, Leonard K. Narod) the Klickitat River Bridge No. 142/9, believed to be the oldest prestressed concrete bridge remaining in the Washington State highway system. In 1964 the Seattle Chapter of the American Society of Civil Engineers honored him as "Engineer of the Year." By the time he retired, Powell was widely recognized as a leading structural designer, and had collected more than a dozen national design awards and eight structural patents.

"It always has been my belief," Powell once said, "that bridges should be beautiful as well as useful, and I have tried to work out my designs with due regard for aesthetic values."[14]

Reginald Heber Thomson (1856–1949)

When R. H. Thomson stepped ashore in Seattle in 1881, the little frontier town of 3,000 was a dismal yet attractive sight—the young man was in search of a place that "needed improving," and he eagerly set about doing just that. Well before his death at age ninety-two, Thomson became something of a legend. Few men have written a larger signature on a city's landscape than Thomson did in Seattle. The accolades he accrued capture the nature of his contributions: "He built Seattle," "He lifted the face of Seattle," "the man who tore down the hills," and "Seattle dreamer" are just a few.

Reginald H. Thomson contributed to more engineering projects in Seattle and Western Washington than any other man.

Washington State Archives, WSDOT Records

Thomson was perhaps Washington's most dynamic and brilliant engineer. A civic leader, politician, and author, as well as an engineer, he was a "builder" in every sense of the word. Tall, strong-minded, and strong-willed, he received master's and doctoral degrees, but not in engineering. Nonetheless, his genius lay in building, and he set out to change Seattle's skyline.

For nineteen years of Seattle's most dramatic growth, from a village of 4,000 in 1892 to a metropolis of more than 200,000 in 1911, Thomson served as city engineer. Under seven different mayors, he managed the design and construction of Seattle's public works projects—some $42 million in bridges, sewers, water lines, and other efforts that significantly changed the city's landscape. While he had many detractors, was called "impractical," and often found himself the center of controversy, Thomson retained a vision of Seattle that gradually became reality.

Among his most notable achievements were the Cedar River gravity water system that brought a bountiful supply of fresh water to the city; Seattle's first sewer line; the city's first paved streets; the Denny Street regrade and other regrades that reclaimed many acres of tideland; the 12th Avenue Bridge; the Grant Street Bridge; the Ballard Locks; and Lake Washington Boulevard. Thomson led the movement that resulted in the establishment of the Port of Seattle and later served as its first chief engineer.

Thomson also served on the Seattle City Council from 1916 to 1921 and again as city engineer from 1930 to 1931. As a consulting engineer for the Washington Toll Bridge Authority in 1937, he also contributed to the design of the first Lake Washington Floating Bridge. In his final years, Thomson wrote his autobiography, *That Man Thomson*, published a year after his death in 1949.[15]

NOTES

1. Quote from "His Expertise Was Expansive," *Seattle Times*, August 21, 1989, F1, F2; *Who's Who in Engineering: A Biographical Dictionary of the Engineering Profession, 1937*, 4th ed. (New York: Lewis Historical Publishing Company, 1937), 28; *Who's Who in Engineering: A Biographical Dictionary of the Engineering Profession, 1941*, 5th ed. (New York: Lewis Historical Publishing Company, 1941), 38.
2. Application for License as Professional Engineer, #233 (1935), Washington State Department of Licenses for Professional Engineers and Land Surveyors; "Cecil Arnold, Longtime Consulting Engineer, Dies" [obituary], *Seattle Times*, January 28, 1963, B20.
3. Washington State Department of Licenses for Professional Engineers and Land Surveyors, License Application #279, Olympia; Washington State University Alumni Office, Pullman; WSU Board of Regents' Distinguished Alumnus Awards, www.regents.wsu.edu/distinguished-alumni; "William A. Bugge, Washington's Highway Builder," *Highway News* 11, no.1 (July-August 1963), 13–14.
4. Clarence B. Bagley, *History of Seattle from the Earliest Settlement to the Present Time*, vol. 3 (Chicago: S. J. Clarke Publishing Company, 1916), 566; Bureau of Vital Statistics, Washington State, Death Index, 1929; "Arthur Dimock's Death Mourned by Civic Leaders," *Seattle Times*, November 15, 1929, 12.
5. "L. R. Durkee Named Engineer of the Year," *Pacific Builder and Engineer*, March 1960, 3; *Who's Who in Engineering: A Biographical Dictionary of the Engineering Profession, 1954*, 6th ed. (New York: Lewis Historical Publishing Company, 1954), 680; Application for License as Professional Engineer, #1308 (1935, 1937), Department of Licenses, State of Washington; Bureau of Vital Statistics, Washington State, Death Index, 1929.
6. Clark Eldridge, manuscript autobiography (1896–1982), loaned by his son, C. W. Eldridge; "Engineer Has Long Career as Bridge Builder," *Seattle Times*, June 1, 1940; Doris Hensel, "Longtime Bridge Engineer Recalls 'Galloping Gertie' with Heartache," *Daily Olympian,* September 3, 1986; "Clark Eldridge, 94, Bridge Builder, Didn't Let Age Dull Desire to Work," *Seattle Times*, November 9, 1990, E9; Application for License as Professional Engineer, #352, November 8, 1935, and Application for Land Surveyor, October 28, 1963, Washington State Board of Professional Engineers and Land Surveyors, Olympia; "Capture and Imprisonment by Japs," *Pacific Builder and Engineer* 51 (December 1945), 44–

49; Clark H. Eldridge, "The Tacoma Narrows Suspension Bridge," *Pacific Builder and Engineer* 46 (July 6, 1940), 35–40; Eldridge, "The Tacoma Narrows Bridge," *Civil Engineering* 10 (May 1940), 299–302.

7. "Last Rites for Admiral Gregory Set" [obituary], *Seattle Times*, September 14, 1960, 57; *Who's Who in America, 1950* (Chicago, IL: International Who's Who), 1066; *Who's Who in Engineering: A Biographical Dictionary of the Engineering Profession, 1948*, 6th ed. (New York: Lewis Historical Publishing Company, 1948), 780.
8. Quoted in "'Hadley's Folly' Turned into Island's Floating Bridge," *Mercer Island Reporter*, January 2, 1991; Mark Higgins, "Floating Bridge to Bear Name of Innovator Hadley," *Seattle Post-Intelligencer*, July 17, 1993, B1, B3; "Designer Honored," *Olympian*, July 19, 1993, C3; "H. M. Hadley Dies; Bridge Designer," *Seattle Post-Intelligencer*, July 7, 1967, 43; Lucile McDonald, "The Inspiration for the First Floating Bridge," magazine section, *Seattle Times*, July 26, 1964, 6; "Hadley, Floating Bridge Proposer, Dies in Lake," *Seattle Times*, July 6, 1967, 55; Application for License as Professional Engineer, #51 (1935), Washington State Department of Licenses for Professional Engineers and Land Surveyors.
9. Application for License as Professional Engineer, #445 (1935), Washington State Department of Licenses for Professional Engineers and Land Surveyors; "Humes Placed among City's No. 1 Citizens," *Seattle Times*, June 26, 1941 [Biography File, Special Collections, University of Washington Libraries].
10. Application for License as Professional Engineer, #559 (1935), Washington State Department of Licenses for Professional Engineers and Land Surveyors; "Seattle Man Helped Build Vital Corregidor Fortress," *Seattle Times*, January 7, 1942 [Biography File, Special Collections, University of Washington Libraries]; "D. W. McMorris, Former City Engineer, Dies," *Seattle Times*, November 10, 1942 [Biography File, Special Collections, University of Washington Libraries]; C. T. Conover, "D. W. McMorris Was Noted for City Engineering Work," *Seattle Times*, January 1, 1960, 24; Clarence B. Bagley, *History of King County, Washington*, vol. 3 (Chicago: S. J. Clark Publishing Company, 1931), 898–903; Clarence B. Bagley, *History of Seattle From the Earliest Settlement to the Present Time*, vol. 3 (Chicago: S. J. Clarke Publishing Company, 1916), 694.
11. [Obituary], *Transactions, American Society of Civil Engineers*, vol. 95 (New York: ASCE, 1931), 1565–1566.
12. *Who's Who in Engineering: A Biographical Dictionary of the Engineering Profession, 1937*, 4th ed. (New York: Lewis Historical Publishing Company, 1937), 990; *Who's Who in Engineering: A Biographical Dictionary of the Engineering Profession, 1941*, 5th ed. (New York: Lewis Historical Publishing Company, 1941), 1275; *Who's Who in Engineering: A Biographical Dictionary of the Engineering Profession, 1948*, 6th ed. (New York: Lewis Historical Publishing Company, 1948), 1427–1428; *Who's Who in Engineering: A Biographical Dictionary of the Engineering Profession, 1954* (New York: Lewis Historical Publishing Company, 1954), 1739.
13. Quote from State Highway Commission Resolution No. 1815, March 20, 1967; "Lacey Murrow, Former Director, Honored at WSC," *Highway News* 8 (May-June 1959), 19; "Ex-State Highway Director Murrow Dies of Gunshot," undated clipping, Lacey V. Murrow biographical file, Washington State Library; *Who's Who in the State of Washington, 1939–1940* (Seattle: Gordon Barteau, 1940), 137–138; *America's Young Men: The Official Who's Who Among the Young Men of the Nation* (Los Angeles: Richard Blank Publishing Company, 1934), 439.
14. Quoted in Robert P. Hammond, "A Bridge Can Be Beautiful," Sunday Magazine, *Seattle Times*, February 17, 1963, 12–13; Application for License as Professional Engineer, #1271 (1935), Washington State Department of Licenses for Professional Engineers and Land Surveyors; "Civil Engineers Honor Powell," *Seattle Times*, May 17, 1964, 9; Tom Paulsen, "Husband's Devotion Unto Death," *Seattle Post-Intelligencer*, June 10, 1991, B1.
15. Reginald Heber Thomson, *That Man Thomson* (Seattle: University of Washington Press, 1950); Application for License as Professional Engineer, #158 (1935), Washington State Department of Licenses for Professional Engineers and Land Surveyors; "R. H. Thomson, Civic Leader, Engineer, Dies," *Seattle Post-Intelligencer*, January 8, 1949, 4; "R. H. Thomson, 80, Hearkens Back to the Days of Cow Paths," *Seattle Times*, March 20, 1932; Lucile McDonald, "R. H. Thomson—He Changed the Face of Seattle," Sunday Magazine, *Seattle Times*, September 6, 1964, 10–11; C. T. Conover, "Just Cogitating: Reginald H. Thomson Was Early Day City Engineer," Sunday Magazine, *Seattle Times*, March 20, 1960, 6; "R. H. Thomson, City Builder and Leader, Dies at 92," *Seattle Times*, January 7, 1949; "Reginald Heber Thomson," in Clarence Bagley, *History of King County*, vol. 2 (Chicago: S. J. Clarke Publishing Company, 1929), 92–100.

Part Two: Premier Bridges

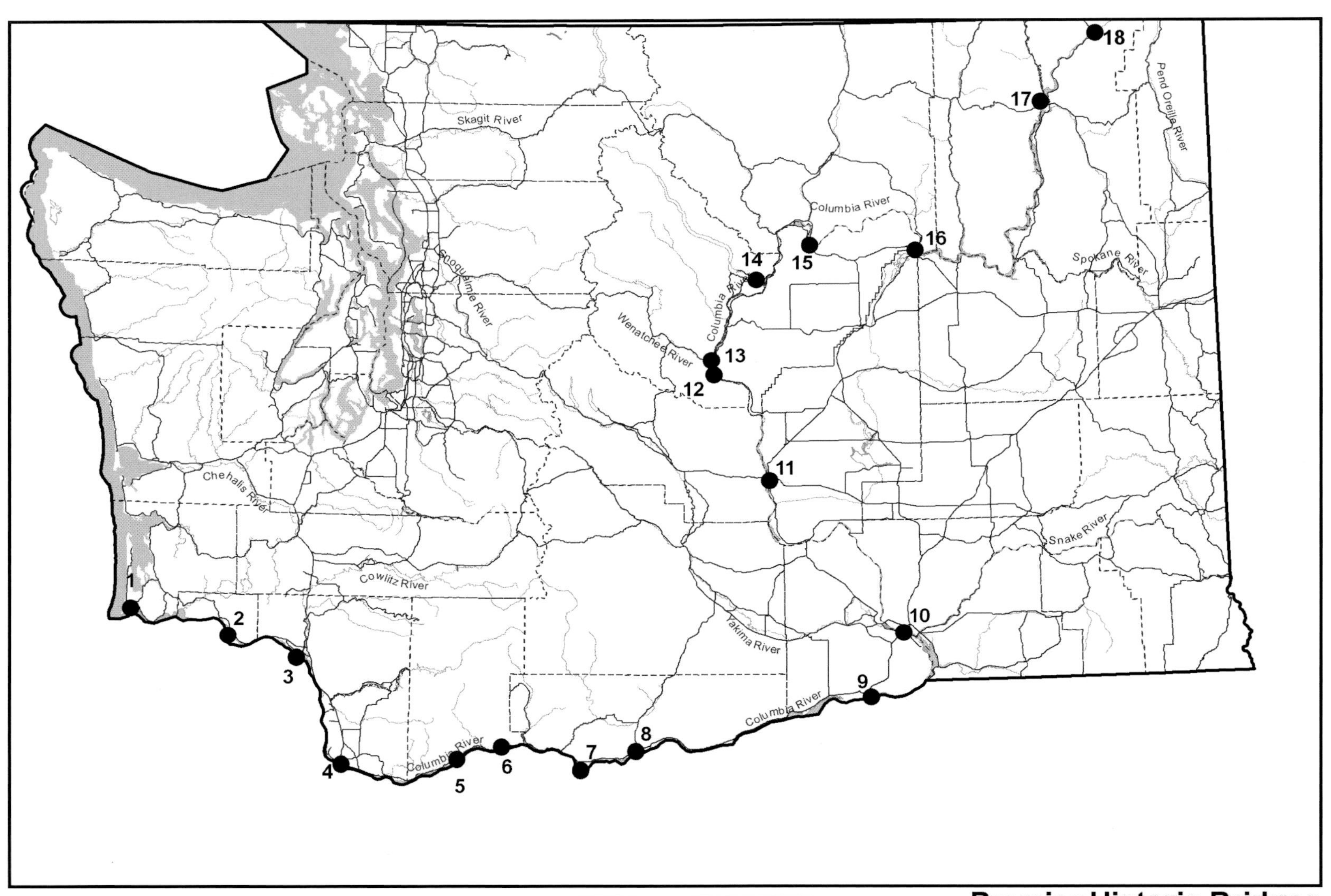

Bridge No.	Bridge Name	Bridge No.	Bridge Name
1	Astoria-Megler Bridge	10	Pioneer Memorial ("Blue") Bridge
2	Julia Butler Hansen Bridge	11	Vantage I-90 Bridge
3	Longview (Lewis and Clark) Bridge	12	Wenatchee Bridge (1950)
4	Vancouver-Portland I-5 Bridges	13	Wenatchee Bridge (1907)
5	Bridge of the Gods	14	Beebe Bridge
6	Hood River - White Salmon Bridge	15	Bridgeport Bridge
7	The Dalles Bridge	16	Grand Coulee Bridge
8	Biggs Rapids Bridge	17	Kettle Falls Bridge
9	Umatilla Bridge	18	Northport Bridge

Premier Historic Bridges of the Columbia River

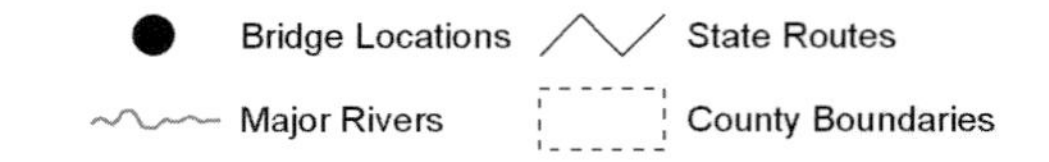

Bridges of the Columbia River

It was called *Nch'i-Wana,* "The Big River," by the region's mid-Columbia Native American tribes. To early explorers, the Columbia River was the legendary "River of the West." Its mighty and revered waters provided the surest, yet a dangerous, route through the mountains to the Columbia Plateau and other interior destinations.

Initially, Europeans searched the rugged coastline for two centuries, hoping to find the fabled "Inland Passage" that would lead to a waterway across the continent. Later, an American sea captain, Robert Gray, discovered the mouth of the river he named for his ship, the *Columbia Rediviva,* in 1792. From the British territories (now Canada) and the newly founded United States came fur traders who relied on the river as the region's highway to and from the interior. As white settlement advanced, ferries were established where the most-traveled roads crossed the river.

As an obstacle to travel, few barriers are more formidable. The Columbia is one of the world's great rivers. Its tributaries in the United States and Canada drain an area of 259,000 square miles. About 680 miles (over half) of the river's 1,210-mile length pass through Washington before emptying into the Pacific Ocean. The Columbia is second only to the Mississippi in volume of water carried by streams in the United States. Where the river crosses the Canadian border, it measures nearly a quarter-mile wide, and at its mouth, it stretches over four miles between Washington and Oregon.

Until dams controlled its flow on all but the "Hanford Reach" and its lower reaches below Bonneville Dam, the Columbia flooded annually each spring, scouring its steeply incised banks in places, while spreading over floodplains elsewhere. For nearly a century after the arrival of the first nonnatives to the region, ferries provided travelers with the only means of crossing its broad, often treacherous waters.

Bridging the great Columbia River called for unique, extraordinary, and massive engineering efforts. The Northern Pacific Railroad first bridged the Columbia near Pasco in 1888, followed by the Great Northern Railway at Rock Island, near Wenatchee, in 1893. Not until the early years of the twentieth century did Washington residents witness the first successful highway bridge construction. In 1907, private businessmen in Wenatchee built the first public road bridge across the Columbia River in Washington. The pin-connected, steel cantilever truss carried irrigation water from the Wenatchee River to arid lands slated for development in East Wenatchee. Pedestrians, horses, and wagons also crossed the bridge, as did automobiles until 1950.

The Spokane, Portland, and Seattle Railway built the first movable bridge (a swing span) across the Columbia between Oregon and Washington in 1908. Highway structures soon followed with completion in 1917 of the Interstate Bridge connecting Vancouver and Portland, publicly funded by tolls and bond sales in both states. The Puget Sound Bridge and Dredging Company erected a steel cantilever highway bridge between Pasco and Kennewick in 1922. In 1924 the Oregon-Washington Bridge Company completed the Hood River Bridge, linking Bingen and White Salmon, Washington, with Hood River, Oregon.

Two years later, the Te Wanna Toll Bridge Company completed the Bridge of the Gods, an elegant steel cantilever. Promotion of a "North Bank Highway" on the river's Washington side helped spur completion of the Bridge of the Gods, crossing near where a massive Columbia Gorge landslide once temporarily blocked the river, an event associated with the gods in Native American legend.

Upriver near Chelan, an agricultural company erected a wooden suspension bridge in 1920 to carry its irrigation water across a lonely stretch of the Columbia. The Beebe Orchard Company charged tolls for local vehicular traffic, as well as for sheep and pedestrians. Between Wenatchee and the Tri-Cities, the state installed a major link in its cross-state highway at Vantage, where a 1,636-foot steel cantilever was completed in 1927. Further upstream, a 2,000-foot-long, timber-decked bridge crossed the Columbia at Brewster as early as 1928. (Forty years later, a spark from a cutting torch ignited newly creosoted deck timbers and the bridge was destroyed by fire.) Just as the Great Depression began to grip the nation, the Longview (Lewis and Clark) Bridge, one of the most spectacular of the early spans across the Columbia, opened to traffic at Longview in 1930.

In the 1930s, when the federal government began building dams on the Columbia, rising waters required modification of some bridges, replacement of others, and construction of new ones. On the lower Columbia above Bonneville Dam, the Hood River Bridge and the Bridge of the Gods were raised to lift their decks above shipping lanes. On the upper river, construction of Grand Coulee Dam resulted in building a new bridge (1935) just below the dam and replacing bridges (1941) at Kettle Falls and Fort Spokane. On the uppermost reaches of Lake Roosevelt behind Grand Coulee Dam, the State Department of Highways built a bridge (1949) at Northport a few miles south of the Canadian border. Construction of Chief Joseph Dam in the early 1950s required a new bridge (1950) at Bridgeport.

The 1950s saw several new outstanding bridges spanning the Columbia. At Wenatchee, construction of a graceful steel arch (1950) garnered the state a national prize for aesthetics in bridge building. In response to population growth in the Hanford area, the Pioneer Bridge (1954) was built between Pasco and Kennewick. (It has since become known as the "Blue" Bridge because of its attractive paint scheme.) Completion of McNary Dam helped bring about construction of the James Sturgis Bridge (1955) at Umatilla. Traffic jams on Interstate 5 between Vancouver and Portland sped installation of a second bridge (1958) parallel to the earlier bridge.

Bridges constructed across the Columbia in the 1960s included some of the most impressive steel through-truss structures in the state: Biggs Rapids (1962), Beebe (1962), Vantage (1962), Vernita (1965), and Astoria-Megler (1965). After a fire destroyed the Brewster Bridge in 1968, a new bridge was built (1970) at the site between Pateros and the mouth of the Okanogan River. Later in the decade, Arvid Grant designed perhaps the most artistically sophisticated bridge in the state, the Ed Hendler cable-stayed bridge (1977), crossing the Columbia between Pasco and Kennewick.

Today, twenty-five highway bridges cross the Columbia River in Washington, including those spanning the border with Oregon. That total does not include railroad bridges or the bridge built in 1907 to carry an irrigation pipe and local traffic between Wenatchee and East Wenatchee. The most recent bridges, the Richard Odabashian (1975) north of Wenatchee, the Glen Jackson (1982) between east Vancouver and Portland, the Lee-Volpentest (1984) connecting Pasco and Richland, and the second Umatilla Bridge (1988), reflect the evolution to posttensioned, reinforced-concrete, box-girder construction.

The river's impact on the region's history, economy, and social life is immense. Construction of six federal dams and five nonfederal dams on the Columbia has brought hydroelectric power, irrigation, and flood control. Irrigation of the fertile river valleys and the dry Columbia Basin allowed for the establishment of orchards and cropland over large areas of Eastern Washington that once were desert. The wealth of electric power from the Columbia's waters also facilitated much of the state's industrial development since World War II. The historic highway bridges spanning the Columbia (discussed below, proceeding upstream from the mouth of the river) have been an integral part of, and vital links in, those dramatic changes.

The 1966 Astoria-Megler Bridge's four-mile length includes a 2,464 foot continuous through truss, the longest of its type in the world.

Washington State Archives, WSDOT Records

Astoria-Megler Bridge, Pacific County, Washington, and Clatsop County, Oregon

As the Columbia River approaches the Pacific Ocean between Point Ellice, near Megler, Washington, and Astoria, Oregon, it is a strong, tidal current some 4.1 miles across. Here, on the well-traveled Pacific Coast Highway, Route 101, officials of both Washington and Oregon dedicated the Astoria-Megler Bridge in 1966. With the bridge's completion, U.S. Highway 101 finally became an unbroken link between the borders with Canada and Mexico.

This great structure is one of the longest bridges in the United States, measuring 21,697 feet in length. The main (south) ship channel near the Oregon shore is crossed by a continuous truss consisting of a 1,232-foot span and two 616-foot flanking spans. U.S. Navy aircraft carriers can sail smoothly at high tide under the spans' 196-foot height above the ship channel. Seven truss and eight girder spans cross the ship channel on the Washington side, where the bridge rises to forty-nine feet of vertical clearance. Desdemona Sands, the shallow mid-section of the river, is crossed by 140 spans of prestressed concrete, each eighty feet long, standing barely twenty-five feet above sea level. The bridge is twenty-eight feet wide, providing two lanes for auto traffic.

The Julia Butler Hansen Bridge, completed in 1939, is the only bridge on the Columbia that crosses only part of the river, connecting Puget Island and Cathlamet, Washington.

Washington State Archives, WSDOT Records

A joint Washington-Oregon effort, the Astoria-Megler Bridge was designed by both states' highway departments, led by William A. Bugee, the Washington State Department of Highways director. The bridge was designed to survive some of nature's fiercest challenges, and can withstand storms off the Pacific Ocean that hurl gusts up to 150 miles per hour. The concrete piers brace against the river's flood speed of nine miles per hour and the occasional assaults of whole trees carried on the rushing waters.

The DeLong Corporation and Raymond International, both of New York, built the bridge's piers and substructure. United States Steel's American Bridge Division constructed the steel trusses. The J. H. Pomeroy and Ben C. Gerwick companies of San Francisco built the Desdemona Sands viaduct portion of the bridge. The $24 million cost was paid for by bonds sold by the state of Oregon and repaid from tolls collected on the Oregon side until 1993. (The state of Oregon owns the bridge.)[1]

Julia Butler Hansen Bridge, Wahkiakum County

This unique historic bridge is located a few miles inland from the Columbia's mouth. It is the only bridge that spans just a portion of the great river, crossing the channel between Cathlamet and Puget Island. A ferry on the other side of the island connects this remote corner of the state with Westport, Oregon.

The first Europeans exploring the area gave the island its name. Lieutenant William Broughton, carrying out Captain George Vancouver's orders, sailed up the river in October 1792 and named the island for a fellow Royal Navy Lieutenant, Peter Puget. Lewis and Clark later called it "Sturgeon Island," but that name failed to stick. At five miles long and two miles wide, it is the largest island in the Columbia River and lies within Washington's Wahkiakum County.

The bridge, designed by Lacey V. Murrow, R. W. Finke, and Clark H. Eldridge of the Washington State Department of Highways, was completed in August 1939. When built, it consisted of four steel spans, including a through-truss cantilever more than 400 feet long, nine timber-deck truss spans of 90 feet each, and 323 feet of timber trestle approaches on the Puget Island side. (The timber trestles have since been replaced by concrete approaches). The Long-Bell Company in Longview supplied much of the 750,000 feet of Douglas fir used in the bridge. The Arch Rib Construction Company assembled the timber trusses for the overall construction contractor, the Parker-Schram Company of Portland. Construction costs totaled $500,000.

The 2,433-foot-long bridge, originally known as the "Puget Island-Cathlamet Bridge," allowed for 60 feet of vertical clearance above the channel. A five-day-long celebration marked the bridge opening, begun by President Franklin Roosevelt's "cutting" of a ribbon by telegraph from the White House. The structure was later named for Julia Butler Hansen, who represented southwest Washington in Congress from 1960 to 1975.[2]

Longview (Lewis and Clark) Bridge, Cowlitz County, Washington, and Columbia County, Oregon

The bridge across the Columbia River at Longview holds a curious distinction. Upon its completion in 1930, the record-setting span with its 1,200-foot central section became, for a time, the largest and tallest cantilever structure in North America. It also had become the most "politically engineered" of the state's spans.

The bridge's tortuous political path began in 1921. That year, the Oregon State Legislature authorized the

When the Longview (Lewis and Clark) Bridge was completed in 1930, it was the biggest cantilever span in North America. Photo 1993.

Jet Lowe, HAER

Oregon State Highway Commission to conduct a site survey below Portland for possible locations for a new span across the Columbia River. When the commission recommended a site opposite the newly founded town of Longview, the Portland Chamber of Commerce and many other Oregonians feared that the bridge, rather than attracting Washington dollars to their area, might instead advance Longview's commercial interests at Portland's expense. When efforts for a joint bridge project by Washington and Oregon failed, private interests took up the cause.

Soon, entrepreneurial dreams of transforming Longview into a thriving port city prompted efforts to construct a bridge between the town and Rainier, Oregon. Robert Long, a visionary midwestern businessman, founded the city bearing his name where the Cowlitz River flows into the Columbia, in close proximity to four railroads and one of the largest timber and logging operations in the country.

Political squabbling between the commercial rivals on both sides of the river reached a high point in 1927. Oregon authorities approved a franchise stipulating extreme bridge design demands and requiring design approval by the secretaries of war, agriculture, and commerce. In November 1927 Congress authorized private construction of the bridge, but required that the span provide a clear channel width of 1,000 feet and vertical clearances of 195 feet at mid-span (to accommodate tall-masted clipper ships) and 185 feet at the channel piers.

Though cost "estimates" for the bridge ballooned to as high as $5.8 million, the project proceeded. Designed by the noted Strauss Engineering Corporation of Chicago, the Longview Bridge is 3,892 feet long. The 1,200-foot steel cantilever central section is flanked by two 760-foot steel anchor spans and five Warren truss approach spans. The 12,500 tons of superstructure were fabricated by Bethlehem Steel in Pennsylvania and erected by J. H.

The 3,538-foot long Vancouver-Portland I-5 Bridge over the Columbia River, with its 275-foot vertical lift span, was completed in 1917 (foreground). The parallel bridge of equal dimension was completed in 1958 (in back). Photo 1993.

Jet Lowe, HAER

Pomeroy and Company of Seattle. The original timber approaches, measuring 2,618 feet on the Oregon side and 1,754 feet on the Washington side, were reconstructed in 1950 and 1963, respectively.

Official dedication ceremonies unfolded on March 29, 1930, with President Herbert Hoover in the White House pressing a golden telegraph key that dropped a blade to cut a long chain of yellow daffodils hanging across the bridge, and the governors of Washington and Oregon met with a handshake. While the span became an important transportation and commercial link for the region, it proved a financial headache. Built at the onset of the Great Depression, the bridge never attracted sufficient auto traffic to garner tolls that would pay for the $588,113 construction cost. Finally, in 1947 the Washington State Toll Bridge Authority purchased the span. Not until 1965 were the last bonds retired and the tolls removed.[3]

Vancouver-Portland, Interstate 5 Columbia River Bridges, Clark County, Washington, and Multnomah County, Oregon

With great joy and satisfaction, residents of Clark and Multnomah counties witnessed completion of a highway bridge across the Columbia River between Vancouver and Portland in 1917. It was a striking engineering feat and an enormous financial accomplishment. The bridge was one of the largest in the world at the time. It represented an unprecedented level of cooperation between Washington and Oregon residents. More importantly, the span became symbolic of a new transportation era, the automobile age, beginning to unfold in the Pacific Northwest.

Building a bridge between the two cities was a dream of local residents for decades. Commerce between the cities relied on a small steam ferry. But a traffic jam at the ferry on June 30, 1905, during Clark County Day at Portland's Lewis and Clark Centennial Exposition, sparked the first widespread popular demand for construction of a bridge.

Over the next dozen years, citizens and businessmen on both sides of the Columbia rallied behind efforts to erect a "Pacific Highway Bridge." Following civic demonstrations by bridge boosters, and lobbying by the Vancouver and Portland commercial clubs, the campaign won approval in 1914 from the Oregon and Washington legislatures for the sale of bonds. Construction began in March 1915.

On February 14, 1917, area residents witnessed the bridge's historic official dedication. At 12:30 p.m. two daughters of Multnomah and Clark county officials untied a ribbon that dropped a rope separating enthralled Portland residents from the equally joyous Vancouver crowd. At the same time, the mayors of the two cities and the governors of the two states clasped hands as four flags were unfurled from the towers of the lift span and a band played *The Star Spangled Banner.* The next day, the Portland *Oregonian* trumpeted, "With brilliant formality, the Columbia River interstate bridge yesterday swung into its niche in the great scheme of commercial and industrial development of the Northwest," replacing the small ferryboat that in a farewell "tooted a mournful salute to the cheering thousands crowded on the bridge."

Built at a cost of $1.6 million, the completed structure was monumental in size. Stretching more than 3,500 feet long, the bridge includes two Parker truss spans 275 feet long, a 531-foot Pennsylvania Petit truss, and a 275-foot vertical-lift draw span with a clearance of 150 feet above high water. Because of the bridge's extraordinary length, plus the difficulty of building falsework in the swift river current, workmen erected all but one of the spans on the shore, then floated them by barges to their piers.

The bridge earned $287 on its opening day, and paid for itself within twelve years. On January 1, 1929, Oregon and Washington purchased the structure from Multnomah and Clark counties and abolished tolls. Although the 1917 bridge was designed to accommodate future traffic growth, it became clear by the 1940s that traffic demands were taxing the bridge's capacity. Also, a dramatic increase in marine activity, requiring more lift span openings, worsened congestion. Oregon and Washington joined in a study of Columbia River crossings, followed by a detailed look at the Portland-to-Vancouver issue. It was determined that a new bridge immediately adjacent to the 1917 bridge was the most feasible solution.

An updated twin of the 1917 structure, the Southbound Interstate 5 Columbia River Bridge was completed in 1958. The Bridge Department of the Oregon State Department of Highways designed the new structure, which today carries three lanes of traffic between Vancouver and Portland. The 3,538-foot-long bridge has sixteen spans and includes a three-span steel Parker through-truss vertical-lift unit, consisting of a 274-foot south tower span, a 278-foot lift span, and a 272-foot north tower span. Each tower rises to a height of 189 feet and supports a counterweight equal to half the lift span's weight. The 279-foot-long lift spans on this bridge and on the adjacent northbound bridge share the state record for lift span length.

Significant engineering features of the 1958 bridge include lift spans and flanking tower spans, its steel Pennsylvania Petit through-truss span, and the unusual precast concrete shell design for the underwater supports of the eleven river piers. This bridge is one of only two lift bridges that were built in Washington in the 1950s. (The other is the southbound crossing of State Route 529 over the Snohomish River.) The 1958 bridge's 531-foot-long Pennsylvania Petit truss span is the tenth longest simply supported steel truss span in North America.

With fanfare patterned after the 1917 ceremony, the new bridge opened to traffic on July 1, 1958, with a notable contemporary addition—the roar of jet aircraft overhead and artillery fire in an opening salute. Following the new bridge's construction, the old bridge was remodeled. The project's total cost was estimated at $14.5 million, including $6.7 million for construction of the new bridge and the Oregon Slough Bridge, and $3 million for remodeling the old bridge.

The Interstate 5 bridges continue to serve a heavy volume of traffic. In 1999, an average of 122,000 vehicles crossed the Columbia River here daily.[4]

Completed in 1926, the Bridge of the Gods straddles the Columbia where a native legend says a great natural stone arch once spanned the river.
WSDOT

Bridge of the Gods, Skamania County, Washington, and Hood River County, Oregon

A native legend described a great, natural bridge arching over the Columbia River just west of Cascade Locks. In the story, a beautiful young woman, Loo Wit, served as guardian of a sacred flame on the bridge. But the two sons of the "Supreme Being," Tyee Sahale, both wanted Loo Wit as a wife, and they quarreled. This angered Tyee Sahale, who destroyed the bridge and, in the process, his sons and Loo Wit. The debris from the bridge became the Cascade Range. The sons and girl are Mount St. Helens, Mount Adams, and Mount Hood.

In the early twentieth century, bridge promoters envisioned a suspension span at this site. Improvements of the so-called "North Bank Highway," the earliest roadway constructed by the Department of Highways linking eastern and western Washington, helped spur completion of the Bridge of the Gods. Congress authorized a crossing, and the U.S. Army Corps of Engineers accepted a bridge plan in December 1920.

The Interstate Construction Company, a consortium of local investors, installed a concrete pier on the river's Oregon side before encountering financial difficulties. The Te Wanna Toll Bridge Company—consisting of the Portland and southwest Washington investors backing the recently built bridge across the Columbia River between Kennewick and Pasco—then acquired the project. In 1926 the Puget Sound Bridge and Dredging Company completed the Bridge of the Gods, an elegant, steel cantilever through-truss, at a cost of $650,000. Motorists paid a toll of fifty cents to cross.

Anticipating rising water levels behind Bonneville Dam, scheduled for completion in 1938, the Bridge of the Gods owners gladly accepted $762,000 from the federal government for needed modifications. One hundred fifty workers added concrete to the tops of the piers, and hydraulic hoists raised the bridge 44 feet, providing 135 feet of vertical clearance for river traffic. When completed in 1940, the modified bridge extended 1,858 feet in length. In 1961 the Port of Cascade Locks acquired the structure, which remains, as it was when first opened to traffic, a toll bridge.[5]

Hood River-White Salmon Bridge, Klickitat County, Washington, and Hood River County, Oregon

Few bridges reflect the evolution in technology, building materials, and of the Columbia River itself like the Hood River-White Salmon Bridge. Spanning the Columbia between White Salmon, Washington, and Hood River, Oregon, today it bears only slight resemblance to the first structure completed in 1924. Originally a fixed-center span, the bridge is now movable, thanks to the addition of towers and mechanisms that lift the Pennsylvania Petit through-truss span vertically to a height of seventy-two feet above the water.

As early as 1886, Captain John Stanley ferried wagons and livestock across the river on a barge and a sailboat at this location. Several decades later, impending completion of the Columbia River Highway on the Oregon side prompted Congress to approve building a bridge here in early 1923.

Amazingly by today's standards, construction began quickly in September of that year after local communities

When built in 1924, the Hood River-White Salmon Bridge was a "fixed" bridge. Its through-truss was converted to a movable lift span in the late 1930s.
WSDOT

and Portland investors raised money to finance the Seattle-based Oregon-Washington Bridge Company. The Gilpin Construction Company completed the bridge on December 9, 1924, making it the fifth highway bridge built across the mighty Columbia. Toll collection started the day after 1,500 citizens gathered for the opening celebration.

Just over a decade later, the impending completion of Bonneville Dam, located downstream, would raise the level of the Columbia. Consequently, the federal government covered expenses for rebuilding the Hood River-White Salmon Bridge. Piers were installed to carry six new steel-deck spans, and existing piers were elevated. The center piers were strengthened to support the lift towers, which were patterned after those of the Hawthorne and Steel bridges over the Willamette River in Portland. Between 1938 and 1940, a lift span and tollbooth were added to the structure at a cost of more than $500,000.

Additional alterations further changed the bridge's appearance. In 1946 the wooden trestles were replaced on both approaches—by steel-deck spans on the Oregon side and by concrete-girder spans on the Washington end. Later, open-steel decking replaced the original wood-plank deck and pedestrian sidewalk. Wood railings and curbs also were replaced, and the tollbooth was moved to the Oregon side. In 1950, the Port of Hood River purchased the structure from the Oregon-Washington Bridge Company for $800,000. The port continues to maintain the bridge and collect tolls.

Today, its most notable attributes are its vertical-lift main span and steel-deck trusses. The 4,418-foot-long bridge is very narrow and has a steel-grate deck with no pedestrian or bicycle facilities.[6]

The Dalles Bridge, Klickitat County, Washington, and Wasco County, Oregon

The name of this outstanding bridge comes from a French term describing the dangerous rapids where The Dalles Dam stands today. By the 1820s, French-Canadian fur trappers dubbed the rocks and treacherous currents here as "les grand dalles." These magnificent rapids on the Columbia had long been a popular meeting, fishing, and trading place for local tribes.

By 1854 a commercial ferry began cross-river service at The Dalles. The year 1865 brought the first of many efforts by enterprising community figures to have a bridge built. Nearly a century of trial, error, struggle, and periods of dormancy followed before success was achieved. In 1947 the Washington and Oregon highway departments supported building a bridge in the vicinity.

Construction of the The Dalles Dam helped move the idea to reality. In early 1951, the U.S. Army Corps of Engineers received approval and $800,000 from Congress to begin planning for the dam. At the same time, Wasco County officials financed a bridge with revenue bonds to be repaid by toll charges, and they awarded a $1.9-million bridge construction contract.

The Dalles Bridge was built in 1953 as part of the The Dalles Dam hydroelectric project.

Craig Holstine, WSDOT

Work on the bridge proceeded until January 1952. Then, another problem appeared, causing another delay.

The Corps completed studies showing that the dam's spillway requirements meant the new bridge would have to be relocated downstream. With construction and steel fabrication well underway, a series of complicated negotiations ensued between county and Corps officials. The new site required a longer bridge than originally planned. In the end, the huge, three-span truss section designed for the old location was incorporated into the new bridge. This enabled bridge builders to use 2,200 tons of steel already fabricated for the main truss.

The Dalles Bridge was finished on December 18, 1953, at a cost of $2.4 million. It connects Murdock and Dallesport, Washington, to The Dalles, Oregon, on U.S. Route 97. This remarkable structure is 3,339 feet long and composed of thirty sub-spans. Its most notable feature is the three-span, riveted steel Warren through-truss section. The longest individual section is the 576-foot main truss span, which is flanked by two 128-foot cantilever anchor spans. The Dalles Bridge was one of only two steel cantilever bridges built in Washington in the 1950s. (The Umatilla Bridge was the other.) It remained a toll facility until November 1, 1974.[7]

Biggs Rapids (Sam Hill Memorial) Bridge, Klickitat County, Washington, and Sherman County, Oregon

Building a bridge across the Columbia River at Biggs Rapids, south of Goldendale, proved more difficult than for most spans on the river. The main problem was a political one. Interstate travel from California to Canada had long depended on a ferry at Biggs Rapids. The ferry operation between Biggs Junction, Oregon, and Maryhill, Washington, had served the needs of farmers, businessmen, and tourists since the early 1920s. By the 1950s, the lack of a bridge on this thoroughfare became a serious hindrance to commerce.

The Washington State Toll Bridge Authority completed plans for a new span in 1959. The proposed bridge would be a two-lane structure nearly a half-mile long. The price tag of $2.5 million would be paid for by vehicle tolls. Bids for construction opened on February 9, 1960.

Then, the project ground to a halt. Problems with the Hood Canal Floating Bridge and the proposed second Lake Washington Bridge garnered both headlines and priorities for funding. Goldendale residents soon understood that the fate of "their" bridge awaited the opening of bids for the second Lake Washington bridge on July 20. Finally, in the autumn of 1960, construction started; $3.5 million in bonds assured completion, much to the relief of local residents. By January 1961, the construction firm, Paul Jarvis Inc. of Seattle, reported that the bridge was on schedule and, weather permitting, would be completed by the end of October 1962. Eight of the fourteen supporting piers stood ready, and steelworkers had hung the steel on the Oregon side between "pier 5" and the approach.

The Biggs Rapids Bridge opened to the public on November 1, 1962, after two years of construction. State and

The Biggs Rapids Bridge opened in 1962.
Washington State Archives, WSDOT Records

Biggs Rapids Bridge is also known as the Sam Hill Memorial Bridge, in honor of the early highway promoter and builder of the nearby Maryhill mansion.

Washington State Archives, WSDOT Records

local political dignitaries proudly cut the ceremonial ribbon on the Oregon side. Then, Washington Governor Albert D. Rosellini, Oregon Governor Mark Hatfield, and highway officials joined local dignitaries for a luncheon at the Maryhill Museum of Fine Arts.

The Biggs Rapids Bridge includes a fixed, single-span, steel through-truss structure. Its two-lane highway is 26 feet wide and 2,567 feet long. The bridge has a vertical clearance of 75 feet and a horizontal clearance of 300 feet for river traffic.[8]

Umatilla (James H. Sturgis) Bridge, Benton County, Washington, and Umatilla County, Oregon

"Sturgis's Folly," they called it in the late 1940s. When Umatilla County judge James H. Sturgis began promoting a bridge across the Columbia River just west of McNary Dam, cynical citizens took a dim view of the project because it looked like a long-term burden for taxpayers.

In the early 1950s, the ferry operation at Patterson and Umatilla hauled more than 250,000 vehicles a year between Washington and Oregon. Long lines and delays faced truckers and travelers, who grew increasingly eager to see a bridge. Meanwhile, dam development in the region held the promise of agricultural expansion and population growth. McNary Dam soon would make irrigation of the nearby Horse Heaven Hills economically feasible. A bridge at Umatilla would give area farmers ready access to markets.

Umatilla County officials decided the time to act was at hand. After a favorable feasibility study and report, the county issued $10 million in bonds to finance the bridge. Tolls collected from bridge users would repay the bonds. Construction began in 1954, and the bridge opened to traffic on April 15, 1955.

Designed by the Tudor Engineering Company of San Francisco, the Umatilla Bridge is 3,308 feet long. The center section of the steel cantilever bridge is a five-span, continuous, Warren through-truss stretching 1,920 feet over the river. Its cantilever and anchor spans are actually a "partial through-truss" design because their lower chords are below the roadway. The two-lane structure is twenty-seven feet seven inches wide, curb to curb, and offers boats and barges a vertical clearance of eighty-five feet.

After the bridge's completion, the area experienced unprecedented agricultural growth and economic development. By August 1974, the bridge bonds were repaid. Three months later, the tolls were removed. Today, it carries southbound traffic on Interstate 82, while northbound traffic uses a parallel bridge just to the east, which opened in 1988.[9]

In the planning stage, the 1955 Umatilla Bridge (foreground) was called "Sturgis's Folly" by those doubting the wisdom of Judge James H. Sturgis, the bridge's staunchest promoter. The second Umatilla Bridge, built in 1988, stands immediately to the east.
Craig Holstine, WSDOT

Pasco-Kennewick Bridges, Benton and Franklin Counties

As late as 1921, a small, six-car ferry carried motorists across the Columbia River between Pasco and Kennewick on the Inland Empire Highway, one of the state's three original trunk routes. At this time, more than 1,400 citizens from seventeen communities in Washington and Idaho joined a public subscription drive, described by the *Seattle Times* as the "greatest community undertaking in the history of the Northwest." The drive resulted in the formation of the Benton-Franklin Intercounty Bridge Company, and subsidized construction of a multi-span steel-truss bridge that was completed in 1922. The *Kennewick Courier-Reporter* expressed sentiments shared by many of its readers: "The day the bridge was opened to traffic, a new era dawned for each community. Pasco

The first highway bridge (foreground) over the Columbia between Pasco and Kennewick was completed in 1922. A half century later the bridge became the focus of an unsuccessful preservation effort.
WSDOT

awoke in possession of thousands of acres of fields and orchards and Kennewick annexed a railroad payroll."

A toll facility until 1931, the bridge remained in private hands until purchased by the state in 1927. By the end of World War II, the old two-lane structure was inadequate for new traffic being generated by expansion of the Hanford Engineering Works outside nearby Richland. Richland, Pasco, and Kennewick (not yet known as the "Tri-Cities") had become one of the busiest centers for the development of atomic energy in the nation. When a farmer lost a load of hay on the bridge during an Army Day parade in April 1948, a traffic jam of unprecedented proportion sparked an investigation by the Army, which concluded that the bridge was obsolete.

Completed in 1954, the graceful "Blue Bridge" (Pioneer Memorial Bridge) is 2,520 feet long.
Craig Holstine, WSDOT

When opened in 1978, the Ed Hendler Bridge, named after a former Pasco mayor, became the second longest bridge of this type in the world.
Craig Holstine, WSDOT

In 1951 a statewide bond issue for road improvements included money for a new bridge, to be located about a mile upstream from the 1922 structure. That year the Washington State Highway Department's Bridge Division designed the new structure. Three Seattle contractors completed construction in 1954 at a cost of $6.5 million. Unity was the theme at the formal dedication ceremony on July 30, 1954, when dignitaries stressed the bridge's role in bringing together the three cities as the hub of southeast Washington's economy.

Its tied-arch main span characterizes the 2,520-foot structure, as does the blue-painted steel spans, the source of its popular name, the "Blue Bridge." A contest sponsored by a local radio station resulted in officially naming it the "Pioneer Memorial Bridge," but the name has yet to stick.

Twenty-four years after the Blue Bridge's completion, yet another monumental bridge was built across this busy stretch of the Columbia River. In September 1978, Governor Dixie Lee Ray arrived to dedicate the first cable-stayed bridge built in the lower 48 United States and the second largest in the world. Olympia engineer Arvid Grant, designer of what may be the state's most innovative and elegant bridge, observed that "one does not build a bridge across a great river like the Columbia but once in a lifetime."

The new bridge was named for Ed Hendler, a former Pasco mayor, who led the fight to have the 1922 bridge removed. After an unsuccessful court challenge by preservationists, the old structure was torn down. In 1998 lights were added to the new bridge, illuminating the nation's only double-span, cable-stayed structure that former Secretary of Transportation Sid Morrison called "one of the world's most beautiful bridges."[10]

Vantage Bridge, Grant and Kittitas Counties

From basalt-cliff overlooks above the Columbia River, motorists crossing Washington via Interstate 90 are usually awed at their first sighting of the bridge at Vantage. The immense structure stands amidst the broad sweep of a dramatic landscape. The graceful tied arch of the main span mimics the high surrounding hills with its parabolic top chord and its color, painted light brown, blends with desert textures.

The present bridge stands about a mile and a half downstream from its predecessor, a steel cantilever structure built in 1927 on the old Sunset Highway. That bridge served cross-state travelers for three and a half decades. When the Grant County Public Utility District announced plans to build Wanapum Dam downstream, the

The Interstate 90 Bridge at Vantage replaced an earlier cantilever structure that was dismantled and rebuilt on the Snake River at Lyons Ferry.

Washington State Archives, WSDOT Records

State Highway Department designed a new bridge to cross over the anticipated new reservoir. Once the new bridge was completed, workers began dismantling the old bridge. Its steel truss components were stored at a railway siding in Beverly, a few miles downstream, until shipped to Lyons Ferry on the Snake River, where the bridge was reerected in 1968 (see Chapter 7).

The J. W. Hardison Company began work on the new bridge's west approaches in mid-1961. In early 1962, the Munson Construction and Engineering Company of Seattle started building the piers and substructure. Finally, in December of that year, the United States Steel Corporation's American Bridge Division began erecting the steel tied-arch superstructure. Highway Department head William Bugge officially dedicated the new bridge on November 1, 1962.

Designed by the Washington State Department of Highways under George Stevens, head bridge engineer, the new Vantage bridge was significantly larger than the older structure. Its 520-foot, tied-arch, main span exceeded the cantilever's suspended center span in length by 320 feet. The older structure's 1,636-foot length paled in comparison to the new bridge's 2,495-foot total length, which includes six steel-girder approach spans and two Warren deck trusses, in addition to the tied-arch main span. Unlike the 1927 bridge, which was a two-lane structure, the present bridge carries four lanes of Interstate 90.[11]

Columbia River Bridge No. 285/10 at Wenatchee, Chelan and Douglas Counties

One of the state's most aesthetically pleasing bridges spans the Columbia near the south edge of Wenatchee and East Wenatchee. When completed in 1950, the 1,208-foot structure was unique, being the first of its kind in Washington. Its main 480-foot span (consisting of a 352-foot span suspended between two cantilever spans) is tied to a steel-arch through-truss, 66 feet above the roadway deck at its crown, and 180 feet above the river.

For many years, the nearby 1907 bridge, a pin-connected steel cantilever structure, was obsolete as a highway bridge. In fact, the structure was never really designed for automobile traffic. As a vital link in cross-state transportation, the crossing here demanded a modern structure capable of carrying increasing traffic loads. The Department of Highways' 1946–48 biennial report described the new bridge as the most necessary structure of all those then under construction in the state. The General Construction Company of Seattle built the bridge using structural steel fabricated by American Bridge, a division of the U.S. Steel Corporation, in Gary, Indiana. When the bridge was completed in 1950, the American Institute of Steel Construction presented it with a National Competitive Award for the most beautiful bridge built that year comprised of one or more spans more than 400 feet in length.[12]

First Columbia River Bridge at Wenatchee, Chelan and Douglas Counties

This graceful, steel through-truss bridge spanning the Columbia River between Wenatchee and East Wenatchee is the only pin-connected cantilever remaining in the state. As the first highway bridge across the Columbia, the Wenatchee span heralded the new era in transportation dominated by the automobile. But this bridge served more than cars, horses, wagons, and trucks. It also carried an irrigation pipe to funnel water from the Wenatchee River that helped fuel East Wenatchee's growth.

With the arrival of the Great Northern Railway in 1890, Wenatchee became the hub of a rapidly developing agricultural mecca. Judge Thomas Burke of Seattle enticed Great Northern tycoon James J. Hill, President George P. Baker of the First National Bank of New York City, and others to invest in a venture combining bridge building, irrigation, and land development in East Wenatchee. In January 1906, these entrepreneurs incorporated the Washington Bridge Company to build

In 1907 the Wenatchee Bridge (background, upstream) became the first non-railroad structure to span the Columbia River in Washington. Today, it is the only surviving pin-connected steel cantilever bridge in the state. The 1950 bridge at Wenatchee (foreground) won an award for its beautiful design.

Washington State Archives, WSDOT Records

a bridge that would carry both a large water pipe and vehicular traffic. They planned to collect tolls from travelers, whose only other choice was to wait for the local ferry. Bridge users faced a steep toll on this first Columbia River crossing—pedestrians and bicyclists paid five cents one way; equestrians and single-horse buggies paid fifteen cents one way or twenty-five cents round-trip; two-horse teams paid twenty-five cents; and four-horse teams, thirty-five cents. Free easement for the water pipe was granted to the Wenatchee Canal Company, which was owned by the same businessmen.

With steel from the American Bridge Company of New York, workers began erecting the superstructure on concrete piers in March 1907. The 1,060-foot steel span was comprised of two anchor arms, each 240 feet long; two cantilever arms, 160 feet each; a suspended span of 200 feet; and one plate girder 60 feet long. Another 565 feet of timber approaches brought the bridge to a total length of 1,625 feet. At the time, this was a monumental structure for the state of Washington. Its roadway width of twenty-six feet six inches was designed to accommodate "ordinary highway traffic," plus a single-track street railway. Brackets on the outer sides of the trusses supported two water lines, each four feet in diameter. Construction costs totaled $170,000 by the time the bridge opened to traffic on January 20, 1908. Even before the opening, the Commercial Club of Wenatchee petitioned the legislature, requesting that the state purchase the bridge. The club declared its goal was enhancement of the developing highway system, but their main motivation was to eliminate the bridge's tolls.

State Highway commissioners who inspected the bridge in June 1909 gave it a glowing assessment. The structure, they reported, would be of "great value" to the state, "inasmuch as there is a very heavy traffic over it, and this traffic will rapidly increase as the orchards on the east side of the Columbia come into bearing." Before the year ended, Governor Marion Hay signed a bill authorizing state purchase of the bridge.

Highway officials rejected the asking price of $190,000, however, and the matter lingered until 1911, when the state finally purchased it for $125,000. The bridge remained a part of the state highway system until a new structure was completed a short distance downstream in 1950. An artifact of early twentieth-century engineering, the 1907 bridge stands as a reminder of how development of irrigated lands spurred construction of the state's first highway bridge across the Columbia River.[13]

First Beebe Bridge, Chelan and Douglas Counties

The Beebe Bridge, like the early structure crossing the Columbia at Wenatchee (1907), was built to transport water to orchards being developed along the river. When irrigation from the east bank terraces above the Beebe Orchard Company's farms proved inadequate, the company purchased the Chelan Bob Allotment west of the Columbia. The allotment contained Indian Camp Springs, a source of abundant irrigation water. Pumps and water pipes laid on the river bottom had failed, so the company decided to build a bridge for two twelve-inch water pipes. Another advantage of having a bridge was the elimination of the ferry that the company had relied on to transport its increasing apple crop.

In the spring of 1919, MacRea Brothers of Seattle began work on a wooden suspension bridge to carry the pipelines, as well as local traffic and fruit trucks, across this isolated stretch of river. Construction was not uneventful. Two workers, John McDonald and Walter Oaks, survived falling 80 feet into the Columbia. In a dedication remembered as "one of the big celebrations of the year" (1920), with a stunt plane flying over and under the structure, the bridge opened on a toll basis to non-company traffic.

Although its 2¾-inch steel cables were designed to support the weight of two twelve-inch water pipes, vehicular traffic, and the bridge deck itself, the structure had a limited capacity for bearing weight. A bridge tender controlled traffic and collected fares, which never totaled half

The Second Beebe Bridge, completed in 1963, is a 1,227-foot long steel through truss.

Robert Jackson Holstine

his annual salary. Flocks of sheep traveling between winter and summer ranges crossed at one cent per head.

The bridge continued in use as the only Columbia River crossing between Wenatchee and Brewster until 1963. By then, the Washington State Department of Transportation had built a new steel truss bridge nearby, while achieving a long-sought goal of completing a water-level route along the Columbia and Okanogan rivers between Wenatchee and the Canadian border. Just downstream from the new bridge, concrete towers remain from the First Beebe Bridge. Also lying unnoticed in thick brush on the west bank are rusting strands of cables that once suspended this key link in the area's early commercial development.[14]

Second Beebe Bridge, Chelan and Douglas Counties

Construction of a highway following the Columbia River's east side from Orondo to Beebe required a modern bridge at Chelan Station. The Chelan County Public Utility District's decision in the 1950s to begin construction on the Rocky Reach Dam accelerated plans for a new bridge. Installation of the bridge piers prior to raising the reservoir behind the dam saved the state considerable expense, eliminating the need to use floating equipment on the west bank.

To meet the tight schedule, the Alton V. Phillips Company of Seattle began construction of pier 3 on July 21, 1958. Working from a barge, the company prepared a coffer dam for pier 2 off the east bank in April 1959. High water destroyed the structure two days after its completion, delaying work on pier 2 until August. Finally, in January 1960, the company finished all concrete work on pier 2.

On October 23, 1961, WSDOT awarded the Troy T. Burnham Company of Seattle a $1.24 million contract to construct the new bridge. Concrete piers and an approach span were in place by June 1, 1962, but a steelworkers' strike delayed erection of the trusses until later in the year. Final riveting was completed on February 5, 1963. Deck pouring and installation of a twenty-four-inch irrigation pipe on the bridge's downstream side occurred shortly thereafter. The nearby forty-four-year-old suspension bridge was then removed.

The Bridgeport Bridge, a three-span continuous deck truss, was completed in 1950 as part of the Chief Joseph Dam project. Photo 1993.
Jet Lowe, HAER

Designed by WSDOT engineers, the Second Beebe Bridge opened to traffic in 1963. The steel through-truss consists of a 520-foot tied-arch center span, flanked by through-truss spans of 262 and 260 feet, with a 180-foot reinforced-concrete box girder on the west approach. The structure's total length is 1,227 feet.[15]

Bridgeport Bridge, Douglas and Okanogan Counties

The 1,150-foot-long span over the Columbia River at Bridgeport, like the Grand Coulee Bridge, was designed and built as part of a federal hydroelectric dam project. Opened to traffic in 1950, it was the only significant steel-deck truss bridge built in Washington from 1941 to 1950.

Bridgeport had its beginnings as a mining town in the 1880s. In the early twentieth century, steamboats plied the Columbia's waters as far north as Bridgeport, but when railroad lines reached the nearby town of Brewster, the end was at hand for commercial river and ferry traffic at Bridgeport. During the severe winter of 1929, heavy ice downed the ferry, terminating service. Fruit from local orchards, as well as wheat and other grain crops, now went to market by train from Brewster.

After World War II, the federal government appropriated funds to build the Chief Joseph Dam at Bridgeport. By 1949 work started on a road between Brewster and Bridgeport, and plans for a bridge moved forward swiftly under the auspices of the U.S. Army Corps of Engineers. The bridge site offered a narrow crossing point, and theoretically a reinforced concrete-arch structure could be built. But the Columbia flowed swiftly here—too fast to allow construction of falsework needed for a concrete bridge.

The result was a three-span continuous deck truss that cost $1.25 million. Like the bridge at Coulee Dam, this one would be strong enough to bear the heavy loads of construction trucks and equipment. Liberal use of silicon steel boosted the load rating while keeping the bridge itself relatively light. The 800-foot-long deck truss has a horizontal top chord. Its lower chords are sloped (which allowed for lower and cheaper pier construction) with a maximum vertical clearance to the water below. The Brideport Bridge is owned by the federal government and maintained by the state.[16]

Grand Coulee Bridge, Douglas and Okanogan Counties

At Grand Coulee, the Columbia River once narrowed to a channel some 600 feet wide. On a blistering-hot August afternoon in 1934, President Franklin D. Roosevelt crossed the Columbia River here on an old ferry to dedicate one of the state's grandest public enterprises under the New Deal—the Grand Coulee Dam-Columbia Basin Reclamation Project. Two months later, work began on the Grand Coulee Bridge, located just a half-mile downriver from the dam site. This location had been a favorite river crossing for many years. Seaton's Ferry provided service to local travelers from 1920 to 1934, but the dam project soon changed that. Starting in 1933, a hoard of workers that later numbered up to 8,000 men began building the dam, which is still one of the largest concrete structures in the world.

Designed mainly by the State Highway Department's team of bridge engineers, the bridge was financed with Public Works Administration funds under the U.S. Bureau of Reclamation (which owned the bridge from 1935 to 1959). Building the span took longer than expected. Work on the piers started in September 1934, but high water and frequent equipment breakdowns soon slowed progress. Then, the clay hillsides proved unstable. The slope next to the east pier slipped, causing the pier to tilt. During much of the pier construction phase in 1935, engineers and workmen labored to stabilize the hillsides and strengthen the east pier.

The bridge is a 1,066-foot-long, single Warren through-truss that stands 152 feet above the river. The steel structure measures 950 feet, with the cantilever and suspended span accounting for 550 feet of that length. Interestingly, both this bridge and the Deception Pass Bridge were designed by the State Highway Department at the same time, and they have similarities. Although the Deception Pass Bridge is a deck truss and the Coulee Dam Bridge is a through-truss, they have identical measurements for their suspended spans and cantilever arms. Both bridges have sloping bottom chords and very similar bracing patterns.

The Grand Coulee Bridge opened to traffic on January 27, 1936. It linked the "engineers' town," Coulee City, on the river's west side, with the dam site and the east side's Mason City, the community built for workers. This hard-serving bridge had to be strong. Engineers designed the span to include as much silicon steel as possible to handle the heavy trucks that daily pounded across during the dam construction.[17]

Opened in 1936, the Grand Coulee Bridge played a vital role in the construction of Grand Coulee Dam from 1936 to 1941. Photo 1993.

Jet Lowe, HAER

Decorative lighting distinguishes the Grand Coulee Bridge. Photo 1993.

Jet Lowe, HAER

With the building of Grand Coulee Dam, rising waters forced construction of a new and higher bridge at Kettle Falls in 1941 (at left in photo), replacing the original 1929 bridge. A railroad bridge stands adjacent (at right) to the highway bridge. Photo 1993.

Jet Lowe, HAER

Kettle Falls Bridge, Ferry and Stevens Counties

In 1941, with the completion of Grand Coulee Dam, the Franklin D. Roosevelt reservoir was formed, extending nearly to the Canadian border. Rising waters changed life for many people—towns, roads, and bridges had to be relocated. One of the two highway bridges that needed replacing was a span over the Columbia River near Kettle Falls.

At this locality, the Columbia originally tumbled over a thirty-foot-high falls, gouging huge, kettle-shaped pockets in the flat bedrock below. French-Canadian traders had referred to these rock formations as "les chaudieres" or "the kettles." Fur trappers came to meet and barter with Indians, who for centuries had gathered here to fish and trade. In the 1860s, white settlers platted the town of Kettle Falls, which throughout the late nineteenth century served area miners, lumbermen, and farmers. Ferries transported travelers here, but by the mid-1920s, local support for a bridge over the Columbia connecting Stevens County with Ferry County gathered momentum.

Completed in November 1929 by J. H. Tillman, the first Kettle Falls Bridge was a 1,219-foot-long structure with a 528-foot steel cantilever main span and deck truss approaches. The bridge served the region for more than a decade, until the rising waters of Lake Roosevelt doomed it for demolition. At Kettle Falls, the reservoir would rise a hundred feet higher than the average level of the free-flowing Columbia. In late 1940 and early 1941, residents of the old town of Kettle Falls watched the construction of the new bridge, while they prepared for the inundation to come by moving their community to higher ground. Local Indians held a "ceremony of tears," mourning the loss of the salmon runs that had sustained their people for thousands of years.

The replacement bridge, which opened on May 3, 1941, is of a modified through-truss design. It has a center span stretching 600 feet between piers. This earned the riveted steel bridge the distinction of having the longest central span of any bridge built in Washington during the 1940s. The structure cost $452,000 (paid for by the U.S. Bureau of Reclamation), achieving a highly cost-effective design for the special features of the setting. The bottom chords on the cantilever and anchor arm spans are sloped, which allowed for a lower height in the main piers, resulting in a substantial savings in pier design and costs. The top chord is nearly horizontal. Total bridge length is 1,266 feet, with the steel portion accounting for 1,050 feet. Nearby, the concrete piers of the old bridge rise out of the lake's placid waters. They are the only visible reminders of the 1929 structure.[18]

Northport Bridge, Stevens County

Some thirty miles upstream from Kettle Falls, and just a few miles south of the Canadian border, stands the Columbia River bridge at Northport in Stevens County. Originally, the Red Mountain Railroad Company built a timber-trestle bridge here in 1897. That bridge later was converted for automobile and truck use. By October 1946, however, it reached the end of its useful life. No longer safe for traffic, it was closed.

Two months earlier, the state had awarded a contract for building a steel replacement span, but the great spring 1948 flood undermined the bridge before it could be completed; swirling water eroded the earth fill around the main pier on the bridge's south end.

The contractors redesigned the structure and resumed work. To remove the weakened pier and make way for a new, deeper bridge foundation, they used special techniques in underwater blasting. Finally, on June 13, 1951, the bridge opened to motorists. Cost for the 1,542-foot Northport Bridge totaled $1.75 million.

The completed bridge is a riveted steel cantilever through-truss, with a 224-foot-long suspended section, two 140-foot-long cantilever spans, and two 168-foot-long anchor spans. The truss sections have a horizontal top chord and lower chords, while the anchor and cantilever arms have sloping bottom chords. Its design, able to withstand extreme flood conditions, makes it not only the largest man-made structure in Stevens County, but also one that residents can rely on for decades into the future.[19]

The 1,542-foot long Northport Bridge, completed in 1951, is a riveted steel cantilever through-truss and the largest man-made structure in Stevens County.
Washington State Archives, WSDOT Records

NOTES

1. "Megler-Astoria: A Bridge for the Columbia's Mouth," *Highway News* 11, no. 1 (July-August), 1963, 7–8; "Astoria Bridge Dedication Program," 1966, brochure in Washington State Department of Transportation Environmental Affairs Office files; "Bi-State Bridge Inventory for Columbia River Crossings," WSDOT, Southwest Region Planning, May 1998.
2. WSDOT Bridge Preservation Office files, Olympia; *Wahkiakum County Eagle*, May 3, 1979; Burwell Bantz, Director, Washington State Department of Highways, to G. S. Paxton, Oregon State Highway Commission, October 30, 1942, copy of letter in WSDOT Environmental Affairs Office; Robert Hitchman, *Place Names of Washington* (Tacoma: Washington State Historical Society, 1985), 242.
3. Robert W. Hadlow, "Longview Bridge," HAER No. WA-89, August 1993; Lisa Soderberg, "Longview Bridge," HAER Inventory, 1980; "Bi-State Bridge Inventory for Columbia River Crossings," WSDOT, Southwest Region Planning, May 1998.
4. Jonathan Clarke, "Vancouver-Portland Interstate Bridge," HAER No. WA-86, August 1993; Lisa Soderberg, "Vancouver Portland Bridge, Columbia River Interstate," HAER Inventory, 1980; Oscar R. "Bob" George, "Southbound Interstate 5 Columbia River Bridge," National Register of Historic Places Nomination, 2001; "Bi-State Bridge Inventory for Columbia River Crossings," WSDOT, Southwest Region Planning, May 1998.
5. Articles from the Stevenson *Skamania County Pioneer*: "Bridge of the Gods Finanaced," July 24, 1925, "Secures Property for Bridge Piers," January 9, 1925, "Work on Bridge Next Week," November 9, 1925, "Work on Bridge of Gods under Way," November 27, 1925, "Bridge Work Being Rushed," April 9, 1926, "Bridge of Gods Progressing Nicely," October 1, 1926, untitled article regarding completion of bridge, October 29, 1926; "Bridge of the Gods Piers Lengthen," *Spokesman-Review,* December 13, 1937; "Port Body Buys Bridge of Gods," *Marine Digest*, December 30, 1961, 24; "Bi-State Bridge Inventory for Columbia River Crossings," WSDOT, Southwest Region Planning, May 1998. The native legend about a great natural bridge was used by Frederic Homer Balch in his *The Bridge of the Gods*, a late-nineteenth century romance novel.
6. "Columbia River Toll Bridge at Hood River, Oregon Now in Service," *Engineering News* 94, no. 1, January 5, 1925, 130, and "Safeguarding Lone Highway Bridge from Fire and Other Damage," *Engineering News* 94, no. 18, April 30, 1925, 722–23; Oregon-Washington Bridge Company, *Annual Reports*, 1934–1942, Hood River County Museum, Hood River, Oregon; Judith A. Chapman and Elizabeth O'Brien, "Hood River-White Salmon Bridge Section 106 Documentation Form," Archaeological Investigations Northwest, Inc., Portland, Oregon, June 21, 2004; Burwell Bantz, Director, Washington State Department of Highways, to G. S. Paxton, Oregon State Highway Commission, October 30, 1942, copy of letter in WSDOT Environmental Affairs Office; "Bi-State Bridge Inventory for Columbia River Crossings," WSDOT, Southwest Region Planning, May 1998.
7. Oscar R. "Bob" George, "Columbia River Bridge at The Dalles," NRHP Nomination, 2001; "Bi-State Bridge Inventory for Columbia River Crossings," WSDOT, Southwest Region Planning, May 1998.
8. "Biggs Rapids Bridge Discussed," *Goldendale Sentinel,* July 21, 1960; "Bid Opening Set Tuesday for Span," *Goldendale Sentinel*, February 4, 1960; "Biggs Rapids Bridge Assured," *Goldendale Sentinel*, August 18, 1960; "Construction Moves Rapidly on Bridge and Approaches," *Goldendale Sentinel*, January 11, 1962; "Biggs Rapids Bridge," *Highway News* 10, nos. 1–2 (September-October), 1962, 16; "Sam Hill Memorial Bridge to Open Thursday Morning," *Goldendale Sentinel*, November 1, 1962; "Rosellini to Attend Opening of Interstate Bridge," *Goldendale Sentinel*, October 25, 1962; "Bi-State Bridge Inventory for Columbia River Crossings," WSDOT, Southwest Region Planning, May 1998.
9. Oscar R. "Bob" George, "Columbia River Bridge at Umatilla," NRHP Nomination, 2001; Pamphlet, "James H. Sturgis Bridge Official Dedication Ceremony, June 1, 1991," WSDOT Environmental Affairs Office files; "Bi-State Bridge Inventory for Columbia River Crossings," WSDOT, Southwest Region Planning, May 1998.
10. Lisa Soderberg, "Pasco-Kennewick Bridge," HAER No. WA-8, 1980; Paul Dorpat and Genevieve McCoy, *Building Washington: A History of Washington State Public Works* (Seattle: Tartu Publications, 1998), 112–13; "The Cable Bridge," *Tri-City Herald*, March 2, 1978; Bob Woehler, "Cable Bridge Opens," *Tri-City Herald*, September 17, 1978; Genoa Sibold-Cohn, "Cable Bridge Brightens the Night," *Tri-City Herald*, September 16, 1998; Peter J. Lewty, *To the Columbia Gateway* (Pullman: Washington State University Press, 1987), 48–49, 116; Lewty, *Across the Columbia Plain* (Pullman: Washington State University Press, 1995), 47, 51–52, 57, 64, 67–70.
11. C. E. Sines, "Five Million Dollar Vantage Bridge Progressing Smoothly," *Highway News* 9, no. 4 (January-February 1961), 32–33; Bill Merry, "Public Pulse of Highway Progress," *Highway News* 11, no. 2 (September-October 1963), 7; Engineering Records, Plans Vault, WSDOT, Olympia.

12. Robert H. Krier, J. Byron Barber, Robin Bruce, and Craig Holstine, "Columbia River Bridge No. 285/10 at Wenatchee," NRHP Nomination, 1991.
13. *Biennial Report of the Highway Commissioner*, 1910, 45; Dorpat and McCoy, *Building Washington* (1998), 111.
14. Leon Cronk, "History of Old Beebe Bridge," *Chelan Valley Mirror*, July 11, 1963; *Wenatchee Daily World*, July 16, 1964.
15. "Beebe Bridge Construction Fulfills Age-Old Need," *Highway News* 9, no. 1 (July-August), 1960, 13–14, 20; "Beebe Bridge," *Highway News* 10, no. 3 (November-December), 1962, 20.
16. Robert H. Krier, J. Byron Barber, Robin Bruce, and Craig Holstine, "Columbia River Bridge at Bridgeport," NRHP Nomination, 1991; Robert W. Hadlow, "Columbia River Bridge at Bridgeport," HAER No. WA-90, August 1993.
17. Robert W. Hadlow, "Columbia River Bridge at Grand Coulee," HAER No. WA-102, August 1993; Robert W. Hadlow, "Washington State Cantilever Bridges, 1927–1941," HAER No. WA-106, November 1993.
18. "Kettle Falls Bridge," *Western Construction News*, August 25, 1929; Robert H. Krier, J. Byron Barber, Robin Bruce, and Craig Holstine, "Columbia River Bridge at Kettle Falls," NRHP Nomination, 1991; Robert W. Hadlow, "Columbia River Bridge at Kettle Falls," HAER No. WA-91, August 1993; Hadlow, "Washington State Cantilever Bridges, 1927–1941," HAER No. WA-106, November 1993.
19. Robert H. Krier, J. Byron Barber, Robin Bruce, and Craig Holstine, "Columbia River Bridge at Northport," NRHP Nomination, 1991.

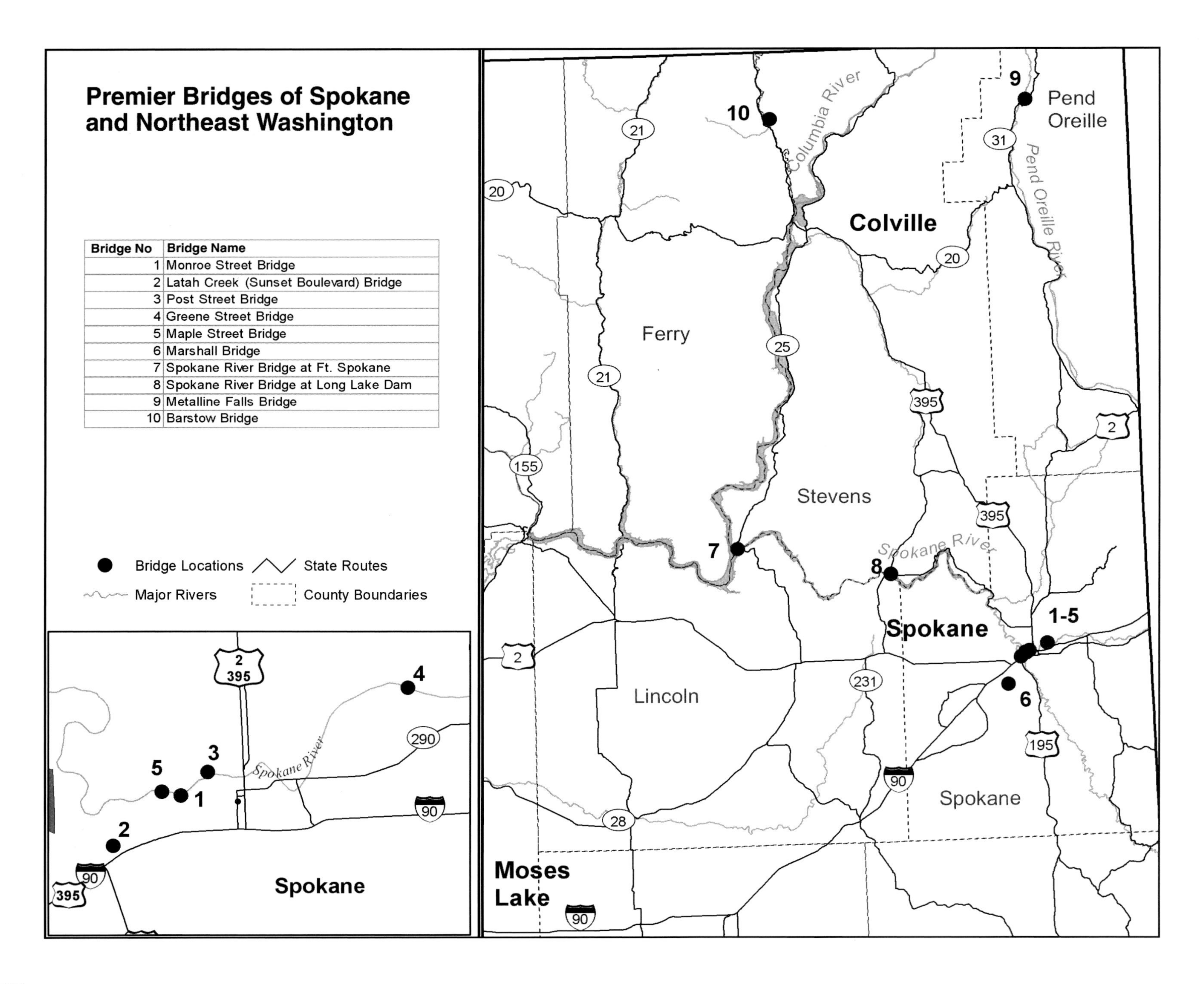

Bridge No	Bridge Name
1	Monroe Street Bridge
2	Latah Creek (Sunset Boulevard) Bridge
3	Post Street Bridge
4	Greene Street Bridge
5	Maple Street Bridge
6	Marshall Bridge
7	Spokane River Bridge at Ft. Spokane
8	Spokane River Bridge at Long Lake Dam
9	Metalline Falls Bridge
10	Barstow Bridge

Bridges of Spokane and Northeast Washington

Some of the Northwest's most rugged terrain is found in Washington's northeast corner. Here, the Spokane, Colville, Kettle, and Pend Oreille rivers flow to the wide Columbia through forested mountain ranges. Mining, milling, logging, lumbering, farming, and ranching fueled the area's economy through the late nineteenth century.

Spokane became the hub of the Inland Northwest's industrial, commercial, cultural, and educational activities. Founded at Spokane Falls to capitalize on abundant waterpower for flour and lumber mills, Spokane is a picturesque example of a community built on a river. The Spokane River bisects the city, and bridges here are a unifying force. To early travelers and settlers, however, the rushing waters represented a substantial barrier.

White settlement in the area dates to 1810, when fur traders erected Spokane House near the confluence of the Spokane and Little Spokane rivers several miles northwest of the present city. After the community's establishment in 1871 and subsequent incorporation ten years later as "Spokane Falls," the city thrived with completion of the Northern Pacific Railroad in 1883. A devastating fire in 1889 destroyed much of the commercial district, but citizens rapidly rebuilt the heart of the town, boosted by rapidly developing regional extractive industries. Soon renamed "Spokane," the city surpassed Walla Walla as the dominant economic and social center of the "Inland Empire."

The first decade of the twentieth century witnessed Spokane's greatest growth. Construction of commercial buildings—from hotels, banks, warehouses, and department stores, to multistory office structures—brought a modern, urban face to the city. By 1910 Spokane boasted a population of some 104,000 inhabitants.

Spokane's Bridges

From its hard-won rebirth after the 1889 fire, Spokane grew rapidly into what many called the "Queen of the Inland Empire," due in part to the city's concentration along the banks of the Spokane River and the falls. In no small way, its rise to prominence was due to the many innovative solutions used to bridge the Spokane River and its tributaries.

The first Spokane area bridges were built outside of what later became the central business district, on thoroughfares connecting Spokane to Walla Walla and southeast Washington, and to the Colville Valley to the north. These early wooden bridges served fortune-seekers scurrying to the region's mining districts, but, for a time, this brought few benefits to the fledgling settlement. Eventually, wooden bridges were built across the Spokane River within the town, but their tenure was brief. In 1890 city officials dynamited a recently completed 260-foot span across the river at Bernard Street when the structure was declared unsafe, even for pedestrian traffic. That year, a timber bridge on Hangman Creek collapsed, killing a team of horses. Seven years later the structure that replaced it also collapsed, and twenty head of cattle were lost. The colossal floods of May and June 1894 swept three more wooden bridges down the Spokane River.

The adoption of reinforced-concrete bridges early in the twentieth century gave the city sturdy, aesthetically pleasing spans that stood the test of time. First in a new wave of concrete spans was the Washington Street Bridge (1908), which directly linked the downtown to the business district north of the river. The attractive, 242-foot,

The Washington Street Bridge was the first of five concrete arch bridges, built 1908–10, over the Spokane River.

Washington State Archives, Eastern Regional Branch, Cheney

Mission Avenue Bridge under construction, February 1, 1909. The Mission, Washington, and Trent bridges have been replaced.

Washington State Archives, Eastern Regional Branch, Cheney

ribbed-concrete-and-steel structure was composed of three 77-foot arches consisting of German cement, reportedly used because of the lack of a domestic supply. Next came the Mission Street Bridge, then the East Olive Bridge (today's Trent Avenue), also concrete arches, effectively securing Spokane's adoption of concrete bridges. Between 1908 and 1910, the city had built five concrete arch bridges. By 1936, thirty-one bridges, twenty of these concrete, were constructed in and around Spokane.[1]

The first Monroe Street Bridge was lost to fire in 1890.

Northwest Room, Spokane Public Library

Monroe Street Bridge

As the primary link between north and south Spokane, the river crossing at Monroe Street has seen three bridges since 1888. A fire in 1890 claimed the first structure—a 1,240-foot bridge consisting of wooden trestle approaches and two wood-and-steel trusses—only a year after its construction. A steel bridge erected in 1891 to replace the wooden predecessor suffered partial collapse and was declared unsafe in 1905. By then increased vehicular and streetcar size, weight, speed, and numbers made the bridge obsolete.

The current bridge had its beginnings in 1908. In December of that year, the city council approved a preliminary design (not the one eventually used) by City Engineer J. C. Ralston and issued $500,000 in construction bonds. Locating the bridge provoked heated debates by residents, who wanted the structure in someone else's neighborhood.

City engineers crafted designs for a new bridge of remarkable aesthetic sensitivity, and construction began in 1910. When a severe wind storm struck the city on July 21, the massive wood falsework collapsed, sending two men to their deaths. Steel trusses replaced the failed wood scaffolding, and work continued. Two of the steel trusses

Built of steel in 1891, the second Monroe Street Bridge represented an advancement in bridge building technology and materials. By 1905, however, it was declared unsafe.

Northwest Room, Spokane Public Library

were later used to build the Howard Street Bridge, which stands today east of the Monroe Street Bridge in downtown Spokane's Riverfront Park (see Chapter 3). Despite the loss of life and construction delays, in the following year workers completed what was then the largest concrete arch bridge in the nation. The structure garnered attention from around the world, and set the course for future spans across the Spokane River.

The bridge measured 791 feet long with a 50-foot roadway and two 9-foot cantilevered sidewalks, and initially included a double streetcar track. The center span rose 115 feet above the river and measured 281 feet long. On each side stood two 120-foot semicircular spans. A single 100-foot semicircular span extended from the south end. Continual problems with the south footing plagued the project. Finally, a 150-foot timber approach had to be constructed. Later, this was replaced with five concrete arches that matched the eight on the north bank.

The architectural firm of Cutter and Malmgren of Spokane designed the bridge's ornamentation. Cutter's original plan included cast concrete profiles of American Indians, in feather headdresses, attached to the sides of the pavilions facing north and south. The figures were to stand in canoes, with bows and torches, protruding from the pavilion walls. In addition, bison skulls were to be mounted under the arches. Cutter's use of Indians and bison were

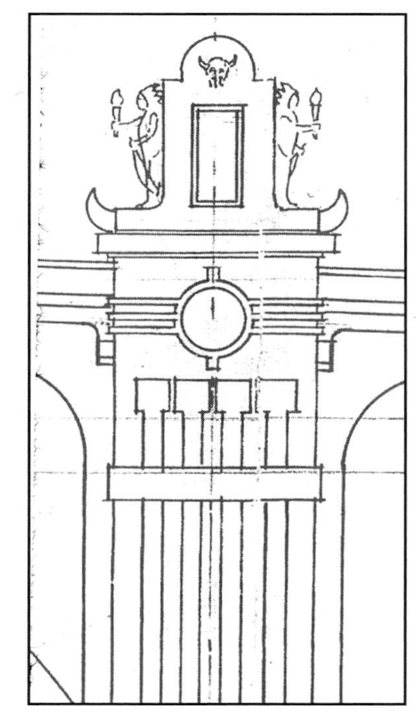

Renowned Spokane architects Kirtland Cutter and Carl Malmgren's initial design for deck ornamentation on the third Monroe Street Bridge evoked native themes: bison heads mounted on pavilions flanked by Indians in canoes. Drawing ca. 1910.

Spokane Public Works and Utilities Department, City of Spokane

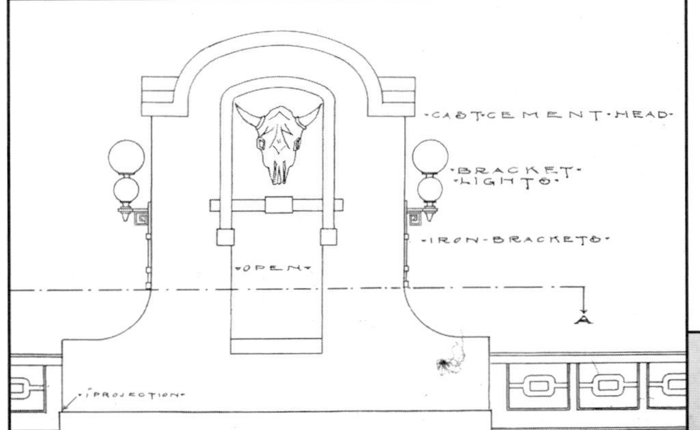

By ca. 1911, Cutter and Malmgren's design for the Monroe Street Bridge deck ornamentation had changed to a "pioneer" theme: ox skulls mounted on covered-wagon-shaped concrete pavilions.

Spokane Public Works and Utilities Department, City of Spokane

apparently meant to evoke a sense of wilderness lost, and to achieve what he called an "effect of age," which he sought in all his building designs (mostly residential and commercial structures). The Indian-and-canoe motif seemed particularly apt for a river crossing. The bison skulls, with their skeletal connotations, provided metaphoric allusion to the exposed structural framework supporting the bridge deck.

For unknown reasons, when Cutter submitted his final design, the Indians were removed, the pavilions reshaped, and bas-relief concrete bands were added to depict Conestoga wagons with oxen yokes over the skulls. (City Engineer Ralston himself referred to them as "ox skulls," but their design remained that of bison.) A "pioneer" theme had obviously replaced the depiction of the region's first

Skulls on the Monroe Street pavilions resembled bison, although bridge designer J. C. Ralston called them "ox skulls."

Northwest Room, Spokane Public Library

Completed in 1911, the main span of the third Monroe Street Bridge was the longest concrete arch in the United States. The Oregon Railway and Navigation Company Bridge, built over the Monroe Street Bridge, was removed in time for Expo '74.

Northwest Room, Spokane Public Library

The Monroe Street Bridge

The early stages of construction, ca. 1910, showing wooden falsework under the concrete rib arches at right.

Northwest Room, Spokane Public Library

For sixty years, the OR&N Bridge crossed over the Monroe Street Bridge, shown here ca. 1920s.

Northwest Museum of Arts and Culture, Eastern Washington State Historical Society, Spokane

A Spokane postcard, showing the Washington Water Power hydroelectric plant near the south end of the bridge.

Northwest Room, Spokane Public Library

The Monroe Street Bridge and the lower falls of the Spokane River in the 1990s.

Craig Holstine, WSDOT

inhabitants, but still allowed Cutter to achieve his "effect of age" with a western motif. The wagon pavilions and skulls are among the most distinctive of any bridge ornamentation in the Pacific Northwest.

On November 21, 1911, more than 3,000 citizens gathered for the opening of one of the largest bridges in the world. At a cost of $488,200, the new monolith attracted attention from the governors of Washington and Idaho and numerous city and county luminaries. Mayor W. J. Hindley compared the structure to those built in ancient Rome, and even invoked a Lincolnian metaphor of bringing "north and south together."

Today, the Monroe Street Bridge, perhaps more than any other natural or man-made feature, serves as Spokane's premier character-defining landmark, appearing as a logo in commercial advertisements and on city government stationery. Designed to achieve maximum aesthetic effect, its supporting arches mimic the tumbling waters of the Spokane River's series of high and low falls. From the bridge deck, there is no finer view of the city and the falls that inspired Spokane's founding. An extensive rebuilding project that was scheduled for completion in 2005 has preserved this landmark bridge as an important transportation link, as well as a source of civic pride.[2]

Crossing a deep canyon west of downtown Spokane, the Latah Creek Bridge served the "Sunset Highway." This photo, taken ca. 1913 as construction neared completion, shows the adjacent wooden trestle still in place.

Northwest Room, Spokane Public Library

Latah Creek (Sunset Boulevard) Bridge

With completion of the Monroe Street Bridge, the stage was set to more readily link the city with communities to the west, including the Big Bend region of Adams, Lincoln, and Franklin counties, which are cradles of dryland wheat farming. Traffic along the Northern Pacific Railroad had grown steadily in the wake of the 1909 Alaska-Yukon-Pacific Exposition in Seattle, as communities west of Spokane promoted settlement and proclaimed the area's great agricultural opportunities.

Latah Creek Bridge, modern view.

Office of Archaeology and Historic Preservation

Spokane lacked a reliable bridge across Latah Creek on the city's west edge, where the stream's deep gorge hindered travel. In August 1911 the city approved plans prepared by W. S. Malony under the supervision of City Engineer Morton Macartney. However, John Lyle Harrington, of Waddell and Harrington, visited Spokane to review the design plans after "it was found impossible to come to an understanding without a personal conference." Harrington "fully approved" Macartney's plans, and the city proceeded to contract with J. E. Cunningham of Spokane. J. F. Greene, an assistant city engineer, supervised construction. The bridge was completed in July 1913 at a cost of $425,000.

The local press took note that the Latah Creek structure was 200 feet longer and required 30 percent more cement than the Monroe Street Bridge, yet the former structure was built for $75,000 less. The *Spokesman-Review* observed that the Monroe Street Bridge was constructed by "day labor" two years earlier, and the Latah Creek Bridge was built by a contractor.

The most striking feature of the Latah span is its series of Roman or semicircular arches. The 940-foot-long bridge is composed of seven arches varying in length from 54 to 150 feet. The *Engineering News* of March 27, 1913, remarked: "In determination of span lengths for the arches, architectural considerations were given much weight and an effort was made to effect a combination which was adapted to the ground line and site.... Each arch consists of four arch ribs, carrying the roadway slab on spandrel columns and arches." The Latah Creek Bridge, like the Monroe Street Bridge, represents one of the state's early examples of a long-span, fixed-end, open-spandrel concrete arch.[3]

Post Street Bridge, built in 1917. This 2005 photo shows it without its original globe lights and standards.

Craig Holstine, WSDOT

Post Street Bridge

The first Post Street Bridge was built in 1893 by the San Francisco Bridge Company. It consisted of a single steel three-hinged arch 244 feet long, two steel trusses, and steel floor beams with a wooden deck. A decade later, it needed replacing. In 1916, City Engineer Morton Macartney prepared plans for a new concrete bridge. The city awarded a contract for $40,000 to the Olson and Johnson Company of Missoula, Montana, and construction began immediately.

Work progressed routinely until the afternoon of February 6, 1917, when part of the falsework collapsed into the Spokane River. When the dust cleared, two workmen lay dead (another died later) and ten injured. After replacing the scaffolding, Olson and Johnson completed the bridge on December 20 of that year. The open-spandrel structure includes two 250-foot clear-span rib arches. In 1936–37 a third rib arch was added to accommodate a widening of the deck. To accomplish the work, the Spokane United Railways steel streetcar bridge located immediately downstream was removed.[4]

The Post Street Bridge falsework collapsed into the Spokane River on February 6, 1917, killing two workmen, and mortally injuring another. An editorial cartoon showed the city belatedly considering the human toll of bridge building—after having experienced two other fatal bridge collapses prior to the Post Street tragedy.

Spokesman-Review, February 8, 1917

The Division Street Bridge collapsed on December 18, 1915, killing five people.

Northwest Museum of Arts and Culture, Spokane

The replacement Division Street Bridge, built in 1917. Harvey S. Rice and Craig Holstine, 1987 photo.

Office of Archaeology and Historic Preservation

Division Street Bridge

The first Division Street Bridge, built in 1882, was a simple wooden-truss structure. A new steel bridge replaced it in 1891. On December 18, 1915, under the weight of an electric streetcar, the twenty-four-year-old structure collapsed into the Spokane River's icy waters. It was the worst bridge disaster in Spokane's history. Five people were killed and twenty injured.

City officials considered two different plans for a replacement, one of steel and the other of concrete. The concrete plan consisted of "three arches of what is known as barrel-type filled with earth and paved, and a number of small arches of 31 foot in the clear, forming a trestle over the Great Northern Railway tracks." City Engineer Macartney proposed that the three arches could be built, while using the existing wooden trestle, and later a new concrete trestle could be built when funds permited. The cost was roughly $100,000.

More than a hundred concerned citizens signed a petition supporting the concrete proposal, and by June 1916, plans were approved for such a structure. The plan called for a fifty-foot roadway and two nine-foot sidewalks—virtually identical dimensions as used in constructing the Monroe Street Bridge. The bridge's total length was to be 600 feet, of which approximately 270 feet represented the main span across the Spokane River. On July 16, 1916, the city awarded a $108,443 construction contract to the Washington Paving Company of Seattle. Although high water swept away half of the falsework in May 1917, the company completed the new bridge later that year. The bridge was replaced in 1992 to enable increased vehicular traffic through the city.[5]

Greene Street Bridge

Traffic congestion in Spokane became an issue in the 1950s, spurring the construction of two notable bridges on Greene and Maple streets. Each crossed the Spokane River and diverted traffic from the downtown business area. The first, spanning the river at Greene Street near the city's east edge, was designed by the Spokane Engineering Department to replace an old, narrow steel-truss bridge unable to carry existing truck traffic loads. The latter had been built in 1916. A steel Baltimore Petit through-truss, it had a 192-foot steel span, concrete piers, reinforced-concrete trestle approaches, a wooden-plank floor, and wooden sidewalks made of creosote-soaked timber. The bridge served area residents for nearly forty years. Over the decades, the steel deteriorated, and by 1946 talk of a replacement span began. The city deferred action as long as possible. Mounting complaints about the bridge's wood-plank surface finally stirred city officials.

In November 1950, a mother and her infant son died in an auto accident caused by the slippery board surface. The gruesome incident prompted a flood of letters to the city council, demanding a solution to the surface problem. When the victims' family filed a $135,000 damage suit against the city, other concerned citizens joined in the

The Greene Street Bridge over the Spokane River, the only multiple-span concrete arch bridge built in Washington since the 1930s. Photo 2002.

Craig Holstine, WSDOT

chorus of complaints. One insisted that "something must be done about the wooden, lateral planking on the Greene Street Bridge, and other wooden surfaced bridges in the city."

By June 1952, Spokane secured state and federal aid for a new concrete arch bridge, with an estimated cost of $425,000. Dismantling of the old structure began in March 1955. Construction on the new bridge followed, supervised by the Henry Hagman Construction Company of Cashmere, Washington. The ribbon-cutting ceremony for the newly named Esmeralda-Greene Street Bridge occurred on July 24, 1956. (The name "Esmeralda" has since been dropped.)

The bridge is 453 feet long and consists of three reinforced-concrete, open-spandrel, rib deck-arch spans. Each approach is flanked by two reinforced-concrete slabs. The bridge's center span rises 27 feet above the water and stretches 133 feet in length. On either side of the center span, asymmetrical, 116-foot-long, parabolic arch spans meet the riverbank. The Greene Street Bridge was one of only two concrete arch bridges built in the state in the 1950s, and the only concrete multiple-arch bridge constructed since the 1930s.[6]

Maple Street Bridge

The Maple Street Bridge was the second bridge constructed in the 1950s to divert north-south traffic from downtown Spokane. The long, steel-plate girder structure was built as a toll bridge. It was designed for the city by the Tudor Engineering Company to carry Maple Street traffic over the river and above Peaceful Valley at the west edge of downtown.

Spokane City Engineer Charles Davis had considered a bridge for this location as early as the mid-1930s. By then, city growth and the increasing dependence on the automobile brought a need for a new bridge to relieve traffic congestion on the Monroe Street corridor. Davis had proposed a concrete span similar to the Monroe Street Bridge, to be located at the Ash-Oak street site, just west of the present Maple Street Bridge. However, funds were unavailable, in part due to the Great Depression. After World War II, attention again focused on a new structure, but

Maple Street Bridge, 2002.

Craig Holstine, WSDOT

money was still unavailable, and outside help was necessary.

In 1953 the city approached the Washington State Toll Bridge Authority, which contracted for site-feasibility and preliminary-design studies. A year later, the authority recommended a toll bridge at the Maple Street site. By February 1956, approval came for the final design plans prepared by Tudor Engineering of San Francisco. The bridge would be financed with a bond, to be paid off by a twenty-cent round-trip toll.

Construction contracts were awarded to Morrison-Knudsen for $675,000 to build the bridge's substructure, and to U.S. Steel Corporation's American Bridge Division in Portland, Oregon, for $1.6 million to erect the superstructure. Several other contracts were let to two Spokane firms for approach work. Begun in 1956, the bridge was completed on July 1, 1958. Scores of Spokane citizens attended the official opening, headed by Governor Albert D. Rosellini.

Initially, the Maple Street Toll Bridge was a welcome addition to Spokane. As time passed, however, the toll increasingly aggravated area drivers. Finally, in July 1990, financial obligations were met, and the Toll Bridge Authority lifted the fee and officially transferred ownership to the city of Spokane. The bridge's significant engineering features include its fourteen-span, 1,713-foot-long, riveted steel-plate girder and floor beam support system using structural low-alloy steel. Its configuration of five multi-span continuous units represented cutting-edge technology in the 1950s. The five 150-foot spans comprise a 750-foot-long continuous unit at the north end, which was the longest in the state when constructed.[7]

Northeast Washington Bridges

Farming, ranching, and logging already had begun when the U.S. Army established Fort Spokane at the confluence of the area's two main rivers, the Spokane and Columbia, in 1882. More than a half-century later, the Army's bridge across the Spokane River had long since vanished when the presence of the federal government again was felt, this time in the form of rising waters behind Grand Coulee Dam. A new bridge of significant proportions was needed to span the wider mouth of the Spokane River.

Upstream, too, the need for a new bridge near Long Lake Dam came suddenly when a wooden structure collapsed under the weight of a sheep herd, in what was perhaps the state's most bizarre bridge failure.

Since the coming of the transcontinental railroads in the 1880s, the village of Marshall southwest of Spokane had been an important rail crossroads, where spur lines branched off to tap the agricultural wealth of the Palouse region to the south. In the 1940s, a modern concrete bridge over busy railroad tracks became the solution to a troublesome road crossing there.

Mining spurred development of the Pend Oreille River country, leading to the incorporation in 1910 of Metalline Falls in the state's northeasternmost corner. A new steel bridge attracted attention from commercial interests to the south, as well as nearby neighbors across the Canadian border.

Further west, mining and subsequent logging activities brought the need for a bridge across the Kettle River between Ferry and Stevens counties.

Spokane River Bridge at Fort Spokane

With the establishment of Fort Spokane in 1882, U.S. Army soldiers were soon operating the first of several ferries crossing the Spokane River here to the Spokane Indian Reservation. Also, in 1883, U.S. Army Lieutenant George W. Goethals, later the builder of the Panama Canal, designed and supervised construction of a 120-foot timber-truss bridge over the Spokane River near the fort. Erecting a bridge across the narrow canyon was the "hardest task I ever had," he later recalled. In the early twentieth century, the Dentillion Bridge, a 200-foot steel-and-timber Howe

Built in 1941, the Spokane River Bridge at Fort Spokane spans reservoir waters behind Grand Coulee Dam. The fort appears on the terrace behind the bridge. Photo 1993.

Jet Lowe, HAER

truss, was erected in the vicinity. A tractor-trailer collision destroyed that bridge in May 1939.

The forming of a reservoir behind Grand Coulee Dam resulted in the construction of the Spokane River Bridge near Fort Spokane—a steel cantilever structure that was the largest bridge erected in Washington in the 1940s, before World War II temporarily suspended bridge building.

The Spokane River's depth near its confluence with the Columbia precluded any bridge designs requiring mid-channel falsework. With plans prepared by the Washington Department of Highways, the Angeles Gravel and Supply Company of Port Angeles constructed the piers and T-beam approaches, and the C. and F. Teaming and Trucking Company of Butte, Montana, built the superstructure and concrete deck. The resulting 953-foot-long structure includes two riveted steel through-cantilever spans and a modified Warren truss suspension span. The Montana firm completed work on December 5, 1941. Construction contracts totaled approximately $280,000, for which the U.S. Bureau of Reclamation, sponsor of the Grand Coulee Project, reimbursed the state.[8]

In 1949, the Spokane River Bridge at Long Lake replaced a wooden structure that collapsed under the weight of sheep in 1942. Photo 1993.

Jet Lowe, HAER

Spokane River Bridge at Long Lake Dam

The wooden Warren through-truss built here in 1911 was designed to carry wagon and light automobile traffic crossing between the timbered and grass-covered highlands on the north and the rolling grain fields south of the Spokane River. Whatever its load-bearing capacity really was, it was not up to the task of supporting a flock of sheep crossing the bridge on September 26, 1942. Apparently, the rhythmic motion of hooves, with the added sheer weight of the flock, collapsed the bridge, sending sheep and shattered timbers into the river (see Chapter 2).

Henry Hagman of Cashmere built a new 486-foot-long bridge at the site in 1949, replacing the destroyed structure. State Department of Highways engineers designed the reinforced-concrete bridge, with a 211-foot, open-spandrel arch, and concrete girder approaches. The arch consists of two parallel arch ribs supporting concrete columns, which in turn support the roadway girders, floor beams, and deck slab. Use of a "considere hinge," causing the arch to act as a fixed arch under live load conditions, was unique among concrete arches built by the state in the 1940s. The bridge, now situated on State Route 231, also represents a relatively late example of an open-spandrel, reinforced-concrete, rib deck-arch, considered a most aesthetically pleasing design for a remote setting of rugged natural beauty.[9]

Marshall Bridge

Keeping automobiles and trains apart was always challenging in Marshall, the important railroad hub nine miles southwest of Spokane. There, the Cheney-Spokane Road crossed several sets of busy railroad tracks, as well as a timber bridge over Marshall Creek. Spokane County turned to the State Department of Highways for help to ease the congestion. In the late 1940s, the department contracted with noted Spokane engineer W. L. "Pat" Malony to design a bridge for Marshall.

After leaving the post of Spokane city bridge engineer, Malony had set up shop in Spokane designing buildings and bridges. In private practice, he designed several build-

W. L. "Pat" Malony, well-known for new college buildings in Pullman, designed the Marshall Bridge (opened 1950).

WSDOT

ings on the Washington State College campus, including Bohler Gymnasium.

For the Marshall crossing, Malony designed a continuous concrete T-beam structure, consisting of ten spans varying in length from forty-one to seventy-five feet. To create a subtle arch appearance, the T-beams were cast with parabolically arched haunches. Constructed for $164,000, the 547-foot-long structure completed in 1950 remains today as an example of the successful integration of highway and rail transportation by one of the area's noted engineers.[10]

The Metalline Falls Bridge, a cantilever deck truss completed over the Pend Oreille River in 1952.
WSDOT

Metalline Falls Bridge

Although the June 14, 1952, dedication ceremony for the new bridge over the Pend Oreille River took place several miles south of the Canadian border, it was an international ceremony. American and Canadian color guards raised flags, and vocalists sang the national anthems of both countries. The city of Trail, British Columbia, furnished the public address system, and that city's "Kilties band... made a decided hit." Obviously counting on improved business from south of the border, the speakers included representatives of the Trail Chamber of Commerce and the Nelson Board of Trade. Also attending was William A. Bugge, the State Department of Highways director, representing Washington Governor Arthur Langlie.

This far-northeastern corner of the state had always maintained connections with communities north of the border, given the significant distance to commercial centers to the south. (Colville is more than forty-five miles southwest, and Spokane approximately ninety miles south.) An earlier bridge, constructed in 1920 immediately downstream of the new structure, had provided the first dependable link between the mines southwest of town and a reduction mill to the northeast. The bridge had been a deck truss, but of inferior dimensions and load capacity. Limited to one lane of traffic and eleven-ton truckloads of ore, the old bridge had outlived its usefulness. When removed, it was reerected at Heron, Montana, over the Clark Fork River, the eastern extension of the Pend Oreille River drainage above Lake Pend Oreille.

Henry Hagman of Cashmere built the new bridge for something in excess of the $548,615.50 contract. With the 1949 legislature earmarking $500,000 for the job, and the Bureau of Public Roads having an "interest" of $282,000 in the project, cost overruns apparently were considerable. At 760 feet in length, the steel cantilever deck truss rises 80 feet above the low waterline of the scenic Pend Oreille River.[11]

Barstow Bridge

Various bridges across the Kettle River at Barstow have been vital to the local timber and agricultural economy. Since the first structure was built here in 1904, however, all but one has met an untimely end thanks to frequent flooding. It took railroad technology, military engineering, and the end of World War II to bring a reliable bridge to the site.

Stevens County solved the perennial problem of bridge loss at the crossing when the Reconstruction Finance Corporation (the former War Assets Administration) announced that surplus, prefabricated railway bridges

The military surplus Barstow Bridge over the Kettle River is the only one of its type in the state.

Washington State Archives, WSDOT Records

were for sale after the war ended. In 1947, a Pratt through-truss railway bridge arrived from a surplus stockpile in Albany, New York. The bridge, one of many similar structures, was intended to replace war-damaged bridges in Europe. Apparently, supply had exceeded demand, and surplus bridges became available on the U.S. market. Designed for rapid field construction by unskilled labor under skilled supervision, the bridges consisted of standard elements, some of which could be reversed end for end. These bridges were designed to be lengthy spans of 90 to 150 feet without intermediate piers.

Because these structures were also designed to carry railroad traffic, the Barstow Bridge's loading rating is higher than for any other Washington highway bridge of like dimensions. Due to the bridge's exceptional weight (131 tons), and anticipated heavy use (such as loaded logging trucks), the contractor installed H-bearing piles under a new concrete pier supporting the contact point of the approach and main span. So-called "T-type" steel trestles were used atop the new pier. On the east bank, the contractor used a concrete abutment remaining from the previous bridge.

The Barstow Bridge is a single-lane structure, sixteen feet wide with a timber plank deck. Its 123-foot-long Pratt truss has a 60-foot approach span consisting of four I-beams. It is the only known prefabricated World War II-era railway bridge in use as a vehicular bridge in Washington. With a total cost of $44,818.58, including shipping and installation, the bridge has proven the economic value of adaptive reuse.[12]

NOTES

1. J. Byron Barber, "The Golden Era of Bridge Building," *The Pacific Northwesterner* 29, no. 1, Winter 1984, 1–10; Jay J. Kalez, "Early Spokane Thrived after Bridge Bug's Bite," *Spokane Daily Chronicle*, July 22, 1967, 10; "Bridge Builder [Morton Macartney] Amazed by City," *Spokesman-Review*, August 22, 1951; "Bridge Designer [J. F. Greene] Tests Spans He Built in City Long Ago," *Spokesman-Review*, August 25, 1953.
2. Craig Holstine, "Monroe Street Bridge," Spokane Register of Historic Places Nomination, 1990; Patsy M. Garrett, "Monroe Street Bridge," NRHP Nomination, 1975; "Historic Spokane Bridge Rebuilt," *Pacific Builder and Engineer*, September 6, 2004, 10–11; "The Monroe St. Bridge, Spokane, Washington; A Concrete Bridge Containing a 281-ft. Arch," *Engineering News*, September 2, 1909, 241–243.
3. "Approve Hangman Plans," *Spokesman-Review*, August 23, 1911; "Sign Big Bridge Contract," *Spokesman-Review*, October 15, 1911; "Bridge Span Progresses," *Spokesman-Review*, August 17, 1912; "The Latah Creek Bridge, Spokane, Wash.," J. F. Greene, *Engineering News* 69, no. 13, March 27, 1913, 616; "Hangman Bridge Opened," *Spokesman-Review*, October 16, 1913; Lisa Soderberg, "Latah Creek Bridge," Historic American Engineering Record Inventory, 1979.
4. "Post Street Bridge Falls into River, Killing Two Men," *Spokesman-Review*, February 7, 1917; Barber, "The Golden Era of Bridge Building," *The Pacific Northwesterner* 28, no. 1, Winter 1984, 1–13.
5. "History of Division Street Bridge Recalled," *Spokesman-Review*, May 10, 1964; "Spokane May Inspect City Bridges," *Pacific Builder and Engineer* 21, January 1916, 34; "Collapse of Spokane Bridge," *Pacific Builder and Engineer* 23, February 23, 1917.
6. Oscar R. "Bob" George, "Greene Street Bridge, Spokane, Washington," NRHP Nomination, June 2001; "New $225,000 Bridge at Greene Street Planned by City," *Spokesman-Review*, February 15, 1949; "State Says Yes on Bridge Plan," *Spokesman-Review*, November 3, 1952; "Greene Bridge Plans Readied," *Spokesman-Review*, November 28, 1952.
7. "Maple Street Bridge Plans and History Reviewed," *Highway News*, July-August 1958, 26–27; "Maple Street Toll Bridge Opened July 1 for Spokane Traffic," *Highway News*, July-August 1958, 26–27; Oscar R. "Bob" George, "Maple Street Bridge," NRHP Nomination, 2001.
8. Walter R. Griffin, "George W. Goethals, Explorer of the Pacific Northwest, 1882–1884," *Pacific Northwest Quarterly* 62, October 1971, 133–37; Robert W. Hadlow, "Washington State Cantilever Bridges, 1927–1941," HAER No. WA-106, 1993; Hadlow, "Spokane River Bridge at Fort Spokane," HAER No. WA-113, 1993; Robert H. Krier, J. Byron Barber, Robin Bruce, and Craig Holstine, "Spokane River Bridge at Fort Spokane," NRHP Nomination, 1991.
9. WSDOT Bridge Preservation Office; Wm. Michael Lawrence, "Spokane River at Long Lake Dam," HAER No. WA-95, 1993; Robert H. Krier, J. Byron Barber, Robin Bruce, and Craig Holstine, "Spokane River at Long Lake Dam," NRHP Nomination, 1991.
10. Robert H. Krier, J. Byron Barber, Robin Bruce, and Craig Holstine, "Marshall Bridge," NRHP Nomination, 1991.
11. *Metalline Falls News*, June 19, 1952; "Metalline Falls Bridge," *Highway News* 2, no. 6, December 1952, 23.
12. Robert H. Krier, J. Byron Barber, Robin Bruce, and Craig Holstine, "Barstow Bridge," NRHP Nomination, 1991.

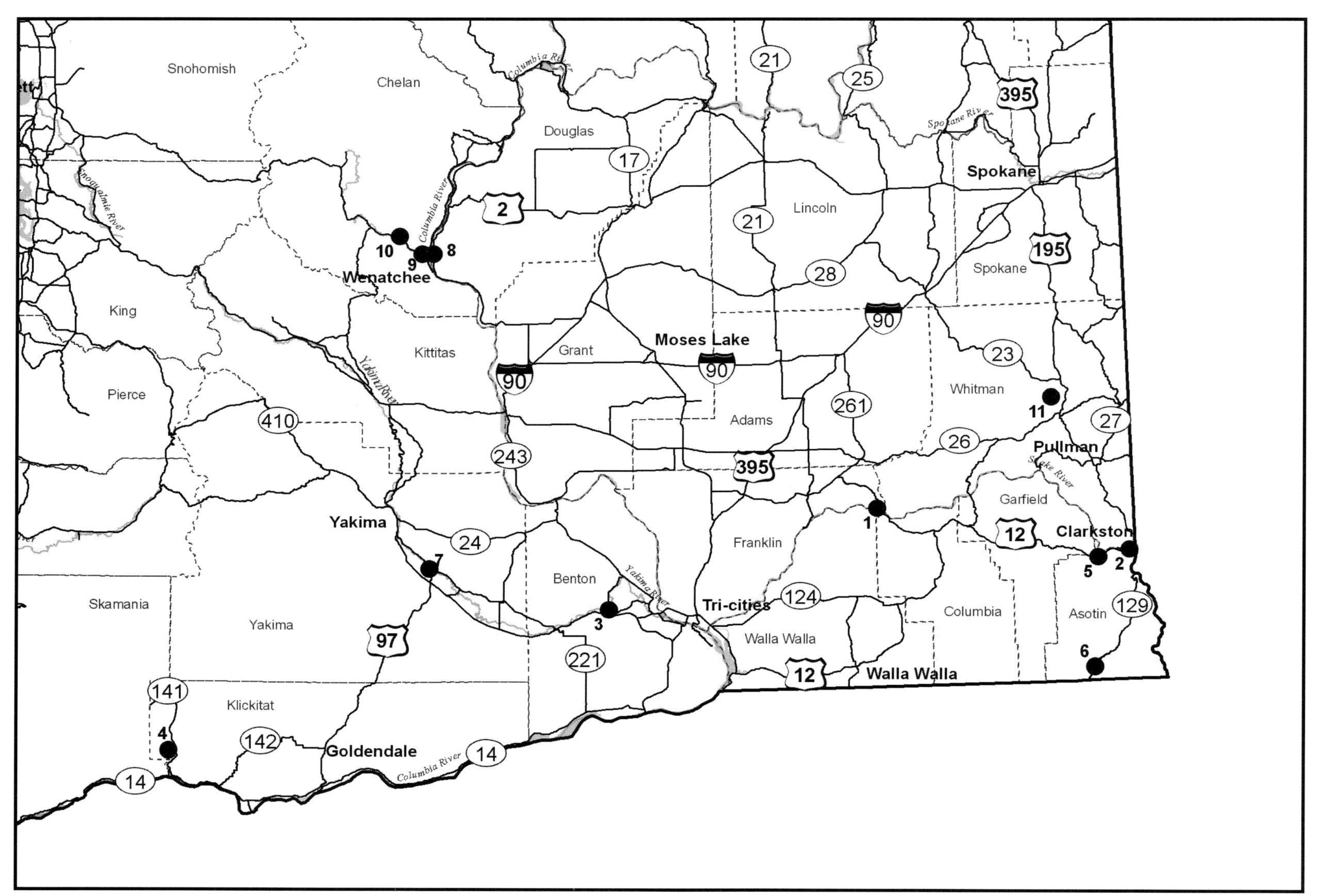

Premier Historic Bridges of Central and Southeast Washington

Bridge No.	Bridge Name	Bridge No.	Bridge Name
1	Snake River Bridge at Lyons Ferry	7	Toppenish-Zillah Bridge
2	Clarkston-Lewiston Bridge	8	Wenatchee Ave Bridge (1933)
3	Benton City-Kiona Bridge	9	Wenatchee Ave Bridge (1955)
4	B-Z Corner Bridge	10	West Monitor Bridge
5	Indian Timothy Memorial Bridge	11	Harpole (Manning-Rye) Bridge
6	Grande Ronde River Bridge		

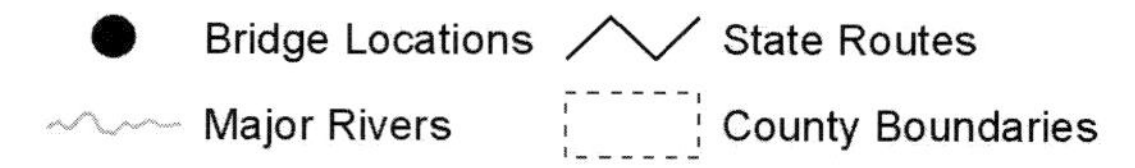

Bridges of Central and Southeast Washington

Central Washington is mostly a broad plateau, but in places deeply incised by the Columbia River and its tributaries, or wrinkled by long east-west trending ridges. The area is semiarid, but supports both irrigated and dryland farming. Much of Washington's agricultural produce is grown here, thanks in large part to the abundant water diverted from the Columbia River and dispersed by extensive canal systems.

Southeast Washington, on the other hand, is exemplified by the Snake River canyon country, barren lava "scablands," and the relatively well-watered Palouse Hills, where expansive dryland grain growing occurs. In the state's extreme southeast corner, the forested, deeply cut Blue Mountains rise to elevations above 6,000 feet.

The Yakima, Columbia, and Snake rivers meet at the Tri-Cities—Pasco, Kennewick, and Richland. Here, premier bridges cross the Columbia (see Chapter 5). Not far away on the Yakima River, a truly innovative bridge connects the small towns of Kiona and Benton City. Further west on the Yakima River, the state's most renowned innovator of concrete bridges designed a concrete-box girder structure, which today remains in service between Toppenish and Zillah. To the north on the lower Wenatchee River, historic bridges represent three distinct eras of bridge building.

Near where the Columbia River begins cutting its way through the Cascade Range, the White Salmon River drains the foothills of southwest Klickitat County. There, in the hamlet of B-Z Corner, a steel deck-arch bridge designed by one of the state's renowned bridge engineers crosses the White Salmon River.

Innovative bridges likewise are found on the Snake River. The Lyons Ferry structure is the oldest remaining steel cantilever highway bridge in the state. It also is the largest bridge to be moved from its original site and reconstructed at a new location. Another bridge standing near the Snake-Clearwater junction is a movable span, to accommodate grain barge traffic. On nearby Alpowa Creek, the concrete through arches of the Indian Timothy Bridge are of equal elegance. Not far south is the Grande Ronde River Bridge, perhaps the first riveted steel-plate girder built on the state highway system.

Finally, one of the state's most photographed bridges, the Harpole Bridge, stands in serene isolation in rural Whitman County. It is one of only three historic covered bridges remaining in Washington. Built to carry rail traffic, it is now a privately owned structure crossed by automobiles and farm vehicles.

Snake River Bridge at Lyons Ferry, Columbia and Franklin Counties

On June 5, 1960, the State Department of Highways celebrated the one hundredth anniversary of commercial ferry operations at this historic Snake River crossing adjacent to the mouth of the Palouse River. Here, the U.S. Army's Mullan Road crossed the river near the Indian village of Palus. Originally known as the "Palouse Ferry," it later became "Lyons Ferry."

On that day in 1960, Mr. and Mrs. N. G. Turner were too busy to attend the ceremony. The Turners were running a twenty-eight-ton, six-car, steam-powered barge

Moved from its original location at Vantage on the Columbia to the Snake River, the Lyons Ferry Bridge nears completion as the suspended span is hoisted into position between cantilever spans. The bridge is now the oldest remaining cantilever highway bridge in the state. Photo 1968.
WSDOT

cable ferry that they had operated here since 1945, and business was too brisk to allow for their attendance. Their absence, too, may have reflected a disdain for the proceedings. Surely they knew that the new bridge being discussed at the festive gathering would end their tenure on the river. A featured speaker at the ceremony, Director of Highways William A. Bugge, proposed the construction of a highway linking to a new bridge spanning the river just downstream of the ferry. Eight years later, this plan finally put the last cable ferry on the Snake River out of business.

To facilitate the project, the State Department of Highways rebuilt here a steel bridge dismantled at Vantage on the Columbia River in 1963. Originally constructed in 1927, the 1,640-foot structure was particularly suited for a two-lane, secondary roadway. (Its replacement at Vantage is the present Interstate 90 four-lane bridge across the Columbia River; see Chapter 5.) The cantilever trusses were hauled from the Beverly railroad siding in Grant County and erected at the Snake River site on new piers with concrete approach spans. The high concrete piers allowed for tugboat and barge clearance on the Lake Herbert G. West reservoir, created in 1969 upon completion of Lower Monumental Dam.

The bridge, now 2,040 feet in length, is a relatively late instance of what was once a common practice—the adaptive reuse of steel bridges. It is the longest bridge in Washington to be rebuilt at a location different from its original construction site, and the oldest steel cantilever highway bridge still in service in Washington.[1]

Clarkston-Lewiston Bridge, Asotin County, Washington, and Nez Perce County, Idaho

When it opened in June 1899, a steel cantilever bridge connecting Washington and Idaho across the Snake River was the largest "wagon" bridge on the West Coast outside of Portland. Lewiston, Idaho, then boasted a modest population, and Clarkston, Washington, was still known as "Concord," a promotional invention of an irrigation company. Tolls were collected until 1913, when the two states purchased the bridge from private owners.

As traffic became too heavy for the narrow structure, locals clamored for a new bridge with a movable span capable of accommodating commercial navigation, i.e., "the eventuality of river shipping service from Lewiston to the sea." With their shares of the Federal Aid to Highways program, the two states financed construction of the steel lift bridge that spans the river here today. The Washington State Department of Highways designed the bridge, and the Idaho Department of Public Works and the U.S. Bureau of Public Roads approved the plans. In May 1939, the Puget Construction Company completed the project at a cost of $750,000.

When completed in 1939, the Clarkston-Lewiston Bridge featured a unique vertical-lift system.
Washington State Archives, WSDOT Records

The 1,423-foot bridge consists of reinforced-concrete approach spans, two 150-foot steel through-trusses, and a 200-foot center lift span. Counterweight towers rise from the center piers at either end of the lift span, with sheet-metal covering the sheaves and drive-machinery atop the towers. The manner in which the cables were attached to the lift span and counterweights was believed to have been unique, never before used on a vertical-lift bridge.

The bridge's overall design reflected what *Western Construction News* described as an attempt to "present as pleasing an architectural appearance as possible," given the structure's "significance as a gateway between the two states." Concrete pillars with inset lighting in Art Deco motif flank the approaches as a testament to that design goal.[2]

Clarkston-Lewiston Bridge drawing.
Washington State Archives, WSDOT Records

Benton City-Kiona Bridge, Benton County

A one-of-a-kind structure completed in 1957 crosses the Yakima River on State Route 225 between the small farming communities of Kiona and Benton City.

The Benton City-Kiona Bridge is considered a prototype for what later would be called "cable-stayed" bridges. Photo 2002.

Craig Holstine, WSDOT

Designed in 1955 by noted engineer Homer M. Hadley, the Benton City-Kiona Bridge appears to be an American prototype for what later would be called "cable-stayed" bridges. Although Hadley is credited with other innovative ideas—such as concrete floating bridges—the idea for securing a bridge deck with stays anchored to towers was anything but new. In 1784 a German engineer by the name of Loscher published a bridge design using timber stays. In 1873, R. M. Ordish used wrought-iron stays radiating diagonally from ornate towers to support the deck of the Albert Bridge, built in Battersea, London.

For the design of the Benton City-Kiona Bridge, Hadley may have borrowed from a contemporary German design (the Stroemsund Bridge in Sweden), or even from the obscure Chow Chow Bridge in Grays Harbor County (see Chapter 2). Hadley used boxed steel girders filled with vermiculate concrete for the cable stays. The resulting "tied-cantilever" (Hadley's term) functioned much the same as the cable-stayed technique, and provided certain advantages for crossing the Yakima River at the site, primarily since a longer span allowed the main piers to be placed away from the deep river channel. The approaches were then high enough to clear the river's record flood stage.

Use of the towers and inclined stays enabled adoption of an extremely shallow cross section for the bridge's superstructure, saving weight and cost of materials. A conventional concrete or steel girder bridge would have provided a depth-to-span ratio in the range of 1/15 to 1/25. On the Benton City-Kiona Bridge, towers and stays support a 3-foot-3-inch-deep, 170-foot main span, consisting of two 55-foot concrete spans cantilevered outward from the two main piers and a 60-foot steel span suspended between the cantilevered spans. Thus, the Hadley bridge provides a depth-to-span ratio of 1 to 52.3. This was an extraordinary engineering achievement for the 1950s. Its innovative use of both concrete and steel components is the only example of its type in Washington. (For an example of the cable-stayed design at its artistic zenith, see the Ed Hendler Bridge connecting Pasco and Kennewick, in Chapter 5.)[3]

B-Z Corner Bridge, Klickitat County

The B-Z Corner Bridge crosses a narrow cataract of the White Salmon River in the small southern Klickitat County community of B-Z Corner. This three-hinged, open-spandrel, steel rib deck-arch was the first of its type in Washington to rely on welded (rather than riveted) connections.

Use of a three-hinged arch design simplified construction at the constricted site. The contractor cantilevered each half of the arch rib from its abutment support, while a crane standing on the bridge approach held it steady. This allowed relatively easy placement of the center pin connection. Welding marked an advancement in steel bridge design and construction, and in a few years (by the mid-1960s), riveted connections would no longer be used by builders.

Two welded, tapered steel-plate girder sections 16 feet apart comprise the 120-foot-long rib arches. Hinged at each end and at the main span's centerline, the structure is free to expand and contract with thermal changes. Steel spandrel columns are connected along the rib arches, and, in turn, support the steel crossbeams under the reinforced-concrete deck. The 24-foot-wide roadway carries two lanes of traffic on the structure's 182-foot length.

The Seattle firm of Harry R. Powell and Associates designed the B-Z Corner Bridge in 1956, and West Coast Steel Works of Portland completed construction in 1958. That year, the American Institute of Steel Construction and the James F. Lincoln Arc Welding Foundation recognized the bridge for its innovative design and construction techniques.[4]

Designed by Harry R. Powell and Associates in 1956, the B-Z Corner Bridge won recognition for design and construction techniques, among them the use of welding instead of riveting on steel components. Photo 2002.
Craig Holstine, WSDOT

Indian Timothy Memorial Bridge, Asotin County

Not often did a state highway department name a bridge for a Native American. This was the case, however, with an attractive half-through concrete arch structure in Asotin County. Born Ta-Moot-Tsoo, Timothy, a Nez Perce Indian, lived in a nearby village at the mouth of Alpowa Creek. According to Washington State Highway Engineer James Allen, Timothy was "very friendly to the early settlers." Legend has it that Timothy served as a guide to an ill-fated U.S. Army expedition led by Lt. Col. Edward J. Steptoe.

In May 1858, Steptoe's command was nearly annihilated by hostile bands of local Plateau tribes near the present town of Rosalia in Whitman County. Timothy reportedly assisted in the soldiers' escape and safe return to Fort Walla Walla, perhaps earning him the blue military jacket he wore into his old age. Certainly, Timothy was held in high esteem by Elgin V. Kuykendall, the director of the Washington State Department of Public Works, which formerly oversaw the Division of Highways. Kuykendall, born and raised in southeast Washington and familiar with local sentiments for naming a monument for the respected Indian leader, ensured that the bridge bore Timothy's title. "I was able to have it designated in the statute authorizing its construction as 'Indian Timothy Bridge,'" said Kuykendall in a local history he authored.

In 1923 the Colonial Building Company of Spokane erected the bridge on the old Inland Empire Highway (later designated State Route 12), linking Clarkston with Walla Walla and the Tri-Cities. State engineers designed the structure to withstand "heavy runoff" in Alpowa Creek, a flow subject to sudden fluctuations resulting from cloudbursts.

The bridge is an example of the "rainbow arch," popularized in the 1910s and 1920s by James Marsh, a Des Moines, Iowa, engineer. It was one of five concrete through-arches constructed by the Department of Highways in the early 1920s—the North and South Hamma Hamma bridges and the Duckabush River Bridge along Hood Canal, and the Goldsborough Creek Bridge in Shelton were the others. Like the Goldsborough Creek Bridge, but unlike the Hamma Hamma and Duckabush examples, the Indian Timothy Bridge is a half-through structure, with the twenty-foot-high ribbed arches lacking overhead cross-bracing ties between the skewbacks.

The Indian Timothy Memorial Bridge honors a Nez Perce Indian respected by local settlers.
Jet Lowe, HAER

Built in 1941, the Grande Ronde River Bridge was probably the first riveted, steel-plate girder bridge in the state highway system.
WSDOT

The arches, which are in two spans of one hundred feet each, are not actually arcs, but rather parabolas whose crowns are half the depth of their springing points at the abutments and center pier. The floor is noncontinuous, with the floor stringers resting upon shelves cut into crossbeams suspended from the arches by vertical concrete hangers. The bridge's aesthetic qualities are enhanced by conical caps on the center pier, the balustrade-style guardrail with arched openings, and grooved, rectangular patterns in the arches' concrete surfaces, giving the structure a vague classical appearance. In designing the Indian Timothy Bridge, engineers used smooth arches evoking the area's nearby high, rolling prairie hills, an effect that may have met with Timothy's approval. No longer in use, the bridge now is a roadside attraction at a highway pullout.[5]

Grande Ronde River Bridge, Asotin County

Nestled at the bottom of one of the state's deepest canyons, the Grande Ronde River Bridge serves State Route 129, three miles north of the Oregon border and fifteen miles west of Idaho. The bridge connects the southeast Washington communities of Clarkston, Asotin, and Anatone with rural Oregon towns to the south.

Although prosaic in appearance, the bridge is significant. When completed in 1941, it was likely the first riveted, steel-plate girder bridge built on the state highway system. The structure is 283 feet long, consisting of concrete T-beam approaches, steel anchor arm and cantilever spans, and a rolled, wide-flange steel center suspended span. The Clinton Bridge Works in Clinton, Iowa, fabricated the steel used in the structure. R. W. Finke was in charge of design for the Washington State Department of Highways. Henry Hagman, builder of many Washington bridges, erected the structure. Few travelers here would ever guess they are crossing what once was the structural prototype for many future bridges in the state.[6]

Toppenish-Zillah Bridge, Yakima County

Homer M. Hadley, one of the state's most creative and prolific bridge engineers, designed this innovative structure in 1947. Spanning the Yakima River between Toppenish and Zillah, it is a single, hollow, concrete-box girder consisting of two cells. The structure differs from another of Hadley's noteworthy concrete-box girder bridges built nearby in 1948, the Donald-Wapato Bridge, also on the Yakima River; two separate, hollow, concrete boxes were used in that structure, which was recently demolished (see Chapter 3).

The Toppenish-Zillah Bridge's components include 34.5-foot cantilevers at each end, flanking four interior spans of 118 feet each. Total length of the bridge is 541 feet. The concrete-box girder is six feet deep and nineteen feet wide, with 4.5-foot cantilevers extending from each side supporting curbs and railings. The lattice-steel railings add aesthetic appeal to the structure. Ramsey and Company built the bridge in 1947.

Known for his innovations in concrete, Hadley undoubtedly used what was, at that time, a new system of mathematical formulations in designing the Toppenish-Zillah structure. Called the "Hardy Cross" method for its creator, the system facilitated accurate computations of "moments" in continuous, indeterminate structures such as this. Later, modern computers would be used in place of the laborious calculations engineers formerly carried out in designing bridges. Hadley's mastery of the engineering techniques of his day is especially represented in the Toppenish-Zillah Bridge. With its sleek lines and efficient use of materials, the bridge represents an advanced method of design analysis for its time, and is the earliest known continuous, indeterminate box-girder bridge in Washington.[7]

Wenatchee Avenue Bridges, Chelan County

Flood and fire claimed the first two bridges crossing the Wenatchee River near the north end of Wenatchee, and fireworks ignited a blaze that destroyed the deck of a third bridge built here in 1916. The deck was repaired, however, and the structure remained in service until the present "east" bridge was constructed in 1932–33.

The "east" bridge is a steel cantilever deck truss consisting of 647 tons of "Structural O.H. Steel." The Wenatchee River's swirling waters, susceptible to high runoff levels due to snowmelt in the nearby Cascades, combined with the adjacent Great Northern Railway line, probably led State

The "east" Wenatchee Avenue Bridge is the older of two structures spanning the Wenatchee River here. Dating to 1932–33, the steel deck-truss cantilever bridge is noted for its decorative metal guardrails. Hilscher, photographer.

Wenatchee Valley Museum and Cultural Center, Wenatchee

When built in 1955, the "west" Wenatchee Avenue Bridge included the longest steel-plate girder span in Washington.

Craig Holstine, WSDOT

Highway Department designers to choose a cantilever-type bridge for this crossing. Cantilevers can span greater distances without center piers, and do not require supporting falsework, which was not practical here given the size of the river and the railroad traffic. The Puget Sound Bridge and Dredging Company constructed the new bridge, removed "dangerous approach curves" on the road leading to the site, and dismantled the older structure.

With two 100-foot anchor arm spans, two 80-foot cantilever spans, a 100-foot suspended center span, plus the approaches, the "east" bridge totals 640 feet in length. Aside from the sweeping cantilever arch over the river, the bridge's most notable features are its ornamental guardrails. Squared metal balusters with spherical finial caps and cylindrical horizontal rails provide a framework for vertical metal bars and circles, the latter arranged below the top rails. The overall effect is one of lightweight transparency. Although the original Westinghouse light poles and 200-watt lamps encased in crown-shaped, glass globes have been removed, the deck nevertheless retains a sense of ornamentation.

When the bridge was completed in February 1933, the *Wenatchee Daily World* reminded readers of the importance bridges had to the community:

> Did you ever stop to think how isolated Wenatchee would be without its two bridges over the Columbia and Wenatchee rivers? Deprived of the two state highway bridges and those of the Great Northern Railway, Wenatchee would be hopelessly cut off from the rest of the world as far as traffic is concerned. Even an aeroplane would have difficulty landing on this [west] side of the [Columbia] river.

Immediately upstream, a companion "west" bridge now carries southbound traffic, leaving its older neighbor to serve northbound vehicles. Seattle contractor Paul Jarvis erected the "west" bridge, fabricated by U.S. Steel's American Bridge Division in Gary, Indiana. Dating to 1955, the newer structure consists of three continuous, riveted, steel-plate girder spans over the river, and two three-cell concrete box girder spans over the railroad. In total, the bridge measures 608 feet.

State Highway Department engineers in the 1950s designed the new bridge with a boast to fame of its own—the "west" bridge included the longest (at 260 feet) steel-plate girder span in Washington. It is still one of the longest such spans in the state. Its length was achieved through the use of structural low-alloy steel (a product new to the market) and a creative erection technique. Together the two adjacent bridges provide a study in the evolution of bridge engineering and technological advancement.[8]

West Monitor Bridge, Chelan County

The West Monitor Bridge over the Wenatchee River is typical of what many smaller bridge crossings once looked like across the country. This 320-foot structure consists of two 140-foot, steel pin-connected Pratt trusses, and two 20-foot timber trestle approach spans. Both trusses are comprised of seven 20-foot panels, braced by two eyebars with turnbuckles. The trusses are supported by two pairs of original, concrete-filled, steel cylindrical piers.

Dating to 1907, the West Monitor Bridge is among the oldest Pratt trusses in Washington.

Zachary Dee Holstine

One of the oldest and least altered pin-connected Pratt trusses remaining in the state, the West Monitor Bridge recently received a new wood-plank deck and has been repainted. Otherwise, it appears just as it did when built by the Puget Sound Bridge and Dredging Company in 1907. In 1977 the Chelan County Board of Commissioners noted, "the West Monitor Bridge is deficient and a candidate for replacement." The bridge still stands today, albeit still in jeopardy, not far from the ever-busier four lanes of State Route 2.[9]

Harpole (Manning-Rye) Bridge, Whitman County

One of the state's most photographed bridges is also one of the most inaccessible. Once a railroad bridge, the Harpole (Manning-Rye) Bridge is now privately owned and used for vehicular access to a rural Whitman County residence. Photographers from around the world are attracted not just by the fact that the bridge once served the Spokane and Inland Empire Railroad. Nor are they drawn by the bridge's simple, single-span, wooden Howe trusses. What photographers find irresistible are the wooden housings enveloping the trusses, making the Harpole technically a "covered" bridge. Contributing to its photogenic appeal is the pastoral setting on a slow-moving river amidst the verdant Palouse hills.

An earlier structure probably crossed the Palouse River here in 1907, when the electrified railroad completed its line between Spokane and Colfax, the nearby county seat. The Great Northern Railway acquired the Spokane and Inland Empire Railroad in 1927. The GN's records show that the present bridge was built in 1922 near its Harpole siding, named for Ed Harpole, then owner of the adjacent lands. Manning and Rye were the two nearest sidings, handling grain and other produce, as well as what must have been only a few local passengers.

Railroad records indicate that the trusses were "housed in 1928." Three additional bridges of this type (in GN's

Harpole (Manning-Rye) Bridge, showing enclosed trusses and planks laid atop railroad ties for vehicular use.

Jet Lowe, HAER

Metal vertical tension rods can be seen in front of the ladder accessing the truss tops on the Harpole (Manning-Rye) Bridge.

Jet Lowe, HAER

Plan No. 115-1562) were supposedly built at that time along the thirty-seven-mile branchline between Colfax and Spring Valley in Spokane County to the north. None of the others survive. Indeed, few if any bridges of this type on former GN lines exist elsewhere in the country.

The housing consists of sawn-board framing attached vertically and built wide enough around the trusses to allow for periodic inspections. Doors beside the tracks/ roadway access the passageways, and wooden ladders climb to the top of the trusses. The bridge itself consists of a 150-foot single timber Howe truss span, constructed with railroad-strength heavy timbers and iron rods on timber piles, with wooden approach spans. Above the roadway, there is no roof, only timbers and iron rods for cross-bracing.

The railroad continued to use the bridge until 1967, when the line was abandoned. Mrs. Ruth Lowe, who had acquired the Harpole site, bought the bridge and had planks laid on it to provide vehicle access to her home across the river from Green Hollow Road. Although the bridge is on private property, photo opportunities abound from the nearby county road.[10]

The Harpole (Manning-Rye) Bridge spans the Palouse River near Colfax in Whitman County.

Steve Hauff

NOTES

1. Robert W. Hadlow, "Snake River Bridge at Lyons Ferry," Historic American Engineering Record No. WA-88, 1993; Hadlow, "Washington State Cantilever Bridges, 1927–1941," HAER No. WA-106, 1993.
2. *Spokesman-Review*, May 27, 1939; "Clarkston-Lewiston Bridge," *Western Construction News*, August 1939, 267–69.
3. David J. Brown, *Bridges: Three Thousand Years of Defying Nature* (St. Paul, Minnesota: MBI Publishing Company, 1988), 128–29; Oscar R. "Bob" George, "Benton City-Kiona Bridge," National Register of Historic Places Nomination, 2001.
4. Oscar R. "Bob" George, "B-Z Corner Bridge," NRHP Nomination, 2001.
5. Wm. Michael Lawrence, "Indian Timothy Memorial Bridge," HAER No. WA-85, 1993; Lisa Soderberg, "Indian Timothy Memorial Bridge," HAER Inventory, 1979; Elgin V. Kuykendall, *Historic Glimpses of Asotin County* (Clarkston, Washington: *Clarkston Herald*, 1954), 17.
6. Robert H. Krier, J. Byron Barber, Robin Bruce, and Craig Holstine, "Grande Ronde River Bridge," NRHP Nomination, 1991.
7. Robert H. Krier, J. Byron Barber, Robin Bruce, and Craig Holstine, "Toppenish-Zillah Bridge," NRHP Nomination, 1991.
8. Wenatchee River Bridge No. 285/20E File, WSDOT Bridge Preservation Office, Olympia; *Wenatchee Daily World*, February 7, 1933, quoted in Robert H. Krier and Craig Holstine, "Wenatchee River Bridge No. 285/20E, NRHP Eligibility Evaluation," Short Report DOT98-20, Archaeological and Historical Services, Eastern Washington University, Cheney, May 1998; Oscar R. "Bob" George, "Wenatchee Avenue Southbound Bridge No. 285/20W," NRHP Nomination, 2001.
9. Lloyd L. Berry, P.E., Chelan County Highway Dept., correspondence to Keith Eggen, Dept. of Highways, March 14, 1977, WSDOT Bridge Preservation Office, Olympia; Chelan County Engineer's response to WSDOT Historic Bridge Inventory correspondence from Craig Holstine, 1993; Robert H. Krier and Craig Holstine, "An Assessment of the Current Status and Condition of Bridges and Tunnels in Washington State Listed in the NRHP," Short Report DOT93-10, Archaeological and Historical Services, Eastern Washington University, Cheney, 1993, 9; Lisa Soderberg, "West Monitor Bridge," HAER Inventory, 1979.
10. Lisa Soderberg, "Manning-Rye Covered Bridge," HAER Inventory, 1979; Richard Sanders Allen, correspondence and accompanying material sent to Eric DeLony, December 21, 1988; Brian T. Berkley, "Harpole Bridge," HAER No. WA-133, 1995.

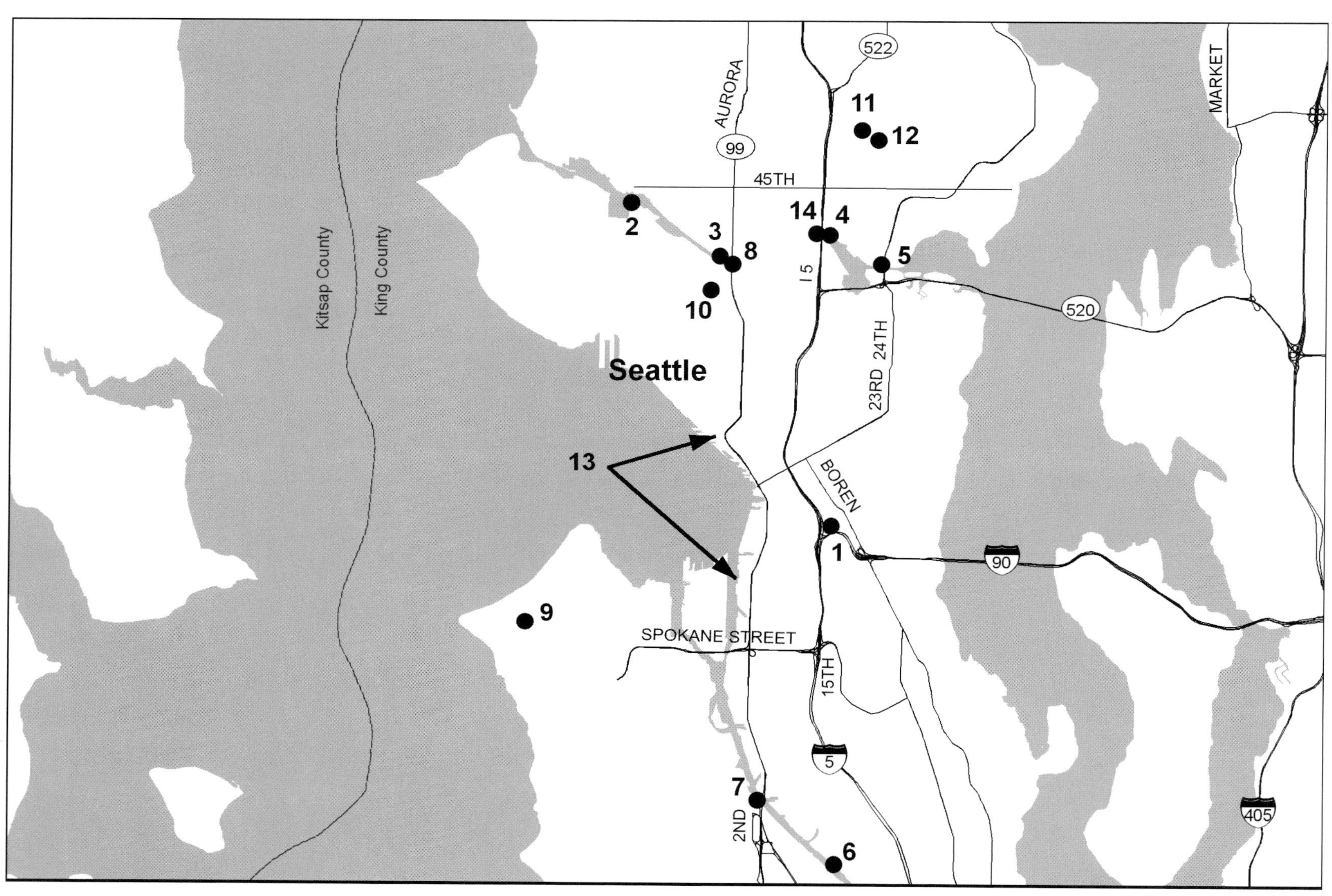

Premier Historic Bridges of Seattle

● Bridge Locations — Local Streets
Major Rivers — County Boundaries

Bridge No.	Bridge Name	Bridge No.	Bridge Name
1	12th Avenue South/Dearborn Street Bridge	8	Aurora Avenue Bridge
2	Ballard Bridge	9	Schmitz Park Bridge
3	Fremont Bridge	10	North Queen Anne Drive Bridge
4	University Bridge	11	Cowen Park Bridge
5	Montlake Bridge	12	Ravenna Park Bridge
6	14th Avenue South/South Park Bridge	13	Alaskan Way Viaduct
7	1st Avenue South/Duwamish River Bridge	14	Lake Washington Ship Canal/Interstate 5 Bridge

Bridges of Seattle

Today, some 150 bridges span Seattle's ravines and waterways. To many "pontists" (bridge enthusiasts), it is the Northwest's premier "city of bridges." Seattle sits between the saltwater of Elliott Bay and freshwater Lake Washington. On its near north side, the downtown straddles Lake Union and the Lake Washington Ship Canal, and to the south, the Duwamish River empties into Elliott Bay through east and west branches around Harbor Island.

In the late nineteenth century, the bustling burg became a Northwest hub of the shipping, fishing, lumbering, dairying, manufacturing, and distribution industries. It was America's nearest continental port to the Far East and Alaska, and by 1888 became a terminus for a transcontinental railroad.

This commerce and a dramatic population growth provided the stimulus for improving the city's transportation network. From a village of under 10,000 in 1890, Seattle grew in the wake of the 1897 Alaska gold rush to a metropolis of more than 237,000 by 1910. By 1920, Seattle's population reached 315,000. As the city developed and the automobile revolutionized transportation, demand increased for bridges across the waterways around Seattle. By the 1890s, new residential and business districts were being developed, and streetcar lines both served and promoted growth in "suburban" neighborhoods, such as Greenlake, Ballard, West Seattle, and Rainier Valley. Bridge building became a key feature of that expansion.

During nineteen years of Seattle's most dramatic growth, from 1892 to 1911, Reginald Heber (R. H.) Thomson served as city engineer. Under seven different mayors, he managed the design and construction of Seattle's public works projects—some $42 million in bridges, sewers, water lines, and other efforts that significantly changed the city's look and feel.[1]

The Grant Street Bridge and the 12th Avenue South Bridge to Beacon Hill were among the many municipal improvements that Thomson promoted. The Grant Street Bridge—a two-mile-long, twenty-four-foot-wide timber span built in the 1890s—provided a vital connection between downtown and the manufacturing district, located across a muddy tidal flat south of the city near Spokane Street. In 1911 Thomson's regrade project on Dearborn Street cut away 112 feet of dirt at 12th Avenue South, eliminating the hilly barrier between downtown businesses and residential areas in Rainier Valley. Next, a bridge was needed here on the thoroughfare between Lake Washington and Elliott Bay.[2]

The Dearborn Street Bridge is the oldest structure of its type in the state.
Seattle Municipal Archives

12th Avenue South, Dearborn Street Bridge

The 12th Avenue South Bridge, erected in 1911, was one of the first permanent steel bridges built by Seattle engineers. Today, the graceful span is the state's oldest steel arch bridge. It features a 171-foot, spandrel-braced arch with Pratt-style web trussing, a 94-foot cantilever span at the south end, and a 96-foot cantilever span to the north. Originally, timber approaches flanked the arch.

A tall-masted ship passes through the opened Ballard drawbridge, March 13, 1918. James P. Lee Collection, image # 294.

University of Washington Libraries, Special Collections, Lee 4023

Regrading work on Beacon Hill continued after the bridge was finished. In the spring of 1917, heavy rains brought a massive mud slide down onto the span, destroying the southern timber approach and shifting the whole structure some thirty inches north, causing it to bow slightly. The city promptly undertook repair work, rebuilding the approaches and replacing damaged bridge parts.

Major renovations undertaken in 1924 included replacing the timber approaches with six concrete bent approach spans, removing the trolley lines, and adding a forty-two-foot-wide concrete roadway. Other changes came in 1966 and 1967, during construction of the Interstate 5 and Interstate 90 interchange. State engineers altered the original north-south alignment by changing three bents at the south end to create a curved roadway. Mindful of historical concerns, they duplicated the decorative steel railings on the new approach span to retain continuity. The bridge was renamed in 1974 after Filipino hero Jose Protazio Rizal y Mercado Alonso.[3]

The Fremont Bridge (seen here), Ballard Bridge, and University Bridge, all built 1917–19, were among the state's earliest bascule bridges.

Office of Archaeology and Historic Preservation

Fremont, Ballard, and University Bridges—Seattle's First Bascules

In the early twentieth century, as automobiles increased on roadways, Seattle began a remarkable three-decade period of bridge building. These projects—spurred by a growing mobile population, shaped by a peculiar geography, shifted by political debates, and influenced by an innovative city engineering staff—left distinctive landmarks still seen today.

The deck and control house of the University Bridge.

Office of Archaeology and Historic Preservation

Under Thomson's urging, the city initiated plans to connect Lake Washington with Puget Sound. The digging of the Lake Washington Ship Canal to join the two large bodies of water opened a new chapter for Seattle and its bridges. City officials outlined an ambitious program of erecting eight bascule bridges across the ship canal, as well as on the two branches of the Duwamish River. Over the next two decades, the city built bridges at six of the sites.

The official opening of the Lake Washington Ship Canal and Government Locks on July 4, 1917, set the stage for completion of Seattle's first three bascule bridges—the Fremont and Ballard bridges, both finished in 1917, and the University Bridge in 1919.[4] These "Chicago-style," double-leaf, trunnion bascules with transverse cross girders were modeled after bridges built in the late 1890s by the City of Chicago's Public Works Department. All three stretched well beyond 200 feet in length, and their 40-foot width accommodated two trolley lines, as well as automobile traffic. The University Bridge cleared the canal by fifty-two feet, the Ballard Bridge by forty-five feet, and the Fremont Bridge by thirty-seven feet.

The bridges' trusses are raised and lowered by huge counterweights, consisting of concrete-filled steel boxes attached on the underside of the decks at the rear of the trusses. One-hundred horsepower motors move each leaf independently. In case of an unexpected motor failure, a leaf can be opened by hand in a mere six hours. In 1928 the city added to the bridges' backup motors. Hailed as "a new venture in movable bridge machinery," the motors were equipped with hydraulic, variable-speed transmissions.

The Fremont Bridge, completed in 1917 for $410,000, is a 502-foot structure with a 242-foot bascule span. The Ballard Bridge, built for $479,000, also dates from 1917, and is 295 feet long. In 1941 the city rebuilt the approaches, and in 1969 its original four towers were replaced by a lone tower. The University Bridge measures 291 feet long with a 218-foot bascule span. Its $825,000 cost was more than double that of the first bridge, owing to foundation costs and inflated prices that came with the nation's entry into World War I. In 1933 two more traffic lanes were added by replacing the wood-block paving with an open-mesh steel deck that weighed considerably less.[5]

Montlake Bridge

A fourth bascule bridge spanning the ship canal is, today, one of Seattle's most recognizable landmarks. The Montlake Bridge's history is somewhat more unusual than that of its three similar predecessors. The structure's foundations were laid at a cost of $40,000 during the ship canal's excavation in 1914, but the bridge was not completed until 1925. The peculiar challenges in constructing this double-leaf, trunnion bascule began with the need to accommodate the design to existing foundations at the canal's edge. The 345-foot bridge—comprised of concrete T-beam approaches and a 182-foot bascule—is distinguished by its

In 1925 the Montlake Bridge became the fourth bascule built over the Lake Washington Ship Canal. Its Gothic ornamentation makes it one of Seattle's most noted landmarks.

Jet Lowe, HAER

two ornate Gothic towers and unique trunnion supports. Of the city's six bascule bridges, it was the only one that employed concrete brackets instead of transverse girders to support the trunnions. Innovations by the Seattle Engineer's office, however, led to patent infringement suits brought by the Strauss Bascule Bridge Company. Fortunately for the city, these court actions were unsuccessful.

The Montlake Bridge stands prominently as the eastern gateway to the University of Washington, and was designed to complement the campus's Gothic architecture. Noted for its rich ornamentation in terra-cotta, the bridge is considered by many to be the city's most beautiful span.[6]

Despite political controversy, public debate, and legal battles that seemed to continually hover around the city's bascule bridges over the years, they remained vital links in Seattle's transportation network. Today, however, the conflicting interests—of boaters needing the bridges to open to permit passage, and car owners dealing with traffic snarls and rush hour traffic—have become acute as the numbers of boats and automobiles have increased.

Duwamish River Bridges

Some of Seattle's oldest and most interesting bridges from the early twentieth century were built south of downtown on the Duwamish River, near where it empties into Elliott Bay—an area with both residential and industrial districts. One of the earliest was a timber-and-steel swing span built in 1902 by King County. The bridge crossed the newly dredged river, connecting Seattle with West Seattle. In 1907 the city built a second bridge, this one a wooden drawbridge with two eighty-foot spans, for both automobiles and streetcars. It also carried water lines, which were temporarily disconnected whenever the draw spans were lifted to allow boats to pass.

In 1917–18, at Spokane Street, city engineers constructed the third drawbridge over the Duwamish. This wood-and-steel structure, costing $90,000, featured a design similar to the 1907 bridge—a Howe truss swing span some 301 feet long with a 20-foot roadway. Finally, after three bascule spans were completed on the Lake Washington Ship Canal, south- and west-Seattle residents demanded that the city build a new bridge across the Duwamish. Bond issues passed in 1920, and construction began.

Seattle's fifth steel bascule bridge, the West Spokane Street Bridge No. 1, was finished in 1924 for $1.2 million. In 1928 the city began building another identical bridge at the same crossing, just eighty-five feet south—the West Spokane Street Bridge No. 2, completed in September 1930. Both structures measured 288 feet between trunnion centers, and carried automobiles and streetcars across the waterway. These "north" and "south" spans were Chicago-style, double-leaf trunnions.

At nearly the same time, the city undertook the construction of another Duwamish movable bridge at 14th Avenue South. This bascule is unique, representing a distinct engineering contribution to Washington's bridge building development. Erected at a cost of $1.1 million, the 14th Avenue South Bridge (or South Park Bridge) is Washington's only Scherzer rolling-lift structure. The span features two 95-foot draw leaves that rotate when rolling on horizontal tracks. Designed by the Scherzer Rolling Lift Bridge Company of Chicago, this type of structure provided

The 14th Avenue South (South Park) Bridge is the only Scherzer rolling-lift bascule in the state. Photo 2005.
Craig Holstine, WSDOT

Lift gears in the 14th Avenue South (South Park) Bridge date to the original construction in 1931. Photo 2005.
Craig Holstine, WSDOT

The 1st Avenue South Bridge over the Duwamish River made engineering history when completed in 1955. Its "floating piers" support a double-leaf bascule that carries four lanes of traffic.

Craig Holstine, WSDOT

greater navigation clearance than a fixed trunnion bascule. The bridge helped relieve traffic congestion, allowing easier passage for some 600 cars a day. As a key part of the Seattle-Des Moines-Tacoma highway, the bridge also contributed to opening up a large part of Seattle's southern residential area, connecting it more closely to the metropolitan heart of the city.[7]

Crossing the Duwamish River elicited other unique engineering answers as well. The next precedent-setting bridge, completed in 1955, made engineering design history. The 1st Avenue South Bridge, like previous spans, was a bascule type. However, it employed floating piers to support the double-leaf, trunnion bascule. The floating concrete foundations were constructed in dry dock, then towed by tugboat to the bridge site. Each pier (with two reinforced-concrete underwater struts connecting them) features a cellular design that prevents the structure from sinking into the river's unstable mud. Portholes allow tide

This cutaway of the 1st Avenue South Bridge shows the cellular construction of its precedent-setting, concrete "floating piers."

WSDOT

waters to flood the counterweight pits, balancing the tendency of the rising water to lift the piers. The four-lane bridge cost $6.5 million and provided a channel width of 150 feet, with a vertical clearance of 40 feet for shipping. To date, this is the only "floating pier" bridge in the world. It was the longest bascule span in Washington, until surpassed in 1996 by a 294-foot adjacent bridge. The two noteworthy structures rank among the longest bascule spans in North America.

The first two bascule bridges at Spokane Street on the lower Duwamish have passed from the city's landscape. First to go was the "north" span. At about 2 a.m. on June 11, 1978, a barge piloted by eighty-year-old Rolf Neslund struck the bridge's east end. Inspectors soon deemed the damage "permanent," abruptly ending a long debate over a new, high bridge that would facilitate the Port of Seattle's growing needs to expand shipping on the waterway. The new $15 million, six-lane West Seattle Bridge opened in July 1984. A decade later, the "south" bascule was replaced by a uniquely designed concrete "pivot-wing" bridge, known as the Southwest Spokane Street Swing Bridge (see the Afterword).[8]

The Aurora Bridge opened in 1932.
Jet Lowe, HAER

Aurora (George Washington Memorial) Bridge

By the late 1920s, the growing automobile traffic crossing the four bascule bridges on the Lake Washington Ship Canal, as well as boat passages requiring bridge raisings, reached unmanageable proportions. As the State Department of Highways developed the four-lane, north-south Pacific Highway (State Route 99) through the city, pressure mounted for a new bridge that would eliminate the delays occasioned by span raisings. In 1930 a coalition of City of Seattle, King County, and state leaders obtained $4.5 million for the project.

The result was the George Washington Memorial Bridge, better known as the "Aurora Bridge," which opened in early 1932. The 2,955-foot span, one of Seattle's longest and highest, became the city's first major highway bridge and the first without streetcar tracks. It symbolized the new dominance of cars and buses over streetcars on Seattle streets. Its 135-foot clearance over the western arm of Lake Union allowed unobstructed passage for most marine traffic (all but the high-mast schooners, or "tall ships," whose day had passed), and its four lanes offered significant relief for motorists from the crowded bascule bridges.

The Aurora Bridge's design is noteworthy. Supervised by the State Highway Department's chief engineer Thomas McCrory and resident engineer Ray M. Murray, the structure's design was completed by the Seattle firm Jacobs and Ober, consulting engineers. A steel cantilever deck-arch truss, it crosses Lake Union between the north side of Queen Anne Hill and the south part of the Fremont district. The deck-cantilever span is 800 feet long, comprised of two 325-foot cantilever arms and a Warren truss suspended span of 150 feet. The Aurora Bridge represented a major advancement in the evolution of cantilever truss design in the twentieth century. Its simple yet bold lines offered a refined merging of functional requirements and aesthetic values that won the bridge acclaim from both professional engineers and citizens.[9]

Schmitz Park, North Queen Anne Drive, and Cowen Park Bridges

The Great Depression brought mixed blessings to Seattle. The economic downturn put many businesses into bankruptcy and thousands of working people into unemployment lines or on the street. On the other hand, President Franklin D. Roosevelt's New Deal programs began to bring federal monies into the area for construction projects, with the Public Works Administration funneling government financing for bridges and other public structures.

In the mid-1930s, City Engineer Clark Eldridge designed some of Seattle's most innovative and picturesque bridges. Particularly noteworthy are the Schmitz Park, North Queen Anne Drive, and Cowen Park bridges, all completed in 1936. They were built by joint funding from the PWA, and local gas tax and highway funds. Each bridge offered a unique expression of minimalist design and Art Deco motifs, which were common in public architecture during the 1930s.

Motorists passing through Schmitz Park in West Seattle during the 1920s and early 1930s crossed a deep ravine on an aging timber-truss bridge. In 1935, as the bridge showed signs of serious decay, the city engineer's office made plans to replace it. Eldridge's design for the new 175-foot span used concrete cells (hollow boxes) for the structural members, which significantly reduced the bridge's dead load, while also creating a rigid frame 60 percent longer than any previously constructed. Built for $134,000, the structure won national recognition at the time as the longest rigid-frame bridge ever built. Carl

The Schmitz Park Bridge in west Seattle won national recognition as the most important innovation in concrete bridge design since the column and slab style.

Office of Archaeology and Historic Preservation

Condit, author of *American Building Art* (1961), hailed it as the most important innovation in concrete bridge design since the column and slab system, and also as one of the most economical.[10]

On north Queen Anne Hill, Wolf Creek Canyon provided Eldridge with another opportunity for a creative bridge solution when replacing an aging timber span. Here, Eldridge designed a dramatic, parabolic steel bridge

Clark Eldridge, Seattle city engineer, designed the stylish North Queen Anne Drive Bridge, built in 1936.

Office of Archaeology and Historic Preservation

Clark Eldridge also designed the Cowen Park Bridge, an elegant, double-rib, open-spandrel, concrete deck-arch, erected in 1936.

Office of Archaeology and Historic Preservation

that stretched 327 feet across the ravine with a 140-foot arch. The solid-rib, two-hinged construction made the light, attenuated steel span both graceful and economical. Built between 1935 and 1936 for $66,118, the North Queen Anne Drive Bridge was the first of its type erected in the state.[11]

The Cowen Park Bridge, constructed by the city in 1936, replaced two temporary wooden structures (one for streetcars; one for pedestrians) at Cowen Park's eastern end on 15th Avenue Northeast. The bridge—an open-spandrel, reinforced-concrete arch some 358 feet in length—provided a roadway 42 feet wide, with two 7-foot-wide sidewalks. Eldridge's design included six concrete-slab approach spans. The 160-foot, double-ribbed arch is braced by concrete lateral struts. Fluted, vertical concrete posts and Art Deco iron lamps accent the bridge's graceful form.[12]

Ravenna Park Bridge

This structure is one of the finest legacies of Arthur Dimock's tenure as Seattle city engineer. Under his guidance, designer Frank M. Johnson created an aesthetically pleasing solution for crossing a deep, wooded gulch in Ravenna Park at 20th Avenue Northeast. Constructed in 1913 to help link rapidly growing residential districts on either side of the park, the bridge stands today as the oldest of the state's two remaining three-hinged, steel lattice-arch spans. The 354-foot-long structure—with a 250-foot arch composed of two ribs that rise 41 feet over the ravine—supports an 18-foot, reinforced-concrete roadway now reserved for pedestrians only.

In 1913, steel arch bridges were less common in the United States than in Europe. Typically, they were more costly to build. While Americans often chose economics over aesthetics, Dimock's design team decided that the arch form would be more pleasing and better integrated into a park setting. The bridge's only ornamentation

Art deco light standards adorn the deck of the Cowen Park Bridge.

Craig Holstine, WSDOT

The graceful, lattice-steel deck arch of the Ravenna Park Bridge complements its setting.

Craig Holstine, WSDOT

appears on the deck underside, where fascia plates are scalloped, creating the effect of a series of small arches.[13]

The Alaskan Way Viaduct, completed in 1953, is a distinctive double-deck roadway that stretches 1.5 miles along Seattle's waterfront.

Washington State Archives, WSDOT Records

Alaskan Way Viaduct

As Seattle's residential districts continued to expand, engineers sought new solutions to enhance traffic flows around and through the crowded downtown corridor. The completion of the Aurora Bridge in 1932 and extension of the Pacific Highway (now State Route 99) helped significantly as motorists moved about in the central metropolitan area. Still, congestion and the pressure to provide better entrance and exit paths for downtown were growing concerns.

One of the state's most creative engineers, Ray M. Murray, who guided the design and construction of the Aurora Bridge, devised an answer. In 1934 Murray proposed building the extensive Alaskan Way Viaduct, the first double-deck bridge in Washington. By 1946, the State

The City of Seattle and WSDOT began studying options to replace the Alaskan Way Viaduct after assessing damage caused by the February 2001 Nisqually earthquake. Photo April 15, 2001.

WSDOT

The Battery Street Tunnel at the north end of the Alaskan Way Viaduct funnels State Route 99 under city streets toward the Aurora Bridge. Shown here are the south portals. Photo 2002.

Craig Holstine, WSDOT

Department of Highways loaned Murray to the Seattle Engineering Department, and joint funding from the city, state, and federal governments launched the project.

The construction area lay along the waterfront near Railroad Avenue. Here, a maze of tracks ran north-south, parallel with a seawall originally built on earth-filled tidal flats. City officials considered the area an underused eyesore. To state engineers, it seemed suitable for a major highway and could be conveniently connected with Aurora Avenue and the Pacific Highway. Work began on the viaduct between Western/Elliott and King streets in December 1949, and on the Battery Street Tunnel in September 1952. The northern viaduct section officially opened on April 5, 1953, the tunnel in July 1954, and the southern section in August. The viaduct sections cost approximately $9.5 million, and the Battery Street Tunnel, $2.4 million.

The viaduct's most distinctive feature is its 1.5-mile-long double-deck roadway that begins near the Pike Place Market and extends south along the picturesque waterfront. Northbound traffic uses the upper deck, and southbound traffic the lower deck. At the southern end of the viaduct, the route connects to a single-level, at-grade roadway. The structure stands on square and rectangular concrete, pier-frame columns on pile foundations, with reinforced-concrete crossbeams and longitudinal beams supporting the reinforced-concrete roadway.

Today the Alaskan Way Viaduct suffers from age and the effects of recent earthquakes. The 6.8-magnitude Nisqually quake on February 28, 2001, inflicted serious damage, requiring a several weeks' closure for repairs. Since then, intermittent closures and weight restrictions have limited use of the roadway, which normally carries 100,000 cars a day, about one-quarter of the city's north-south traffic. Its longevity is limited. On June 26, 2001, a month-long study of options by an independent team of five engineers recommended that the structure be torn down and replaced.[14]

Lake Washington Ship Canal Bridge, Interstate 5

As the 1950s unfolded, Seattle grappled with steadily mounting traffic problems. The opening of the first section of the Alaskan Way Viaduct in April 1953 was emblematic—and a sign of things to come. Only eighteen minutes after city officials enthusiastically snipped the ribbon, motorists eager to try out the new highway were crowded bumper to bumper in one of the worst traffic jams Seattle had yet seen. At the time, the area's two main arterials, State Route 99 and Marginal Way, handled most north-south traffic. East-west travel continued via the first Lake Washington Floating Bridge, and communities on both sides of Lake Washington were deeply engaged in controversy over a second floating bridge.

By then, state engineers were busily planning on a grand scale. In 1953, political leaders and engineers proposed a sixty-six-mile, multilane, toll road between Tacoma and Everett. At an estimated cost of $178 million, the idea readily garnered opposition, and the Washington State Supreme Court declared the project unconstitutional. Then, in 1956 the federal government stepped forward with one of the most far-reaching pieces of transportation

The Lake Washington Ship Canal Bridge under construction, 1961.

WSDOT

canal's sixth crossing. The smaller, earlier bridges were, of necessity, movable structures, but the new freeway bridge would stand far above all navigation on the canal. Construction began with the pouring of concrete piers just west of the University Bridge. Erection of the steel superstructure started on May 10, 1960, and was completed eight months later on January 14, 1961. The remaining features were finished by early fall 1961.

Construction of the bridge and the new Seattle freeway suffered several delays. On three occasions labor-management disputes slowed progress. Dredging work on the Duwamish River to the south hit snags from time to time, as did the relocation of utility lines on various portions of the route. A long-running public debate over the siting of the Evergreen Point Bridge interrupted work on some sections of connecting freeway. And, various proposals for covering the downtown freeway took months of public discussion to resolve.

Despite delays, workmen finished the ship canal bridge a year before the opening of its connection to Interstate 5. Initially, completion coincided with planning for the Seattle World's Fair, and the city proposed using the idle

legislation ever passed—the Federal Aid Highway Act. Under this law, a federal gasoline tax provided funds to pay for 90 percent of a superhighway network across the nation. Washington qualified for 726 miles of interstate roadway. Engineers began detailed plans for a twenty-three-mile stretch of what would become Interstate 5 through the center of Seattle, including a new span across the Lake Washington Ship Canal.

By August 1958, state officials awarded the first contract for the new Lake Washington Ship Canal Bridge. The massive $16 million span would be the

The Lake Washington Ship Canal Bridge, built 1958–61, measures 4,429 feet long. Photo 2002.

Craig Holstine, WSDOT

Lake Washington Ship Canal Bridge on Interstate 5, north approaches.
WSDOT

structure as a parking lot for visitors. Those plans, which would have made the bridge the largest raised parking lot in the world, were abandoned when the city yielded to using private parking facilities located nearer to the fairgrounds. Putting the final touches on the bridge was a picturesque scene, aptly described by Seattle historian Paul Dorpat: "Twelve painters, with nerves that matched the truss material, applied 10,000 gallons of undercoat and green linseed oil topcoat with brushes so that nearby houses and cars would not be inadvertently decorated" by spraying machines.

The Lake Washington Ship Canal Bridge is the largest structure of its kind in the Northwest. The span stretches 4,429 feet over the ship canal, from Capitol Hill north to the University District. The bridge's six steel truss sections support a double-deck roadway 2,294 feet long, with eight traffic lanes on an 82-foot-wide upper deck and another four reversible express lanes on the lower deck. Three of the steel sections are simple Warren truss spans, and the others are a flanking, three-span, continuous unit, the center of which was erected as a cantilever over the waterway. With a vertical clearance of 135 feet between the upper and lower deck, and a distance of 191 feet from the ship canal to the upper deck, the bridge soars over the canal with an immensity reflecting its significance in the city's transportation network.[15]

Seattle's Historic Bridges

Seattle has one of the state's richest bridge building traditions, and a long-standing commitment to maintenance and preservation. Historic bridges, however, often have been lightning rods for political gamesmanship and civic activists. Seattle's bridges are beloved, yet vulnerable. Age and operational stresses continue to threaten this heritage. Seismic retrofitting of Seattle's bridges was supported by federal funds, but has required a huge investment of state monies as well.[16]

Whatever challenges lie ahead for Seattle's historic bridges, they will remain among the most remarkable of the city's public works. These spans have helped define the city over the years. They remain an integral part of Seattle's transportation system and its identity.

NOTES

1. See R. H. Thomson, *That Man Thomson* (Seattle: University of Washington Press, 1950); "R. H. Thomson, Civic Leader, Engineer, Dies," *Seattle Post-Intelligencer*, January 8, 1949, 4; "R. H. Thomson, City Builder and Leader, Dies at 92," *Seattle Times*, January 7, 1949; "Reginald Heber Thomson," in Clarence B. Bagley, *History of King County*, vol. 2 (Chicago: S. J. Clarke Publishing Company, 1929), 92–100; various newspaper clippings in "R. H. Thomson," Biography File, Special Collections, University of Washington Libraries. Thomson's personal papers are in Special Collections, University of Washington Libraries.
2. Paul Dorpat, "Now and Then: The Latona Bridge," Northwest Magazine, *Seattle Times*, January 13, 1991, 10; Myra L. Phelps, *Public Works in Seattle: A Narrative History: The Engineering Department, 1875–1975* (Seattle: Seattle Engineering Department, 1978), 35–46; "Latona Bridge," *University Herald*, January 10, 1979, 8; "Freeway Span Towers over Site," undated newspaper article, clipping file, Seattle Public Library; Paul Dorpat, "Tampering with Lake Union," *Seattle: Now and Then,* 2nd ed. (Seattle: Tartu Publications, 1984), 66; D. V. Corsilles, ed., *Rizal Park: Symbol of Filipino Identity* (Seattle: Magiting Corporation, 1983), 3, 14.
3. Lisa Soderberg, "12th Avenue South over Dearborn Street Bridge," HAER Inventory, 1980; Phelps, *Public Works*.
4. A. H. Dimock, "Bridging of Inter-City Waterways at Seattle," *Pacific Builder and Engineer*, March 8, 1913, 153–55; A. H. Dimock, Report of City Engineer to Seattle City Council, April 20, 1914, and other correspondence in "Seattle Bridges," Bridge Files, Office of Archaeology and Historic Preservation, Olympia; Wm. Michael Lawrence, "Montlake Bridge," HAER No. WA-108, 1993; James Warren, "Looking Back: Links to Seattle's Growth—When a Ship Canal Cut a City in Two, Placement of Bridges Became a Hot Issue," Northwest Magazine, *Seattle Post-Intelligencer*, February 15, 1981, 14.
5. F. A. Rapp, "Three Double-Leaf Bascule Bridges at Seattle, Wash.," *Engineering News-Record* 84, April 8, 1920, 718–722; John Olson, "Bridges to the 20th Century: Four Drawbridges over the Ship Canal," *Seattle Weekly*, November 20, 1996, 21–27; Lisa Soderberg, "University, Fremont, and Ballard Bridges," HAER Inventory, 1980; J. A. Dunford, "Seattle Remodels Its University Bridge," *Engineering News-Record*, October 12, 1933, 439; "Service Performance of Grid Deck on University Bridge, Seattle," *Engineering News-Record*, September 20, 1934, 376.
6. Wm. Michael Lawrence, "Montlake Bridge," HAER No. WA-108, 1993; "Bridge Tenders Have Their Ups and Downs," *Seattle Post-Intelligencer*, August 20, 1972, A16; Lisa Soderberg, "Montlake Bridge," HAER Inventory, 1980.
7. W. D. Barkhuff, "City of Seattle's Bridge Program," *Pacific Municipalities*, March 1929, 107; W. C. Morse, City Engineer, letter to *Western Construction News*, April 26, 1928, copy in "Spokane Street/Duwamish" OAHP File; Phelps, *Public Works*, 41–42; Thomas D. Hunt, "Fourteenth Avenue South Bridge, Seattle," *Western Construction News*, June 10, 1931, 287.
8. Kevin A. Palmer and Christine Savage Palmer, "14th Avenue Bridge South," King County Landmark Registration Form, 1996; "Duwamish Span to be Semifloating," *Seattle Times*, October 23, 1953, 16; "Job to Set Engineering Precedent," newspaper clipping in "Spokane Street/Duwamish," OAHP Bridge Files; Phelps, *Public Works*, 49–50; Oscar R. "Bob" George, "Duwamish River Bridge at First Avenue South," National Register of Historic Places Nomination, 2001; John H. Clark, "West Seattle Swing Bridge, Seattle, Washington," *Structural Engineering International*, January 1995, 23–25.
9. Robert W. Hadlow, "Aurora Avenue Bridge," HAER No. WA-107, 1993; Lisa Soderberg, "Aurora Avenue Bridge," HAER Inventory, 1980; "Lake Union Bridge, Seattle, Washington," *Western Construction News*, May 10, 1930, 226–229; City of Seattle Landmark Nomination Form for George Washington Memorial Bridge, January 2, 1980, OAHP Bridge Files; "Lake Union Bridge Provides New Traffic Link in Seattle," *Engineering News-Record*, March 8, 1932, 313; Phelps, *Public Works*, 48–49.
10. Lisa Soderberg, "Schmitz Park Bridge," HAER Inventory, 1980; J. A. Dunford, "Record Rigid-Frame Bridge," *Engineering News-Record*, June 24, 1937, 939–42.
11. Lisa Soderberg, "North Queen Anne Drive Bridge," HAER Inventory, 1980.
12. Lisa Soderberg, "Cowen Park Bridge," HAER Inventory, 1980.
13. "Ravenna Park Bridge," HAER Inventory, November 1981; Frank Melvin Johnson, "The Ravenna Park Steel Arch Bridge" (master's thesis, University of Washington, 1916).
14. R. W. Finke, "Seattle's Long-Planned Alaskan Way Viaduct Scheduled for Construction," *Pacific Builder and Engineer*, August 1949, 52–55; Oscar R. "Bob" George, "Alaskan Way Viaduct, Battery Street Tunnel," NRHP Nomination, 2001; George Foster, "Discussion Starts on What to Do with the Viaduct," *Seattle Post-Intelligencer*, June 20, 2001; Tom Madden, Engineering Manager, Alaskan Way Viaduct, WSDOT, personal communication, 2004.

15. Paul Dorpat and Genevieve McCoy, *Building Washington: A History of Washington State Public Works* (Seattle: Tartu Publications, 1998), 131; William Dugovich, "Seattle's Superfreeway," *Washington Highways* 14, no. 2, May 1967, 2–5; George Andrews, "Steel Bridge—May to January," *Pacific Builder and Engineer* 67, June 1961, 86; Kay Conger, "Seattle Freeway Bridge to Be Sixth Ship Canal Crossing," *Highway News*, January-February 1958, 10–13; "Priority Urged for High Level Traffic Span," *Seattle Times*, March 16, 1952; Oscar R. "Bob" George, "Lake Washington Ship Canal Bridge," NRHP Nomination, 2001.
16. Peggy Andersen, "Time and Nature Take Toll on State Bridges," *Seattle Post-Intelligencer*," February 20, 2001; Peyton Whitely, "Seattle's Traffic Is 2nd-worst: City Trails only L. A. in Study," *Seattle Times*, May 8, 2001.

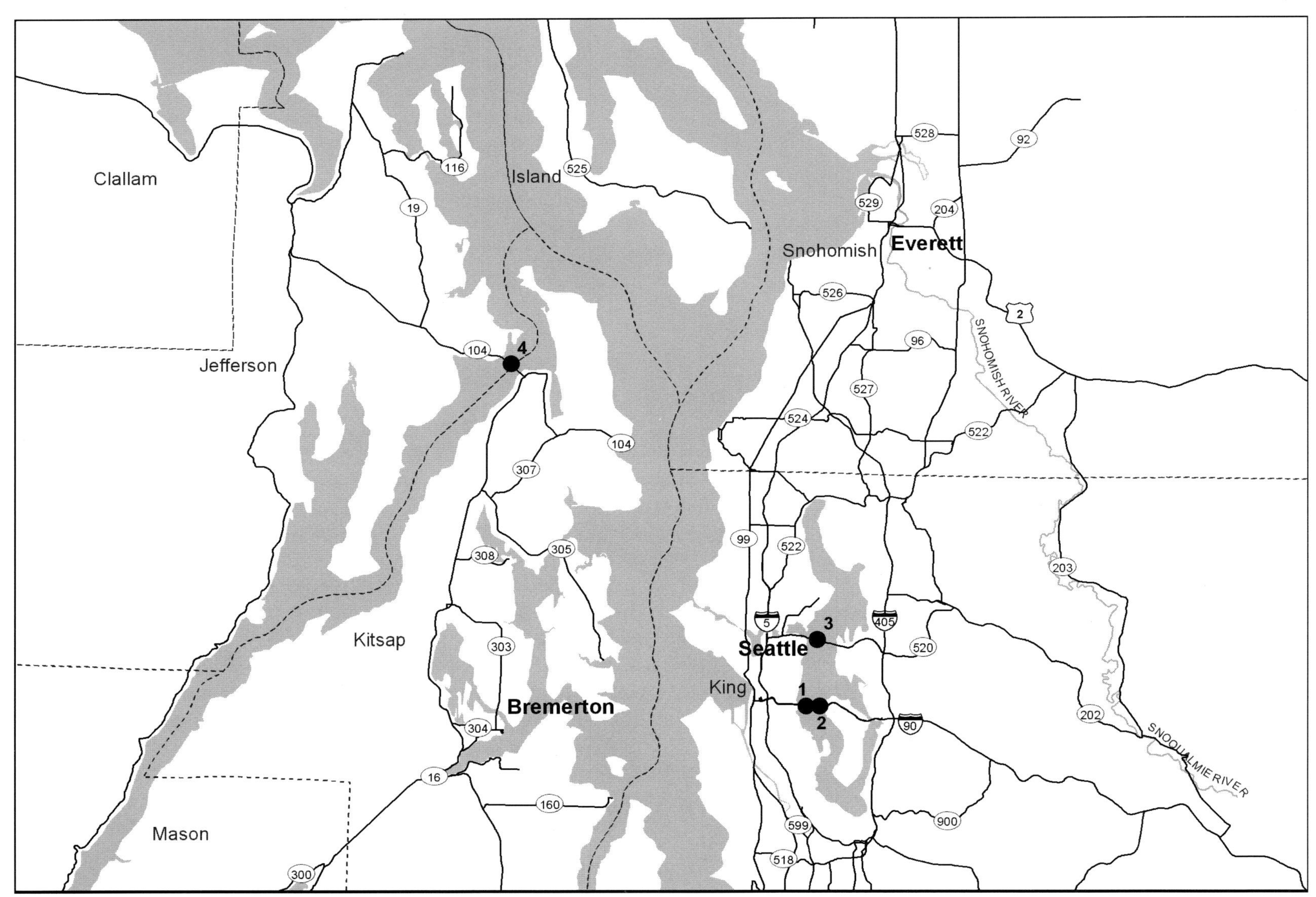

Bridge No.	Bridge Name
1	Lacey V. Murrow Bridge
2	Homer M. Hadley Bridge
3	Evergreen Point (Albert D. Rosellini) Bridge
4	Hood Canal (William A. Bugge) Bridge

Floating Bridges of Washington

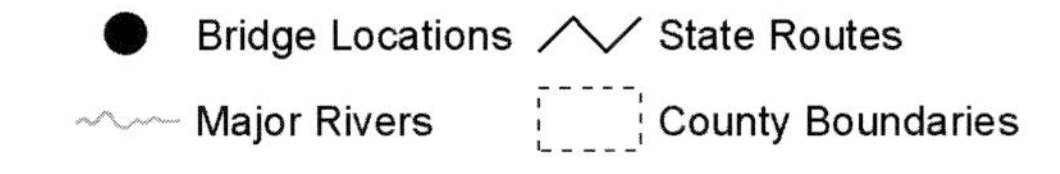

FLOATING BRIDGES

Lake Washington Floating Bridge—1940

The story of the first bridge across Lake Washington is a fascinating saga—a curious mixture of peculiar geography, intriguing personalities, power politics, community activism, and brilliant engineering. It is also a story of patience, dogged persistence, intuition, and sweat and hard work over the course of two decades.[1]

Lake Washington on Seattle's eastern edge is some twenty miles long and four miles across at its widest point. For many years in the late nineteenth and early twentieth centuries, the lake had been a scenic destination for area residents. By the 1920s, however, as the automobile began reshaping how people traveled and lived, the large body of water came to be seen as an obstruction. Mount Baker Ridge on the west shore likewise stood as a barrier for travelers between Seattle and points to the east; roads circled north or south of the lake and ridge.

To bridge builders, the lake's width—one and a half miles at its narrowest—posed major challenges, as did the lake bottom, some 100 to 200 feet deep and composed of soft mud. The building of piers across Lake Washington would be difficult and expensive. Thus, geographic circumstances seemed to augur for the viability of a pontoon bridge, but pontoon spans were vulnerable to high winds, currents, ice, debris, and changes in water levels.

In a sense, Mercer Island too was an obstacle. Settled as early as 1850, the area remained remote and grew slowly. Pioneer residents traveling to Seattle took a boat to the nearby east shore, then traveled overland north or south around the lake. Steamer ferry service began in 1890 and remained the only link west across the lake to Seattle for a half century. In the late 1930s, Mercer Island was a community of a mere 1,200 citizens, with some farms, a scattering of summer cottages, and hardly a dozen businesses. Financial firms viewed the construction of any bridge as a far greater investment than the low traffic volume would ever justify.

By the 1930s, however, there was a shift in thinking. Residents of Mercer Island and other communities, as well as cross-state travelers, faced a twenty-six-mile drive

The tug *Patricia Foss* tows pontoon "L-1" through the Lake Washington Ship Canal en route to the Lake Washington Floating Bridge construction site, January 27, 1940. Alfred Simmer, photographer.
Washington State Archives, WSDOT Records

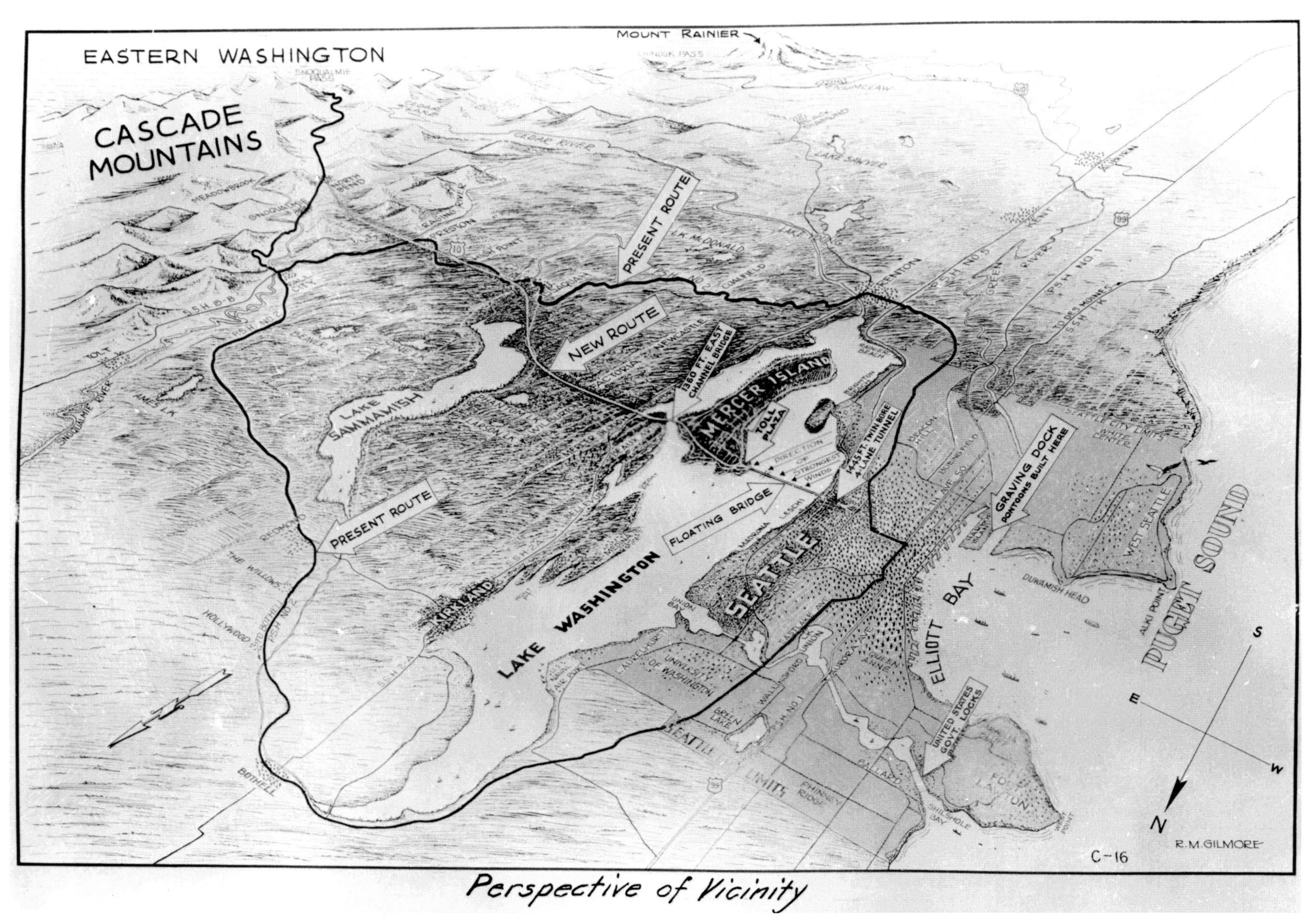

A bird's-eye view showing how the Lake Washington Floating Bridge would change highway traffic into and out of Seattle. The map also shows the locations of the Mount Baker Ridge Tunnels, the new highway, and the "Graving Dock" where the concrete pontoons would be built.

Office of Archaeology and Historic Preservation

around the lake's southern end, partly through residential neighborhoods. The route was tedious and, increasingly, time consuming.

An idea for bridging the lake had surfaced at the start of the 1920s. On a spring morning in 1920, a young engineer named Homer Hadley, then working in the Seattle School District architect's office, stood at his bathroom sink, shaving and looking out the window across Lake Washington. Suddenly, he realized how to bridge the lake—with hollow concrete barges connected end to end and supporting a roadway deck. It was the simple association of two things in his personal experience—first, hollow concrete barges like the ones he had helped build in Philadelphia during World War I; and second, the familiar and rather unsightly war surplus wooden hulls then moored on Lake Union.[2]

Pontoon bridges were no novelty. Under the right circumstances, they were a feasible option, especially when a body of water was too deep to support piers for a traditional span. And, pontoon bridges had been used since ancient times. Persian armies crossed the Hellespont on a pontoon bridge to invade Greece in the late fifth century B.C.

Hadley pointed to more recent wood and steel examples—two railroad pontoon spans maintained since 1874 at Prairie du Chien, Wisconsin; another railroad pontoon draw span on the Missouri River; a 300-foot pontoon railroad and highway bridge across the Panama Canal; a 1,000-foot pontoon on the Rhine in use since 1819; a 1,770-foot pontoon bridge over Russia's Dvina River; a pontoon bridge in service since 1874 in Calcutta; and a 1,530-foot pontoon bridge completed in Turkey in 1912 across the Golden Horn, the latest in a succession of bridges at that location.[3]

Still, pontoon bridges were most unconventional. Hadley's idea to build the longest one ever attempted seemed outrageously ambitious. As one observer has noted, "Floating concrete played like a vaudeville joke in 1921."[4] When Hadley approached local bankers with his concept, they scoffed. Impractical, impossible, a screwball idea, they said. The tag "Hadley's Folly" hung on the floating bridge idea and dogged the young engineer for years. Hadley refused to give up and continued to enter public debates on a bridge across the lake, but all his efforts failed to gather support.

As seen from Lake Washington Boulevard, construction progresses, December 4, 1939. Alfred Simmer, photographer.
Washington State Archives, WSDOT Record

When Hadley took a job with Portland Cement as the regional structural engineer, it seemed to sound the death knell for his proposal. "From then on," he later recalled, "whenever I mentioned my idea, someone would express the fear that it was part of a nefarious plot of the cement companies to desecrate Lake Washington for profit."[5] The effort to build a pontoon span met with legal barriers, financial hurdles, political problems, competition from alternative plans, and other delays. As time passed, the effort became a series of intense struggles. Other proposals came forward to compete with Hadley's.

In 1921, two engineers, Robert Montell and J. H. Dirkes, suggested that a pontoon bridge could replace ferry service across the lake. Their idea resurfaced a year later, when a Mercer Island citizens group urged King County commissioners to establish a fund to buy World War I surplus wooden hulls at $100 each. But, the proposal had little support, and the commissioners refused. Other interested parties advocated tunneling under the lake, but the $12 million price tag made the idea very unattractive.[6]

In 1924 the county completed the East Channel Bridge, a timber-and-steel span linking Mercer Island's east side to the east mainland. This fueled the enthusiasm and determination of islanders and east siders to have a bridge built west across Lake Washington. The South Mercer Island Community Club, led by George W. Lightfoot, became one of the most ardent advocates. A core of community activists on both sides of the lake persisted in this dream, devoting endless hours to lobbying, reviewing proposed plans, and garnering support wherever they could.

View of the floating bridge, looking east toward Mercer Island from atop Mt. Baker Ridge Tunnel, October 2, 1940. Alfred Simmer, photographer.
Washington State Archives, WSDOT Records

In 1931 the Seattle City Council decided to hold hearings on four proposed bridges across the lake, including one suggested by Hadley and his supporters. The debate dragged on as Seattle newspapers took sides. The *Post-Intelligencer* backed Hadley, at least until too many citizens criticized his affiliation with Portland Cement. Colonel C. B. Blethen, the *Seattle Times* publisher, righteously fought against a pontoon structure, calling it "unthinkable."[7]

However, 1937 proved to be a turning point, when the Public Works Administration pledged federal funds for up to 40 percent of construction costs, but the offer was available only through the end of the year. With a December 31 deadline, and no definite plans or commitment from the state, the window of opportunity was rapidly closing. Fortuitously, that spring the State Legislature created the Washington State Toll Bridge Authority, giving it the power to finance, construct, and maintain bridges with bonds and tolls.

When Hadley learned about the Toll Bridge Authority's creation, he visited Lacey V. Murrow, director of the State Department of Highways since 1933. On June 10, the two spent an hour discussing Hadley's bridge idea and the most economical site for its construction. Murrow emerged from the meeting in full agreement, and became the leading proponent of a concrete pontoon span, just like Hadley had suggested in 1920.[8]

Murrow immediately gathered a project team of some of the state's best bridge engineers. Notable among them were Clark Eldridge (who soon became the bridge engineer for the first Tacoma Narrows Bridge), Charles E. Andrew (principal engineer in direct charge of the project), Ray M. Murray, Luther E. Gregory, and L. R. Durkee (resident engineer for the PWA).

Murrow and the Toll Bridge Authority also established a board of consulting engineers comprised of Andrew, who served as chairman, Gregory, Ray B. McMinn, and R. H. Thomson to study options and make recommendations. In February 1938, the board approved a concrete pontoon span as the most economical, and selected the Mount

Baker Ridge route (via tunnels) as the best approach to Seattle. They also approved plans for a bond issue and toll revenues, then ordered immediate preparation of detailed drawings and specifications.[9]

Residents of the Mount Baker Ridge and Seward Park neighborhoods objected to the plan, as did many others. Squabbling continued in local newspapers and the Seattle City Council chambers. The Lake Washington Protective Association called the notion of a concrete pontoon bridge a mere "experiment," predicting that "a chain of skows [sic] across Lake Washington would stand out as a municipal eyesore and a desecration of the city's greatest natural asset." The debate became so intense that the PWA declared it would withdraw its offer to finance 40 percent of the bridge's cost unless the Seattle City Council endorsed the proposal.

First-term Congressman Warren G. Magnuson stepped forward to try and settle the matter. "If Seattle will stop arguing over what it wants and decide on something," he said, "funds will be available. There is no objection whatever in Washington, D.C. to a pontoon bridge." A final council vote of five in favor, four against, sealed the issue. The PWA approved funding, and the Toll Bridge Authority accepted bids for construction. On December 29, 1938, two days before the PWA deadline, Governor Clarence D. Martin and Lacey Murrow presided at groundbreaking ceremonies. The struggle for the bridge finally was over.[10]

Construction began on January 1, 1939. The 6.5-mile project employed 3,000 people for eighteen months. About 1,200 workers handled the actual construction, and the remaining 1,800 provided logistical and administrative support. The project's main features included the 6,620-foot floating span with a sliding draw pontoon, twin Mount Baker Ridge tunnels, west- and east-side approaches, transition spans between the approaches and pontoons, and highways on Mercer Island. The 45-foot-wide roadway, with two four-foot sidewalks, was supported by the pontoons, which were connected end to end, forming a rigid box-girder span across the lake.

The deck and "bulge" slip span on the Lake Washington Floating Bridge.
WSDOT

Near the center, a 378-foot draw span allowed a clear channel opening of 200 feet. The sliding pontoon moved in a slot between two flanking pontoons. Here, at the "bulge," the roadway split onto guide pontoons. Two electric motors (seventy-five horsepower each) opened the massive draw pontoon in only ninety seconds. The transition sections, connecting the pontoons and the column-supported approach spans, featured a system of universal joints and rockers that enabled the structure to withstand the stresses of pontoon movements from virtually any direction. A 960-foot-long steel-and-concrete approach bridge stood at the east end.[11]

The deck truss and tied-arch approach linking the floating pontoons with Mercer Island. The east portals of the Mt. Baker Ridge Tunnel are visible in the distance, September 5, 1940. Alfred Simmer, photographer.

Washington State Archives, WSDOT Records

Costs totaled $8.85 million. The PWA contributed about $3.8 million and the State Toll Bridge Authority financed the remainder with a bond issue. The toll schedule mainly applied to personal autos and commercial trucks, but included fares for horse-drawn vehicles—thirty-five cents when pulled by one or two horses, and fifty cents for three or more horses.[12]

On July 2, 1940, the Lake Washington Floating Bridge officially opened with dazzling fanfare. Some 2,000 people gathered for the highly publicized "christening" festivities to celebrate the occasion. National radio coverage, hosted by Richard Keplinger, carried the event live. Governor Martin paid the first toll, and in twenty-four hours, some 11,600 cars crossed the bridge. Newspapers hailed the "seven-minute travel time" from Seattle to Mercer Island, calling the bridge "the eighth wonder of the world."[13]

At its completion, the Lake Washington bridge was the largest floating structure in the world and the first pontoon bridge constructed of reinforced concrete. The pontoon section and sliding draw span were unprecedented. The innovative bridge received honors from around the nation and the world. It served as a model for subsequent spans across Lake Washington and Hood Canal, as well as for bridges overseas. Washington State became the international leader in floating bridge technology.

The bridge proved so popular that it was paid for by its ninth anniversary on July 2, 1949. The state removed the tolls and dismantled the toll plaza—a full nineteen years ahead of schedule. In 1950, the bridge's success resulted in the demise of ferry service on the lake. Most significantly, the floating bridge became the catalyst for a dramatic population growth in Bellevue, Mercer Island, Redmond, Kirkland, and other east-side communities. Completion of the Evergreen Point Floating Bridge in August 1963 relieved much of the congestion on the first Lake Washington bridge. Still, by 1975 the bridge was handling an average of 52,000 vehicles per day.

Recognizing the key role played by former Highway Director Lacey V. Murrow, the State Legislature approved an official name change. In early 1967, just months after Murrow's death, the span became the "Lacey V. Murrow Memorial Bridge." Homer Hadley, it seemed, would remain the unknown "father" of the floating bridge.

In 1989 increasing traffic loads prompted completion of an adjacent floating bridge, parallel to the original structure and just twenty yards to the north. With the fifty-year-old bridge closed to traffic, the state began renovating its pontoons in late spring 1990. On November 22–23, 1990, a Thanksgiving weekend storm flooded one pontoon, and on November 25 seven pontoons sank. The entire floating section had to be scrapped. Today, all that remains of the original bridge are the approaches. These pier-supported, 215-foot, steel tied-arch spans provide a vertical clearance of thirty-five feet for the passage of small craft. They are the bridge's most visually distinctive features.[14]

Pontoons slip beneath the surface of Lake Washington, November 25, 1990. Water-tight hatches not yet installed were stacked on the deck, and were lost, along with the vehicles shown here.

WSDOT

Tugboats hold the Homer M. Hadley Bridge in place after sinking pontoons from the Lacey V. Murrow Bridge severed some of the anchoring cables.

WSDOT

A unique component of the Lake Washington project lay at the west end. The route into Seattle needed to burrow through one of the city's seven hills, Mount Baker Ridge, which rises steeply 260 feet above the lake. The ridge's geology was unusual. Virtually rock free, it was composed of heavy, glacial-blue clay, which dictated a horseshoe-shaped tunnel design. Engineers decided to bore two tunnels for separate traffic lanes. When finished, these were the world's largest-diameter, soft-earth, twin tunnels. Their centers were spaced 60 feet apart, and measured 1,466 feet long. Twenty-four-foot wide roadways, with vertical clearances of 23 feet from pavement to arch crown, passed through each.[15]

One of the most distinctive features of the Mount Baker Ridge Tunnels appears at the east entrance. The east portal, greeting travelers headed toward downtown Seattle, includes Northwest Coast native motifs in a flowing Art Deco style. Artist James Fitzgerald and architect Lloyd Lovegren created the decorative treatment, which was hailed at the time as "strikingly beautiful." The arch entrance includes a series of three tiered setbacks. On either side of and between the two tunnels, three relief sculptures—of swirling figures and geometric shapes—present a visual portrayal of Seattle as the "Portal of the North Pacific."

Artists design the Mt. Baker Ridge Tunnel portals, May 9, 1940. Alfred Simmer, photographer.

Washington State Archives, WSDOT Records

A night view of the east portals, Mt. Baker Ridge Tunnels, October 22, 1940. Alfred Simmer, photographer.

Washington State Archives, WSDOT Records

In 1982, as efforts moved forward for the parallel Interstate 90 bridge, plans were adopted for a new tunnel. This one superseded the first tunnel as the "world's largest soft soil (or, 'stack drift') diameter tunnel." Completed in June 1989, the new $200 million Mount Baker Ridge Tunnel was the first of its kind. It is a single tunnel 1,476 feet long, with an interior diameter of 63 feet. It features a compression ring of twenty-four separate, concrete-filled drifts, each 9 ½ feet in diameter, lining the core. In 1987 the tunnel's designers received the American Consulting Engineers Council's accolade for excellence. In 1990 the tunnel earned an American Society of Civil Engineers award for "Outstanding Civil Engineering Achievement."

The snow-topped Olympic Mountains rise in the distance beyond Seattle, with the University of Washington (mid-photo), and the Evergreen Point (Albert D. Rosellini) Bridge on Lake Washington (foreground).

Washington State Archives, WSDOT Records

When completed in 1963, the Evergreen Point Bridge was the longest (1.4 miles) floating bridge in the world; and, costing over $24 million, the most expensive. Aerial view to the east, ca. 1963.

Washington State Archives, WSDOT Records

Evergreen Point (Albert D. Rosellini) Bridge—1963

As early as 1946, citizens began calling for a second Lake Washington Floating Bridge. Little did they imagine that seventeen years of struggle and controversy would elapse before the span was built.[16]

Pressure for a new bridge came from the mushrooming population on the east side. The post-World War II years witnessed unprecedented expansion of greater Seattle's urban core and the suburbs. The first floating bridge helped the east side to become Puget Sound's fastest growing area. The engineers who built the bridge expected that its capacity to handle up to 20,000 cars a day would be sufficient for a long time, but it reached that volume in just over a decade. Between 1950 and 1960, while Seattle's population increased by 26 percent, the suburban east side grew by nearly 88 percent. Vehicular use of the first floating bridge rose from 3,476 cars per day in 1940, to 17,884 in 1950, and topped 50,000 by 1960. Traffic delays and accidents reached alarming levels.

Choosing a location proved one of the most contentious issues. On both sides of the lake, advocates wrangled over their preferred sites. Numerous population and traffic studies considered the advantages and disadvantages of the options. In all, six different sites were reviewed. The determining factor was saving time for commuters. After years of public hearings, lawsuits, and debates, the State Toll Bridge Authority announced a compromise plan in 1957. It would build two bridges, one on the Evergreen Point-Montlake route, and a second parallel to the original Lake Washington Floating Bridge. After more hearings, debates, and legal battles, work finally began in 1960.[17]

The bridge's most prominent advocate was Governor Albert D. Rosellini, the presiding official who drove a bulldozer at the bridge's groundbreaking in August 1960, and who would snip a ribbon at its opening ceremony, August 28, 1963. In 1988 the bridge was named in Rossellini's honor, though it continues to be known today as the Evergreen Point Bridge.[18]

More than two years (837 days) of construction brought the bridge to completion. At 1.4 miles in length, it was the largest floating span in the world. With a $25 million price tag (the floating section alone cost $10.9 million), it also was the most expensive. Located some four miles north of the first floating span, it was the central segment of a 5.8-mile project to connect two main north-south highways, Interstate 405 on the lake's east side and Seattle's Interstate 5.

The Evergreen Point Bridge features a floating section of 7,578 feet, comprised of thirty-five pontoons, the largest of which measure 360 feet long by 60 feet wide, and 14.8 feet deep with a 7-foot draft. The smallest pontoon weighs 4,700 tons, and the heaviest, 6,700 tons. The reinforced-concrete anchors, 62 in total and each weighing 77 tons, are connected by 2¾-inch steel cables to the pontoons. The roadway's 54-foot width (compared to 45 feet for the first bridge) provides four lanes—each a spacious 13-feet wide—plus a 2-foot-wide median and 3-foot walkway.

A notable improvement in the new bridge was its sophisticated and unique "no-bulge" lift-draw span, opening to 200 feet for shipping. The lift spans are raised seven feet, allowing retraction of the movable pontoons. At each end of the floating section, elevated steel truss spans with fixed piers connect to the shore and provide enough vertical clearance to accommodate large pleasure craft.[19]

The bridge proved successful beyond anyone's imagination. Motorists gladly paid the thirty-five-cent toll to cut valuable minutes from their commute. The State Toll Bridge Authority financed the bridge with a $30 million, forty-year bond. Traffic rose so dramatically that the debt was paid off by June 1979, more than twenty years ahead of schedule.[20]

Like the first floating bridge, the Evergreen Point Bridge eased traffic for a time, but also stimulated suburb development on the east side, especially for north Bellevue, Redmond, and Kirkland. As these areas grew in the 1970s and 1980s, commuting across the two bridges could be almost nightmarish. State Route 520's popularity with motorists eroded into frustration; by 1988, the Evergreen Point Bridge was handling 100,000 cars a day—twice its intended maximum capacity. A newspaper headline told the story in agonizing brevity: "At 25 Years Old, Span of Route 520 Is Beset by Chronic Congestion...This Drive Requires 'Commuter's Lobotomy.'"[21]

Homer M. Hadley Memorial Bridge—1989

The second floating bridge across Lake Washington had hardly opened in 1963 when the fight started over the third span. One of the first plans proposed a new ten-lane bridge for interstate traffic, while relegating the original floating bridge to serving only Mercer Island residents.

This proposal outraged many Seattle residents, as well as other various neighborhood, community, and environmental groups. As one observer noted, "They viewed the 14-lane I-90 corridor as a monster, a broad snake that would sever Mercer Island and wipe out low-income housing in the Mount Baker neighborhood—while delivering more cars into a city already beset by traffic and pollution woes." East-side and west-side interests would wage political battles for another two decades as plans were modified, and construction finally proceeded.[22]

The new Lacey V. Murrow Bridge (left) and the Homer M. Hadley Bridge (right) on Interstate 90. Seattle provides the backdrop. *WSDOT*

After 1,050 days (almost three years) of construction, the third Lake Washington Floating Bridge was completed, and officially opened on June 4, 1989. At a price of nearly $65 million, the span ranked first in expense. The bridge stood just sixty feet north of the first Lake Washington Floating Bridge. Charles Stewart Gloyd, one of the world's leading experts in constructing floating-concrete structures, guided design for the WSDOT. The new span had many similarities to the first floating bridge—features that have become standard in floating bridge technology. It also benefited from years of engineering experience and included some modern improvements.

The floating section, 5,811 feet in length, is comprised of eighteen pontoons, each 354 feet long by 75 feet wide. Ten of them support a cantilevered roadway 105 feet wide that carries five lanes of traffic and a 10-foot sidewalk. The other eight pontoons are the same length and width, but are deeper at sixteen feet, and support an elevated superstructure of columns and crossbeams that connect at each of the bridge's approaches.

The pontoons, like those for previous floating bridges, are cellular. The new design featured twelve watertight compartments of five cells each, all encased by nine-inch-thick reinforced-concrete walls. The average pontoon weighs roughly 9,000 tons and is held in place by cables attached to anchors weighing up to 90 tons. A total of fifty-two anchors hold the bridge in place.[23]

West-end spans of the second Lacey V. Murrow Bridge (below) and the Homer M. Hadley Bridge (above).
WSDOT

A feature of earlier floating bridges, the draw-span, was not included. At either end of the floating section, a transition span provides an opening 195 wide and a vertical clearance of 35 to 39 feet to accommodate small watercraft. Ships, barges, and other large vessels would use the East Channel on the other side of Mercer Island, where there is a 65-foot vertical clearance under the Interstate 90 bridge spans.

Engineers took full advantage of technological advances. "The really big difference in modern floating-bridge design," said the state's chief bridge engineer, Al Walley, "is that the computer allows us to do dynamic analysis during the planning stage. We factor in the wind and waves, tidal changes and currents that potentially can push the bridge aside or dislodge the anchorages."[24]

After years of effort by Hadley's family and friends, the State Legislature finally agreed to acknowledge his contributions. In July 1993, the parallel span was officially renamed and dedicated as the Homer M. Hadley Memorial Bridge.[25]

Fixed Interstate 90 spans cross the East Channel, connecting Mercer Island to Bellevue.
WSDOT

The Homer M. Hadley Bridge (right) carries traffic across Lake Washington, as the second Lacey V. Murrow Bridge is being completed.
WSDOT

Lacey V. Murrow Bridge—1940/1993

Just two weeks after the pontoons of the first Lake Washington floating bridge sank (late November 1990), WSDOT engineers launched plans for replacement of this 1940 structure. Recommendations by a Governor's Blue Ribbon Commission and WSDOT's own engineering team were incorporated into the design of the replacement structure (as well as in the new, adjacent Homer M. Hadley Bridge). Most important were the regular monitoring and maintenance of the pontoons' watertight capacities, combined with use of bilge-pipe water pumping systems.[26]

As part of the planning process, WSDOT set a new standard for community involvement. At the direction of Transportation Secretary Duane Berentson, the agency conducted an unprecedented 1,300 public hearings on the seven-mile Interstate 90 link between Bellevue and Seattle.

Completed in 1993, the 1.5-mile-long replacement bridge stands about sixty feet south of the Hadley Bridge and in approximately the same alignment as the original span. The floating section is comprised of 20 concrete pontoons bolted together to form a rigid structure some 6,620 feet long. A typical pontoon is 360 feet long, 60 feet wide, and almost 17 feet deep, with a water draft of just under 10 feet. Like the pontoons of the first floating bridge, those of the new bridge are a cellular design, with watertight compartments to minimize flooding. To prevent the domino effect that sank the original bridge, each new pontoon contains forty-eight separate cells. The pontoons are individually fastened to an anchor by eight galvanized steel cables. The bridge employs two types of anchors, one weighing 95 tons, the other 300 tons.

The bridge was constructed with a new high-performance microsilica concrete that has very low levels of permeability and shrinkage. In other words, the pontoons absorb virtually no water and keep their shape. Another improvement is the water-entry warning system mounted in the pontoons. Sensors in each compartment automatically activate alarms monitored by WSDOT emergency response personnel. The bridge opened to traffic nearly a year ahead of schedule, on September 12, 1993. Completion of the seven-mile, $73.8 million construction project finished the final western segment of Interstate 90, which stretches 3,000 miles from Seattle to Boston.

The well-being of Lake Washington's floating bridges is always of public concern. On July 29, 2000, during one of the year's busiest travel weekends, a sleepy boat captain plowed his tug and barge into the Evergreen Point Bridge, crippling three of six columns under the span and snarling traffic for weeks. The word "gridlock" appeared daily in newspapers and on radio and television for the rest of the summer. Of greater concern is the continuing traffic growth, as population expands in the flourishing suburbs north and south of Seattle and east of Lake Washington. The Evergreen Point Bridge carries 102,100 vehicles a day; the Interstate 90 parallel floating bridges (Murrow and Hadley) count 114,800 per day.

In cooperation with King County, other local governments, and the region's citizens, WSDOT has implemented an extensive study of the State Route 520 corridor. The Trans-Lake Washington Project identifies the Evergreen bridge as "the heart of traffic congestion" in Puget Sound, and posts a provocative banner on its Web site: "If you think getting to the moon was tough, try crossing Lake Washington during rush hour." Through its Internet site, community forums, and other means, a host of alternatives are being considered, including bridge replacement, and building a tunnel or tube below the lake.

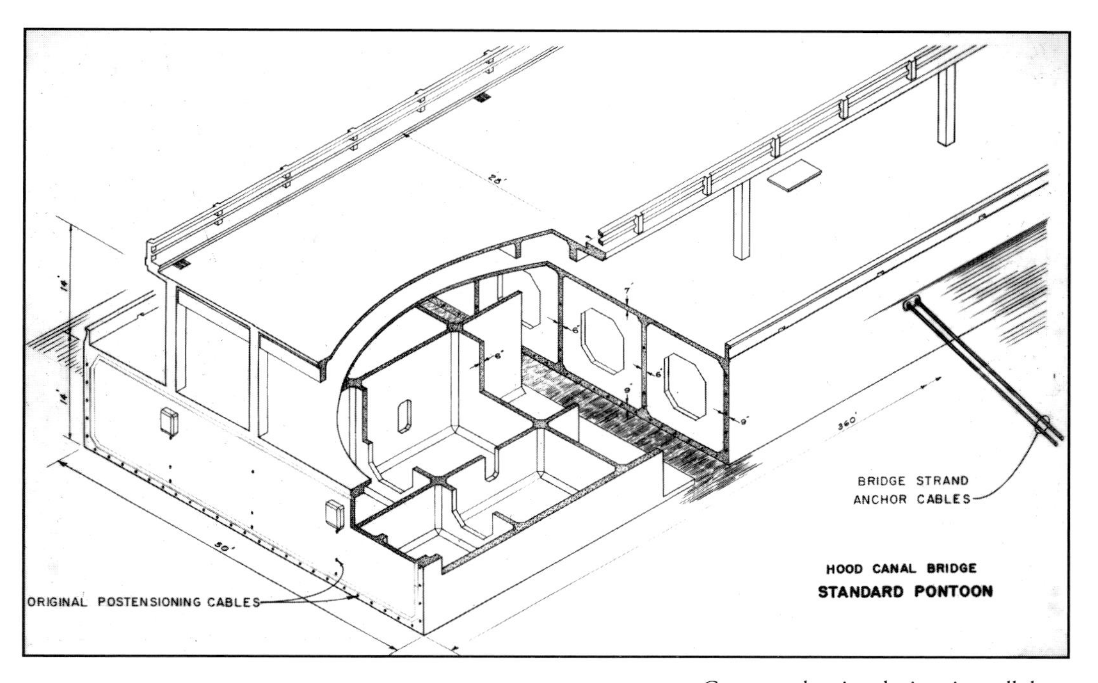

Cut-away showing the interior, cellular construction of a Hood Canal Bridge floating concrete pontoon, with raised roadway.

Washington State Archives, WSDOT Records

Hood Canal (William A. Bugge) Bridge—1961/1982

The Hood Canal Bridge is one of Washington's four world-class concrete pontoon floating bridges. It straddles the entrance to Hood Canal, connecting north Kitsap Peninsula with the Olympic Peninsula on State Route 104. At a total length of 7,967 feet and a floating length of 6,521 feet, it is the longest floating bridge on salt water in the world.

Engineers began planning for a bridge across Hood Canal in 1950. They faced formidable obstacles and limited alternatives. Still fresh in the minds of engineers and the public, however, was the successful Lake Washington Floating Bridge, which opened in 1940. A similar pontoon structure on Hood Canal seemed feasible. But it would be a risky undertaking as the first bridge of its type built on an arm of the ocean, due to the locality's strong currents, frequent high winds, and tidal fluctuations of up to eighteen feet.

Tug towing a pontoon section, with raised roadway deck, from a Duwamish River construction site, Seattle, ca. 1961.

Washington State Archives, WSDOT Records

Artist's rendition of the Hood Canal Bridge, showing "bulge-style" slip spans. View is to the northeast.

Washington State Archives, WSDOT Records

Constructed in three and a half years, the Hood Canal Bridge opened in the summer of 1961. It consisted of twenty-three concrete pontoons—each measuring 360 feet long by 50 feet wide, and weighing 500 tons—held in place by cables attached to forty-two concrete anchors and bolted together to form two separate, rigid spans meeting at the canal's center. There, two retractable pontoon draw spans opened to form a 600-foot-wide ship passage. Concrete columns elevated the roadway fourteen to twenty feet above the pontoons to keep saltwater spray off vehicles. Special precautions were taken to protect electrical and mechanical equipment exposed to seawater in the "splash zone" beneath the roadway.

A "storm within a storm," generating 120 mph winds, struck the Hood Canal Bridge on February 13, 1979, destroying the west half.

Washington State Archives, WSDOT Records

As the February 13, 1979, storm subsided, pontoons had either sunk or floated away, leaving the east half in place and the west truss approach collapsed (seen on the distant shoreline).

Washington State Archives, WSDOT Records

Local newspapers likened the official opening on August 12, 1961, to a "carnival." Governor Albert D. Rosellini cut the ceremonial ribbon, as some 6,000 spectators cheered, and a caravan of automobiles began streaming across the new span. More than 13,000 vehicles crossed the first weekend, celebrating the new connection between the Olympic and Kitsap peninsulas. Sixteen years later, in 1977, the Washington State Highway Commission named the bridge in honor of William A. Bugge, Director of the Washington State Department of Highways during the project's development.

On the evening of February 12, 1979, a severe storm struck the area. The next morning, the west half of the bridge ripped free, scattering or sinking half of the pontoons and thus destroying much of the bridge.

After several studies in the months following the disaster, WSDOT decided to replace the bridge's demolished west half. The new section was to be three to four times

The Hood Canal Bridge approach truss collapsed with the destruction of the structure's west half, February 13, 1979.
Washington State Archives, WSDOT Records

stronger than the original structure. The replacement components were constructed in Tacoma and towed seventy miles to the site. Twelve new pontoons, measuring 60 by 369 feet, were substantially bigger than the previous pontoons. Each weighed 8,300 tons, was as long as a football field, and as tall as a two-story building. Each pontoon was divided into thirty-six watertight compartments to add buoyancy. For added strength, the pontoons were posttensioned vertically, transversely, and longitudinally, then joined together with prestressed cable strands.

Moving twenty-six bucket-shaped concrete pontoon anchors from the assembly site in Tacoma to Hood Canal proved as challenging as towing the pontoons. Each anchor measured 56 feet wide and 29 feet high, and weighed 1,500 tons. The anchors were first shipped by rail, then

The rebuilt Hood Canal Bridge included a "lift/draw" span to allow passage of U.S. Naval ships and other vessels.
WSDOT

Aerial view of the Hood Canal Bridge to the northeast, showing Puget Sound and the snow-covered Cascade Range.
WSDOT

barged, to Hood Canal. At the construction site, specially designed giant winches on barges lowered the ballast-filled anchors (now weighing 2,200 tons each) into place in the deep channel. Three-inch-thick cables (nearly twice the thickness of the cables that had failed during the 1979 storm) were threaded through ducts around the anchors, later to be connected to the new pontoons.

The restored Hood Canal Floating Bridge opened to traffic on October 25, 1982. It consisted of two concrete and two steel truss approach spans, twenty-nine floating concrete pontoons, a concrete roadway elevated above the pontoon decks, two draw spans, and forty-two submerged concrete anchors. The pontoons were arranged longitudinally, connected end to end, and held in place laterally by steel cables attached to the anchors on the canal bottom to either side of the floating structure. Five pontoons floated in a transverse configuration—three supported the newer west-half lift/draw span, and one each supported the floating piers under the Warren truss transition spans.

The new lift/draw span consisted of a complex assembly of ten pontoons, including three placed transversely, and a 300-foot draw pontoon that retracted into a U-shaped receiving structure beneath three raised steel decks. Hydraulic lifts lowered the decks to the roadway level, at the time a new use of that technology. When the new draw span opened, the navigational channel reached the 600-foot width required by the U.S. Navy for the passage of Trident submarines and other vessels. Costs of replacing the bridge's west half and completing all subsequent refurbishing (Stages 1 through 3) eventually totaled $143 million.[27]

Further structural deterioration eventually led WSDOT to determine that the bridge's older, east half needed replacing. On August 16, 2003, the department launched a $205 million project to upgrade the Hood Canal Bridge. Work on the so-called "graving dock" where the concrete pontoons were to be built soon halted, however, when remains of a prehistoric village, known as *Tse-whit-zen* to the Lower Elwha Klallam people, were unearthed at the Port Angeles construction site. The incident served as an example of how bridge construction and rehabilitation can be unavoidably influenced by tribal sovereignty, cultural resources laws, and the ever-present shadow of the past.[28]

NOTES

1. General sources for the first Lake Washington Floating Bridge story include: C. S. Gloyd, "History of Floating Bridges in the State of Washington" (unpublished manuscript, WSDOT Library, Olympia, n.d.); Paul Dorpat, *Building Washington: A History of Washington State Public Works* (Seattle: Tartu Publications, 1998), 120–22; Lake Washington Floating Bridge Scrapbook, Special Collections, University of Washington Libraries; "Lake Washington Floating Bridge," official dedication booklet, Lake Washington Floating Bridge Committee, 1940.
2. Lucile McDonald, "The Inspiration for the First Floating Bridge," *Seattle Times*, July 26, 1964, 6; John Engstrom, "Our Hearts Sank," *Seattle Post-Intelligencer*, December 20, 1990, C8.
3. Wm. Michael Lawrence, "Lacey V. Murrow Memorial Bridge," HAER No. WA-2, 1993; "Pontoon or Floating Drawbridges," *Engineering News* 62, April 30, 1908, 474–79; "A Pontoon Bridge with Submerged Pontoon," *Engineering News* 68, 1912, 148; F. C. Coleman, "New Pontoon Bridge over the Golden Horn at Constantinople," *Engineering News* 70, November 20, 1913, 1018–20; Homer Hadley, "Report on Proposed Pontoon Bridge at Seattle, Wa., Oct. 1, 1921," in OAHP File "Lacey V. Murrow Bridge"; Wm. Michael Lawrence, "Montlake Bridge," HAER No. WA-108, 1993.
4. John Engstrom, "Our Hearts Sank," *Seattle Post-Intelligencer*, December 20, 1990, C8–11.
5. Hadley quoted in Lucile McDonald, "The Inspiration for the First Floating Bridge," *Seattle Times*, July 26, 1964, 6; John Engstrom, "Our Hearts Sank," *Seattle Post-Intelligencer*, December 20, 1990, C8–11; "'Hadley's Folly' Turned into Island's Floating Bridge," *Mercer Island Reporter*, January 2, 1991 (recollection by Hadley's wife, Margaret Hadley, assisted by daughter Eleanor Hadley).
6. "Pontoon Bridge Proposed to Relieve Seattle Ferries," *Engineering News-Record* 87 September 22, 1921, 511; "South End Mercer

Island Improvement Club," unpublished history 1920–54, folder 1, Gertrude Pool Papers, Mss 061, Washington State Library, Olympia.

7. James Warren, "The Idea of a Floating Bridge Surfaced in 1920," *Seattle Times*, February 27, 1983, E3; Lucile McDonald, "The Inspiration for the First Floating Bridge," *Seattle Times*, July 26, 1964, 6; John Engstrom, "Our Hearts Sank," *Seattle Post-Intelligencer*, December 20, 1990, C8–11.
8. Lacey V. Murrow, "A Concrete Pontoon Bridge to Solve Washington Highway Location Problem," *Western Construction News*, July 1938, 249–52; John Engstrom, "Our Hearts Sank," *Seattle Post-Intelligencer*, December 20, 1990, C9.
9. "Sliding Span for Proposed Pontoon Bridge," *Western Construction News*, February 17, 1938, 248; Lacey V. Murrow, "Early History of the Lake Washington Floating Bridge," *Pacific Builder and Engineer*, July 1940; Lacey V. Murrow, "A Concrete Pontoon Bridge to Solve Washington Highway Location Problem," *Western Construction News*, July 1938, 249–52; Homer M. Hadley, "Concrete Pontoon Bridge—A Review of the Problems of Design and Construction," *Western Construction News*, September 1939, 293–98; "Experts Approve Pontoon Bridge," *Engineering News-Record*, July 14, 1938, 38.
10. John Engstrom, "Our Hearts Sank," *Seattle Post-Intelligencer*, December 20, 1990, C8–11 (Magnuson quote on page C9); "'Hadley's Folly' Turned into Island's Floating Bridge," *Mercer Island Reporter*, January 2, 1991; "Public Works—Lake Span," *Argus*, January 7, 1939, 3; Dorpat, *Building Washington*, 120–22.
11. The July 6, 1940 (vol. 46), issue of *Pacific Builder and Engineer* was the special "Bridge Dedication Number" for the Lake Washington Floating Bridge and Tacoma Narrows Suspension Bridge. Journal issues over the next six months included articles on the project by many of the leading engineers: Charles E. Andrew, "The Lake Washington Floating Bridge" (July 6, 29–33); L. R. Durkee, "Floating Bridge More Stable than Fixed Spans" (August 3, 38–41); Richard Barber, "Massive Concrete Pontoons" (August 3, 41–43); Champ E. Corser, "The Transition Section" (August 3, 44); B. Wooliscroft, "World's First Draw Pontoon" (August 3, 46); Lacey V. Murrow, "Early History of the Lake Washington Floating Bridge" (October 5, 44–48); Harold V. Judd, "Twin Bore Tunnels: Construction Methods Used to Bore 30 Tunnels through the Glacial Blue Clay of Mount Baker Ridge" (October 5, 46–48); Cecil C. Arnold, "Field Problems and Construction Methods" (November 2, 28–30). See also: E. H. Thomas, "Approaches to Lake Washington Bridge," Bridge Files, Environmental Affairs Office, WSDOT; "Building Lake Washington Pontoons," *Engineering News-Record*, August 3, 1939, 50; "Moorings for Lake Washington Bridge," *Engineering News-Record*, July 18, 1940, 46–48; Charles E. Andrew, "Building the World's Largest Floating Bridge," *Civil Engineering* 10, January 1940, 17–20; "Bridge that Floats," *Scientific American* 162, February 1940, 75–77; Lucile McDonald, "The Floating Bridge: How It Grew," Pacific Parade, *Seattle Times*, January 29, 1947, 2.
12. Lawrence, "Lacey V. Murrow Memorial Bridge," HAER No. WA-2, 1993; Lisa Soderberg, "Lacey V. Murrow Floating Bridge/Lake Washington Floating Bridge," HAER Inventory, 1981.
13. *Seattle Times,* July 2, 1940, 1, 30; *Seattle Post-Intelligencer*, July 3, 1940, 1.
14. "The Fall of the Bridge," Special Report section, *Seattle Post-Intelligencer*, December 20, 1990; *Seattle Times*, November 26 and 27, 1990; "Panel Finds Holes Helped Sink I-90 Bridge," *Seattle Times*, May 2, 1991, A1; M. Myint Lwin and Donald O. Dusenberry, "Responding to a Floating Bridge Failure," *Public Works,* January 1994, 39–43; Donald Dusenberry, "What Sank the Lacey Murrow?" *Civil Engineering* 63, no. 11, November 1993, 54–59.
15. Sources for the Mount Baker Ridge Tunnel include the following: Jonathan Clarke, "Mount Baker Ridge Tunnel," HAER No. WA-109, 1993; Lisa Soderberg, "Mount Baker Ridge Tunnel," HAER Inventory, 1980; Harold V. Judd, "Twin Bore Tunnels: Construction Methods Used to Bore 30 Tunnels through the Glacial Blue Clay of Mount Baker Ridge," *Pacific Builder and Engineer* 46, October 5, 1940, 46–48; "Twin Tunnels Driven through Clay for Lake Washington Bridge Project," *Western Construction News* 15, July 1940, 246; Edgar B. Johnson, Lee J. Holloway, and Georg Kjerbol, "Unearthing Mt. Baker Tunnel," *Civil Engineering* 55, no. 12, December 1985, 36–39; Harvey W. Parker and Robert A. Robinson, "The World's Largest-Diameter Soil Tunnel," *Underground Space* 7, November 1982/January 1983, 175–81; Charles M. Gordon, "World's Largest Stack Drift Tunnel," *Pacific Builder and Engineer* 91, July 22, 1985, 8–9; Rita Robinson, "The Stacked-Drift Tunnel," *Civil Engineering* 60, no. 7, July 1990, 40–42; Paul Dorpat, *Building Washington*, 124–25.
16. Principal sources for the Evergreen Point Floating Bridge include: Evergreen Point Bridge Scrapbook (ca. 1953–58; newspaper clippings, letters, reports, misc.), Special Collections, University of Washington Libraries; Stephen J. Hitch, "The Evergreen Point Floating Bridge in Crisis: An Approach to Evaluating Alternative Solutions," (Master's thesis, University of Washington, 1999); Tudor Engineering Company, *Lake Washington Bridge Crossings*, Legislative Reconnaissance and Feasibility Report, December 1968; "New Seattle Bridge," *Washington Highway News*, November-December 1960, 16–19; "The Evergreen Point Bridge, 1960–1963," *Washington Highway News*, July-August 1963, 1; Paul Dorpat, *Building Washington*, 125.
17. See, for example, "Second Lake Span Endorsed," *Mercer Islander*, August 11, 1949, 5, and September 8, 1949, 3; "Second Float-

ing Bridge Seattle Need in 1955," *Engineering News-Record* 144, February 16, 1950, 28; "Legislature OKs Second Bridge Bill," *Bellevue-Mercer Islander*, March 12, 1953, 1; various articles on the controversy in *Seattle-Post Intelligencer*, February-April 1957; "Six Lake Bridge Sites Considered before Montlake Was Chosen," *Seattle Times*, August 5, 1957, 16; "Green Light for Second Bridge at Evergreen Point," *Seattle Times*, November 7, 1957, 1; Ross Cunningham, "Bickering in Seattle Area Blamed for Bridge Delays," *Seattle Times*, June 25, 1958, 1; Ross Cunningham, "State High Court Voids Lake-Bridge Financing Plan," *Seattle Times*, July 31, 1959, 1; "Lake Bridge: No Span in Sight after Six Years of Controversy," *Seattle Times*, September 6, 1959, 2.

18. Douglas B. Mauldin, "Washington's Wondrous Highways that Float," undated newspaper clipping, WSDOT Environmental Affairs Office.
19. "Lake Span to Be Started Immediately," *Seattle Times*, August 9, 1960, 1; "Pontoon Work Begins April 1," *Bellevue American*, March 2, 1961, 1; "Evergreen Point Floating Bridge to Open to Traffic August 28," *Seattle Times*, July 24, 1963, 46; "$25,792,000 Floating Span Is Most Expensive in State," *Seattle Times*, August 25, 1963, 7; "Lift Decks Will Be Featured on Draw Span," *Seattle Times*, January 29, 1959; "New Seattle Bridge," *Washington Highway News*, November-December 1960, 16–19; "The Evergreen Point Bridge, 1960–1963," *Washington Highway News*, July-August 1963, 1.
20. "Evergreen Bridge Toll to End June 22," *Seattle Post-Intelligencer*, March 21, 1979, 1; McDonald, *Lake Washington Story*, 143.
21. "A Bridge Too Small," *Seattle Times*, August 28, 1988, A1, A14; Hitch, "Evergreen Point Floating Bridge," 3–4; Tudor Engineering Company, *Lake Washington Bridge Crossings*, Legislative Reconnaissance and Feasibility Report, December 1968.
22. Quote from Don Carter, "Fight: The Story of a Bridge over Troubled Waters," *Seattle Post-Intelligencer*, June 2, 1989, B3; "Rapid-Transit Plan for 3rd Bridge Pushed," *Seattle Times*, March 13, 1964, 5; Robert Barr, "Plan Told for 3rd Lake Bridge," *Seattle Times*, November 22, 1966, 1; "Ten Years of No Progress on I-90 and Bridge," *Seattle Times*, March 5, 1973, A12; "I-90 Bridge Progresses," *Seattle Times*, July 14, 1984, A11; Scott Handley, "Home of Floating Bridges Soon to Open New Span...First Response to Idea Was Far from Positive," *Construction Data and News, Supplement to Engineering News-Record* 230, no. 22, May 31, 1993, P3–19.
23. Frederick Case, "Bridging Technology," *Seattle Times*, August 21, 1989, F1–F2; Hitch, "Evergreen Point Floating Bridge," 48; Dorpat, *Building Washington*, 125–126.
24. Quoted in Frederick Case, "Bridging Technology," *Seattle Times*, August 21, 1989, F1.
25. "Man Who Inspired Floating Bridge Gets His Due," *Journal American*, July 13, 1993.
26. M. Myint Lwin and Donald O. Dusenberry, "Responding to a Floating Bridge Failure," *Public Works,* January 1994, 39–43; Andrew G. Wright, "Buoyed by a Water-tight Design; Despite Two Failures, State Engineers Retain a Pioneering 1940s Concept," *Engineering News-Record* 230, no. 22, May 31, 1993, 22; "Pontoons Redesigned for Seattle Crossing," *Engineering News-Record* 228, no. 3, January 1992, 40–41.
27. Craig Holstine, "Hood Canal Floating Bridge," National Register of Historic Places Determination of Eligibility, 2002; Richard Hobbs, "Hood Canal Floating Bridge (William E. Bugge Memorial Bridge)," HAER Report, June 2004; A. D. Andreas, "SR 104 Hood Canal Bridge: Task Force Preliminary Report, July 1979, and Summary Report, October 1979," WSDOT, 1979; C. S. Gloyd, "Concrete Floating Bridges," *Concrete International*, May 1988, 17–24; "The Hood Canal Floating Bridge," *Highway News*, May-June 1961, 18–19; Thomas R. Kuesel, "Floating Bridge for 100 Year Storm," *Civil Engineering*, June 1985, 60–65.
28. Brenda Hanrahan and Jim Manders, "Graving Work at Least Week from Restart," *Peninsula Daily News*, September 4, 2003; Andrew Garber, "Finding Indian Remains May Sink Bridge Project," *Seattle Times*, September 12, 2003; Lynda V. Mapes, "Tribe's Letter Deepens Dilemma over Project," *Seattle Times/Seattle Post-Intelligencer*, December 12, 2004.

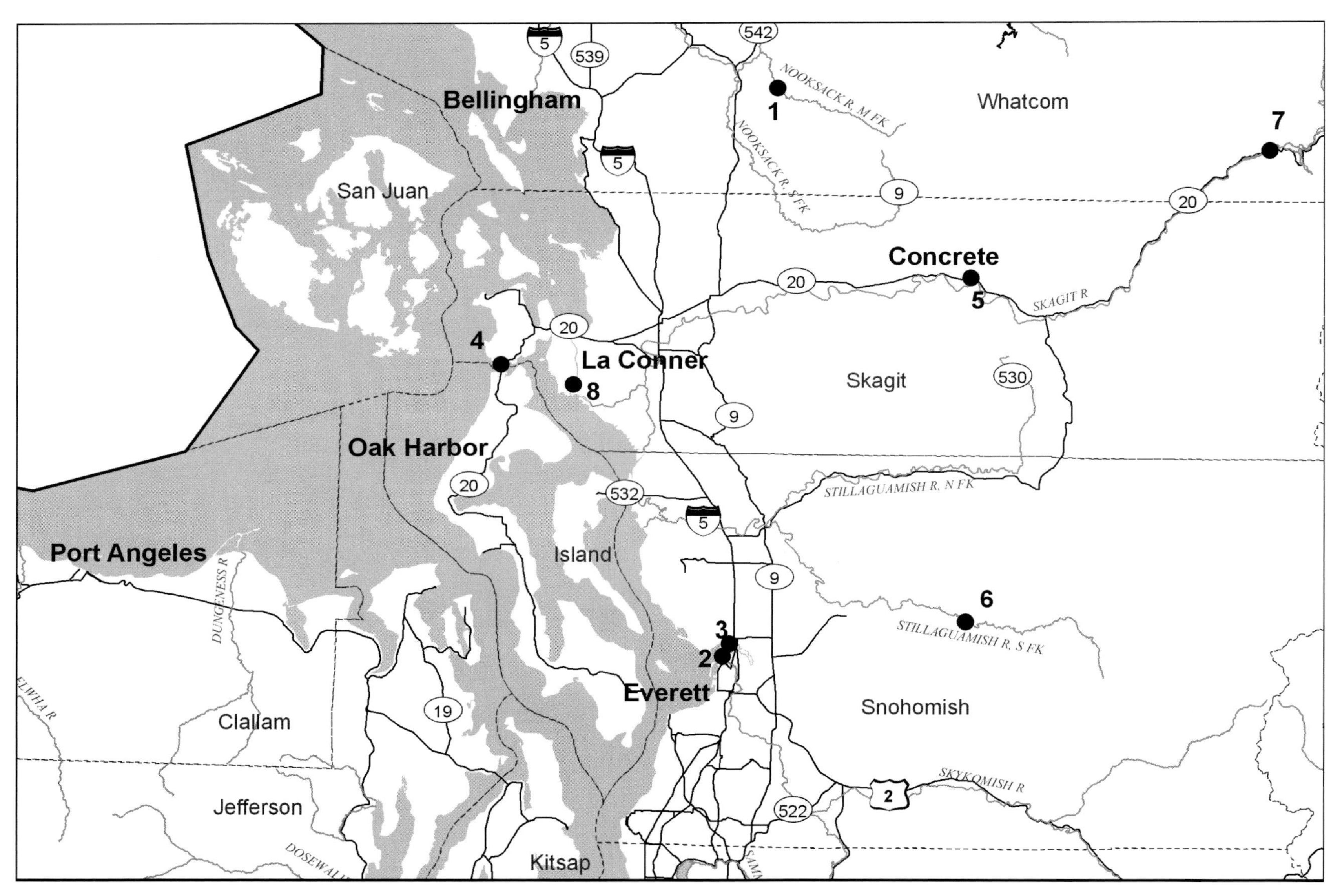

Premier Historic Bridges of Northwest Washington

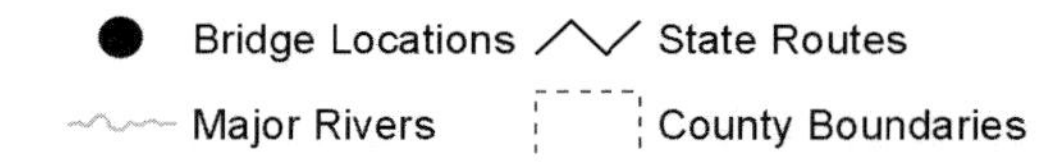

Bridge No.	Bridge Name	Bridge No.	Bridge Name
1	Middle Fork Nooksack River Bridge	5	Dalles Bridge
2	Snohomish River Bridges	6	Red Bridge
3	Steamboat Slough Bridges	7	Gorge Creek Bridge
4	Deception Pass and Canoe Pass Bridges	8	Rainbow Bridge

Bridges of Northwest Washington

Residents drawn to Northwest Washington have developed thriving industries in lumbering, fishing, farming, and other commerce. Essential to progress have been the bridges on the many substantial streams that drain the state's northwestern section, among them the Skagit, Stillaguamish, Nooksack, and Snohomish rivers. Here, bridge engineers devised increasingly sophisticated solutions to the problems posed for transportation in the region.

One of the state's rarest and oldest (built in 1915) steel truss structures crosses the Middle Fork Nooksack River on a remote, secondary road in the North Cascades foothills. Two Snohomish River bridges just north of Everett, erected in 1926 and 1954, are among Washington's most noted vertical lift spans. Barely a mile to the north, two of the state's last highway swing bridges span Steamboat Slough.

The Deception Pass and Canoe Pass bridges, constructed in 1935 to connect Fidalgo and Whidbey islands, present a striking refinement of the steel cantilever truss design in one of the region's most picturesque maritime settings.

In the North Cascades, the Dalles Bridge, a one-of-a-kind steel through-truss structure, crosses the Skagit River south of the town of Concrete. The Red Bridge stands on the South Fork Stillaguamish River, one of the last Pratt/Parker trusses built in the state. And in Whatcom County, the Gorge Creek Bridge is a steel deck-arch span built in 1955, initially serving local traffic and dam workers. In the 1970s, it became a tourist attraction along the new North Cascades Highway.

Across the Swinomish Channel near La Conner stretches the state's longest steel through-arch span, the 791-foot Rainbow Bridge. This award-winning structure stands in the scenic Skagit County farm country, where each spring a "rainbow" of flowers blossom to attract thousands of tourists.

Middle Fork Nooksack River Bridge, Whatcom County

This bridge was built in 1915 by the Toledo Bridge and Crane Company and the Weymouth Construction Company. It was originally on the Guide Meridian Road (now State Route 539) over the Nooksack River near Lynden. In 1951 the steel through-truss was moved to its present location across the Middle Fork Nooksack River, when a modern structure replaced it on the Guide Meridian Road. When relocated, two truss panels were removed, reducing

Middle Fork Nooksack River Bridge. Constructed in 1915 on the Guide Meridian Road, and moved to its present location in 1951, the Pennsylvania Petit truss is the longest pin-connected highway bridge in Washington.
Whatcom County Public Works, Bellingham

the 380-foot span to its present 338 feet. Nevertheless, the bridge remains the longest pin-connected highway bridge in Washington and a rare example of a modified Pennsylvania Petit truss.

Despite its narrow, one-lane roadway, the old bridge serves log-truck traffic on the Mosquito Lake Road, connecting the North Cascade foothills via State Route 542 to Bellingham, Sumas, and Sedro-Woolley. To maintain this service, Whatcom County is planning to rehabilitate the bridge by replacing the laminated wood deck, the steel pins in the web members, and both approach slabs. When restored in 2007, the bridge will be completely repainted.[1]

Snohomish River Bridges, Snohomish County

Two bridges stand side-by-side at the Snohomish River crossing of what was originally State Road No. 1, the 310-mile "Pacific Highway" connecting Blaine on the Canadian border with Vancouver, Washington. At first glance, the two structures—both vertical lift movable bridges with fixed metal truss and concrete approach spans—appear to be siblings. In fact, they are from different generations and reflect the evolution in bridge engineering technology between the 1920s and 1950s. The elder bridge dates from Pacific Highway construction in 1926, the younger to 1954.

Snohomish River Bridges—the lift towers of the 1926 bridge (background) differ from those on the 1954 bridge (foreground).
WSDOT

Work began on the Everett-Marysville segment of the Pacific Highway as early as 1914. In June 1925 the J. A. McEachern Company of Seattle broke ground on the construction of four bridges that eventually would connect Everett and Marysville. These were the Ebey Slough and Steamboat Slough bridges, both steel swing spans; the Union Slough Bridge, a concrete T-beam structure; and the Snohomish River Bridge. Together, they constituted what was characterized as one of the "largest construction jobs yet undertaken by the state highway department."

The 1926 Snohomish River structure was designed by State Highway Department engineers under the direction of Charles E. Andrew. Renowned bridge designer J. A. L. Waddell is listed on the construction drawings as "Consulting Engineer, New York." The Department of Highways chose to use his vertical lift patent (dating to 1909) for the movable section of the bridge. The bridge spans not only the river, but also the Great Northern Railway. In total, the bridge consists of a 145-foot long vertical lift span, a 140-foot steel fixed span, seven 180-foot steel fixed spans, and 1,135 feet of concrete girder approaches.

The 1926 bridge's most notable feature is its "span drive" system. This now antiquated technology relied on machinery centered on the lift span, powering a drive shaft that raises and lowers the movable span via cables attached at either end of the shaft. The sheaves (or pulleys), cables, and counterweights carry the additional weight of the operating machinery. This arrangement is also rather unattractive by modern standards, as the sheaves, cables, and trunnion bearings are clearly visible and open to exposure in northwest Washington's ever-changing weather.

In contrast to the older bridge, the operating machinery on the 1954 bridge is enclosed atop the lift towers. The newer bridge employs a "tower drive" mechanism representing 1950s technology. The bridge's centerpiece is a 141-foot-long steel, Warren through-truss vertical-lift span. It is flanked on each side by two 180-foot-long tower spans, four 240-foot steel through-truss spans, and fourteen concrete T-beam spans on each end. All through-truss spans have a Warren configuration and a polygonal top chord.

The story of the latter bridge began in 1950, when the Washington State Department of Highways' Bridge Division started designing it for this crossing. Lead designers for the lift and tower spans were Larry Robertson and Willis B. Horn, while Derby Livesay and Carl West designed the operating system. George Stevens, chief bridge engineer, approved the final plans on September 24, 1952. With a bond issue of $1.9 million, Manson Construction and Engineering Company began work in 1953 and completed the bridge in 1954.

The two adjacent lift spans share a common operators' control house, located above the lift-span truss on the 1926 bridge. Both bridges are excellent examples of vertical-lift engineering, and they are among the last of their kind in Washington. The Snohomish River bridges are not exactly "dinosaurs," yet people may come to see them that way. Since the end of the 1950s, only one lift structure has been built in Washington, the Hoquiam River Bridge at Riverside, constructed in 1970. With modern fixed spans being higher and longer, and with the diminishing need to accommodate river traffic, it is unlikely we will see many more new lift bridges in the future.[2]

Steamboat Slough Bridges, Snohomish County

Two of Washington's seven swing, or pivot, highway bridges stand together on Steamboat Slough just north of Everett. Built in 1926 and 1953, the bridges accommodate two lanes each of State Route 529. The northbound lanes cross the newer bridge, while the older swing span standing 300 feet to the west carries the southbound lanes.

Beginning in 1926, motorists traveling on Primary State Highway 1 between Everett and Blaine crossed the first swing bridge constructed over Steamboat Slough. The J. A. McEachern Company of Seattle built the structure, along with three others on adjacent waterways. As traffic congestion increased over the years, especially in and around the thriving city of Marysville, the need for a second bridge carrying a parallel highway became apparent. Willlis B. Horn of the State Department of Highways

Swing bridge built in 1926 on Steamboat Slough between Everett and Marysville. Harvey S. Rice, photographer.
Office of Archeaology and Historic Preservation

The Steamboat Slough Bridge, built in 1953 as one of the state's last swing, or pivot, highway bridges, differs little from its adjacent neighbor, built in 1927. They are among the state's seven remaining swing highway bridges, two of which no longer are movable structures.
WSDOT

designed the new swing span, and Carl M. West drew up plans for its operating machinery. In February 1953, the state awarded a $930,000 construction contract to M. P. Butler of Seattle. Gunderson Brothers Engineering of Portland, Oregon, fabricated the structural steel. The bridge was completed in 1953.

Both bridges consist of riveted steel Warren through-truss spans that swing open as a single unit, pivoting about a center-bearing pier. When in the open position, the swing unit acts as two balanced cantilevers, hanging over what is termed a "drawrest." The swing spans do not actually "rest" on the drawrest. That misnamed component simply protects the cantilevered unit from being damaged by passing navigation. As marine commerce has diminished over the years, the likelihood of collisions has decreased also. The bridges opened only seventy-five times for marine traffic in 2000, considerably fewer annual openings than in previous years.

In 1944, sixteen highway swing bridges existed in Washington. Today there are seven. Besides the two on Steamboat Slough, they include the Ebey Slough Bridge north of Everett (built 1925); Riverside Bridge (built 1938) and Skagit River Bridge (built 1953), both located near Mount Vernon; and the Wishkah River Bridge in Aberdeen (built 1949). The 1953 Steamboat Slough Bridge was the final "steel" swing span built in the state. It will likely be the last. The Duwamish Waterway Bridge, a concrete swing structure, was built in 1991.[3]

Deception Pass and Canoe Pass Bridges, Skagit and Island Counties

Few bridges stand in a more picturesque setting than the two magnificent structures at rugged Deception Pass connecting Fidalgo and Whidbey islands. The Deception Pass waterway, with small Pass Island in mid-channel, connects Rosario Strait on the west to Skagit Bay to the east. Also located here is one of the state's most popular recreation areas, Deception Pass State Park.

By the early twentieth century, local inhabitants envisioned bridges connecting Fidalgo and Whidbey islands. In 1907, a Swinomish Channel bridge linked Fidalgo Island and its only city, Anacortes, with the mainland to the east, but only ferries served Whidbey Island. Ferry

Canoe Pass Bridge (left foreground) and Deception Pass Bridge (right background). Photo 1993.

Jet Lowe, HAER

operators opposed a new bridge, fearing they would lose their livelihood. Not until federal funding became available under President Franklin Roosevelt's New Deal did work begin in earnest. In 1933 the Department of Highways surveyed the Deception Pass site as money became available—$245,000 from the Washington Emergency Relief Administration, and $150,000 from Island and Skagit counties, with $87,000 to be repaid to the state from the Public Works Administration.

O. R. Elwell of the State Highway Department served as chief engineer on the project. Because the highway passed through Deception Pass Park, the National Park Service played a role in project development, and the park's landscape architect shared in the bridge design. The NPS also supervised Civilian Conservation Corps crews, which built the highway within the park and the log-and-stone guardrails that line the roadway north and south of the bridges. The bridges, together with State Route 20 and the many recreational facilities built by the CCC, are the most visible components of the park's Depression-era historic landscape.

In June 1934, the Department of Highways awarded the Puget Construction Company of Seattle a $304,755 contract to build the two bridges. With assistance from

CCC workers, pier excavation began in August. Because of the crossings' extreme height and width, falsework could not be built to support the steel structures. Using the cantilever method, workers first installed the bridge across the north channel from Fidalgo Island to little Pass Island. Erection of the 511-foot, spandrel-braced, steel-arch "Canoe Pass Bridge" depended upon a cable traveler derrick operated by a gasoline engine that transported cement and silicon structural steel across the chasm.

Then, a light railroad was constructed on the Canoe Pass Bridge to move materials onto Pass Island and building began on the second, or "Deception Pass," bridge. Again using the traveler derrick, workers constructed the south channel bridge's 175-foot north anchor span and the 175-foot north cantilever span. Rails were then laid on the Whidbey Island side, and the traveler derrick was put back into service, completing the 175-foot south anchor arm and 175-foot south cantilever span. Only the 200-foot suspended span remained to be installed.

Sweating in the heat of a June afternoon, the crane operator lowered the center span, to be pinned between the cantilever spans. To everyone's amazement, the span was three inches too long, and failed to nestle into place. Paul Jarvis, the Puget Construction Company's founder, hurriedly calculated the span's expansion coefficient. His estimation that a thirty-degree temperature drop would shrink the steel and allow the span to fit proved correct. Early the next morning, the span again was lowered into place, a perfect fit in the cool, predawn darkness.

The result was stunning. The Deception Pass Bridge is an elegant, 976-foot steel deck-truss representing innovation, evolution, and refinement of a cantilever-type bridge perfectly suited to its dramatic setting. A picture postcard for some, a vital transportation link for all, together the two bridges symbolize the marriage of beauty and technology.[4]

The Dalles Bridge on the Skagit River, a one-of-a-kind, three-hinged steel through-truss.
WSDOT

Dalles Bridge, Skagit County

In north central Skagit County, State Route 20 winds along the scenic Skagit River. Familiar to cross-state travelers as the North Cascades Highway, it passes through the small town of Concrete in the west Cascade foothills. A mile and a half south of Concrete, the unique Dalles Bridge straddles the Skagit and is the only three-hinged steel through-truss in Washington. It consists of a 300-foot riveted span, flanked by 75-foot riveted through-truss anchor spans at each end.

Two ferries were put out of business the day that the bridge opened in August 1952. Designed by the Cecil Arnold and Raymond G. Smith firm of Seattle, it was built by the General Construction Company of Seattle from steel fabricated by the U.S. Steel plant in Memphis, Tennessee. The contractor used the cantilever method for construction, which was made easier by the three-hinged design. It was the first known use of this method in the state for arch bridge construction. The trussed arch ribs in the design, rather than solid ribs, minimized deflections at the crown caused by traffic loads and temperature fluctuations.[5]

Red Bridge, Snohomish County

One of the last Pratt/Parker through-truss bridges built in Washington crosses the South Fork Stillaguamish River between Silverton and Verlot on the Mountain Loop Highway. The road is well named, with snowcapped peaks of the Boulder River and Henry M. Jackson wilderness areas towering above the forested valley. The bridge lives up to its name as well. It is painted red in honor of its predecessor, a steel railroad bridge constructed on the Everett and Monte Cristo Railway that hauled ore out of remote Monte Cristo (now a ghost town). The original steel bridge remained unpainted so long that it developed a thick coat of rust and became known as the "Red Bridge."

The Mountain Loop Highway follows the route of the railroad that was constructed in 1892 to haul gold, copper, lead, and iron ores to smelters in Everett. In 1936 the rails were removed, and the present road and bridge were built on the right-of-way.

The Red Bridge is a fine example of a steel riveted Pratt/Parker through-truss erected with 1950s technology. The Bureau of Public Roads office in San Francisco designed the structure in 1953. The next year Peter Kiewit Sons Company built it using steel fabricated by the U.S. Steel plant in Gary, Indiana. The 160-foot-long bridge was one of three Pratt/Parker trusses built in Washington in the 1950s. None have been erected in the state since.[6]

The Red Bridge over the South Fork Stillaguamish River. This structure is painted red in honor of its predecessor, whose rusted patina provided the name for the crossing.

Craig Holstine, WSDOT

Gorge Creek Bridge, Whatcom County

Gorge Creek Bridge, built in 1955, is a 250-foot-long steel deck-arch span. Its prosaic beginnings barely hinted at its transformation into part of a scenic tourist attraction by the 1980s.

The bridge was designed in 1954 for the City of Seattle Lighting Department by engineers Cecil C. Arnold and Raymond G. Smith. It was part of the construction of the Newhalem-Diablo Highway to provide access to new hydroelectric facilities. Near the bridge, three dams—Gorge, Ross, and Diablo—harness the Skagit River, providing water and electricity for Seattle.

However, completion of the North Cascades Highway and the North Cascades National Park in 1968 brought a horde of travelers through the Newhalem and Diablo

Gorge Creek Bridge is the last riveted steel deck-trussed arch bridge and the only three-hinged deck-trussed arch remaining in Washington.

WSDOT

Canyon areas. By the 1970s, increased traffic had created safety concerns. Pedestrians were drawn to the bridge deck and its narrow curbs for a view of the gorge and the creek below.

Gorge Creek Bridge is a 16-foot-wide, riveted steel deck-truss, with a central 180-foot trussed arch span, flanked by a 30-foot anchor span at each end. The bridge carries one lane of traffic in each direction on a 22-foot roadway. In 1985 sidewalks five feet wide were cantilevered from each side of the bridge to safely accommodate pedestrians. Aluminum-alloy grating and railing were used to minimize the added weight, and a viewing area was constructed to the south of the structure.

The arch span itself is a three-hinged Pratt truss, with eight 22-foot panels and pin hinges at the arch supports and at the center of the span. The arch's lower chord rises 27 feet above its elevation at the end hinge support. The Gorge Creek Bridge was the last riveted steel deck-trussed arch bridge to be built in the state. It is the only three-hinged deck-trussed arch bridge still standing in Washington, and the only remaining one of its type to use a Pratt truss design for the arch span.[7]

Rainbow Bridge, Skagit County

Each spring bright tulip, daffodil, and iris blossoms splash over 700 acres along the Swinomish Channel, spreading color across Skagit County's fertile farmlands. This world-renowned display provides the backdrop for the Rainbow Bridge.

The bridge's graceful, rainbow-shaped arch is painted a vibrant red. It is one of the state's most photographed spans. Located along the Pioneer Parkway, a busy highway between Anacortes and Conway, the bridge crosses the Swinomish Channel near the picturesque town of La Conner. Swinomish Channel, separating Fidalgo Island from the mainland, is a busy waterway for small boats. The oldest town in the county, La Conner was settled in the mid-1860s and rapidly became a prospering hub for the local logging, farming, and fishing industries. After the 1940s, the old-fashioned community became a mecca for artists and writers. Since the 1970s, the area has drawn thousands of tourists annually.

At a total length of 791 feet, the bridge is the longest steel arch span in the state. Its centerpiece is a 550-foot fixed steel through-arch span—the first constructed in Washington—that at its peak measures 116 feet above the channel. The arch ribs rise from the spring line to 92 feet at the centerline. A 30-foot-wide roadway carries two traffic lanes and sidewalks. The arch span provides for 75 feet of vertical clearance above the mean high water level in the channel.

The span is widely recognized for its pioneering use of high-strength, low-alloy steel in highly stressed areas, where the higher-capacity

Designed by Harry R. Powell, the Rainbow Bridge over Swinomish Channel won national honors in 1958.
WSDOT

steel could reduce the bridge's weight and cost. The new steel lowered the weight by 250,000 pounds and reduced the final cost by 10 percent. The Rainbow Bridge was designed by Harry R. Powell and erected in 1957 at a cost of $633,000. Its clean, sweeping lines and economical construction brought immediate national acclaim. In 1958 the bridge received a prestigious honor from the American Institute of Steel Construction—a First Honorable Mention Award for "Bridges with Spans Greater than 400 Feet Costing More than $500,000." At the time, the Rainbow Bridge held the distinction of having the lowest construction cost of any bridge to win this class since the competition began in 1928.[8]

NOTES

1. Lisa Soderberg, "Middle Fork Nooksack River Bridge," HAER Inventory, 1979; Steve Dillon, Whatcom County Public Works Department, personal communication, November 1, 2004.
2. WSDOT, 10th *Biennial Report*, 1924, 36, and 11th *Biennial Report*, 1926, 43; Oscar R. "Bob" George, "Snohomish River Bridge [1954]," National Register of Historic Places Nomination, 2001; Homer Wassam, "Movable Bridges of the Snohomish," *Everett Herald*, March 23, 1974, 6–7; Craig Holstine, "Snohomish River Bridge [1926]," NRHP Determination of Eligibility, 2005.
3. Oscar R. "Bob" George, "Steamboat Slough Bridge No. 529/20E," NRHP Nomination, 2001.
4. Robert W. Hadlow, "Washington State Cantilever Bridges, 1927–1941," HAER No. WA-106, 1993; Robert W. Hadlow, "Deception Pass Bridge," HAER No. WA-103, 1993; Robert W. Hadlow, "Canoe Pass Bridge," HAER No. WA-104, 1993; Lisa Soderberg, "Deception Pass, Canoe Pass Bridges," HAER Inventory, 1979; Dorothy Neil, *A Bridge Over Troubled Water: The Legend of Deception Pass* (Langley, WA: South Whidbey Historical Society, 2002).
5. Oscar R. "Bob" George, "Dalles Bridge," NRHP Nomination, 2001.
6. Oscar R. "Bob" George, "Red Bridge," NRHP Nomination, 2001.
7. Oscar R. "Bob" George, "Gorge Creek Bridge," NRHP Nomination, 2001.
8. Oscar R. "Bob" George, "Rainbow Bridge," NRHP Nomination, 2001.

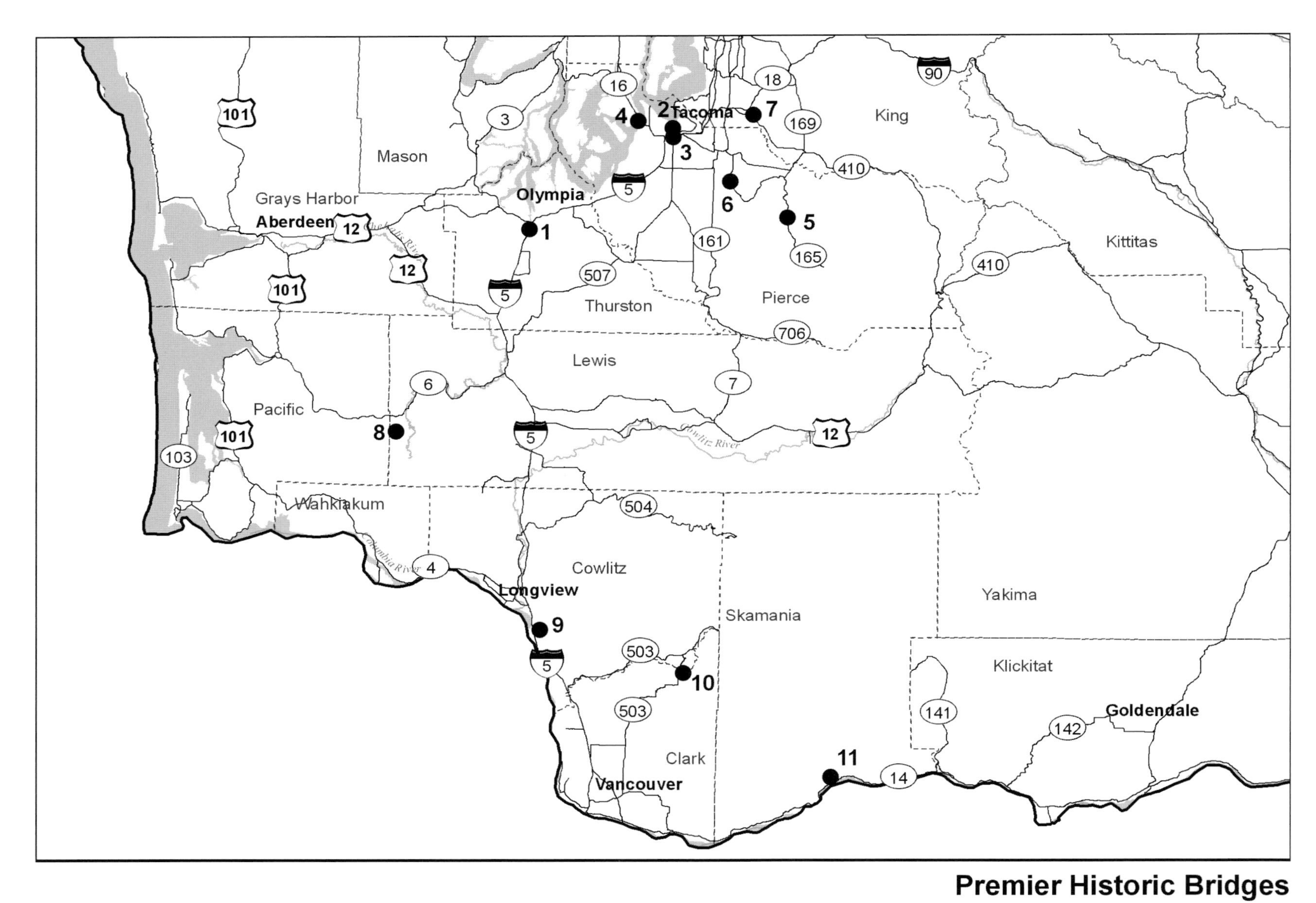

Premier Historic Bridges of Southwest Washington

Bridge No.	Bridge Name	Bridge No.	Bridge Name
1	Capitol Boulevard Bridge	7	Patton Bridge
2	City Waterway (Murray Morgan) Bridge	8	Pe Ell Bridge
3	East Thirty-Fourth Street Bridges	9	Modrow Bridge
4	Tacoma Narrows Bridge	10	Yale Bridge
5	Fairfax Bridge	11	Conrad Lundy Jr. Bridge
6	McMillin Bridge		

Bridge Locations
State Routes
Major Rivers
County Boundaries

Bridges of Southwest Washington

One of Washington's most distinctive bridges welcomes travelers on the old Pacific Highway into south Olympia. The lavish Art Deco ornamentation of the Capitol Boulevard Bridge demonstrates how artistic expression can be achieved without a complicated structural design.

In the Commencement Bay industrial area in Tacoma, one of the nation's leading engineering firms designed a movable bridge to accommodate shipping on the Thea Foss Waterway, close to the city's heart. The City Waterway (also known as the Murray Morgan, or 11th Street) Bridge, completed in 1913, is among the state's most significant historic structures. Drivers on nearby Interstate 5 also are attracted by views of two sweeping concrete arches built in the 1930s and 1940s in the nearby hills on Tacoma's 34th Street.

Motorists leaving west Tacoma for Gig Harbor and the Kitsap Peninsula via State Route 16 cross the majestic Tacoma Narrows Bridge. Completed in 1950, the graceful suspension bridge will soon have an equally elegant neighbor when another bridge is constructed immediately adjacent.

Tourists en route to Mount Rainier National Park on State Route 165 cross the three-hinged steel lattice arch of the Fairfax Bridge over the scenic Carbon River gorge, where miners and loggers were once the main travelers. Also in rural Pierce County, along the meandering Puyallup River, the McMillin Bridge built in 1934 was the longest reinforced-concrete truss span in the nation at the time. It also was notable for its hollow-box design, contributed by renowned engineer Homer M. Hadley. Sixteen years later, Hadley again designed an innovative box-girder bridge, this time of both concrete and steel construction. Known as the Patton Bridge, it spans the Green River in south King County.

One of the state's three remaining covered bridges crosses the Chehalis River at the western edge of Lewis County. To the southeast, the Modrow Bridge over the Kalama River is the second steel, open-spandrel, ribbed, deck-arch bridge built in Washington. On the border between Clark and Cowlitz counties motorists cross the Lewis River on the state's only steel short-span suspension structure, the Yale Bridge, built in 1932. Eastward, the Wind River Road heading north out of the Columbia River Gorge National Scenic Area and the town of Carson crosses the Conrad Lundy Jr. Bridge. Unique in the state, the deck-truss is supported by an immense, steel-lattice tower that rises up from the bottom of the deep Wind River canyon.

Capitol Boulevard Bridge, Tumwater, Thurston County

At the north edge of Tumwater, where Capitol Boulevard crosses the Deschutes River just before entering Olympia, a long, concrete-girder bridge offers an eye-catching entranceway to the state capitol. It is one of Washington's finest examples of Art Deco influence in bridge architecture. Designed by Clark Eldridge, the 1,100-foot-long, four-lane, Capitol Boulevard Bridge was built in 1936–37 by the U.S. Bureau of Roads as a New Deal federal works project. The clean lines of this common bridge allowed Eldridge an opportunity to experiment with modernist treatments on the nonstructural elements.

Strung along the low concrete walls on the outer edges of the sidewalks are a number of Art Deco touches,

Clark Eldridge designed the Capitol Boulevard Bridge over the Deschutes River in north Tumwater, 1936.

Washington State Archives, WSDOT Records

Capitol Boulevard's totem poles, light standards, and chevron-festooned guardrails greeted travelers on the old Pacific Highway. Photo 2002.

Craig Holstine, WSDOT

including molded chevrons, zigzags, rectangular forms, and geometric curves in a series of low-relief setbacks. Cylindrical concrete lampposts rise from rectangular piers in the walls. The most striking vertical elements are the four ornate polychrome totem poles flanking both ends of the bridge. After dark, lamps installed under the thunderbirds' elongated beaks light up the roadway and sidewalks. Tumwater's claim to being the state's earliest pioneer settlement is emblazoned in bas relief on one of the concrete walls. Raised concrete letters adorning the other three walls adjacent to the totem poles remind visitors that this is a starting point for venturing to the Olympic Mountains, Canada, and Alaska, as well as the state's capitol.

Completion of Interstate 5 in the late 1950s, less than a city block away, initiated a change in the bridge's ownership when officials transferred title to the city of Tumwater. Close by, the Lower Custer Way Bridge, a three-span Luten arch built in 1915, stands just below Tumwater Falls, and 300 feet downstream from there, the Upper Custer Way Bridge, an open-spandrel, concrete arch constructed in 1956, was extended across Interstate 5. (See photos in the Introduction.)[1]

City Waterway (Murray Morgan) Bridge, Tacoma, Pierce County

When Tacoma civic leaders opened the City Waterway Bridge on 11th Street in February 1913, there was much to be excited about. The structure was designed by the renowned engineering firm of Waddell and Harrington of Kansas City, Missouri. Moreover, three features made it one of the most remarkable bridges of the times. First, its lift span was the first ever built on a grade. Second, its deck rose 60 feet over the waterway, making the span one of the highest of this type ever constructed. And third, a light overhead truss between the two towers carried a 16-inch water pipe.

Waddell and Harrington designed the City Waterway (Murray Morgan) Bridge, completed in 1913. Drawing 1993.

Wolfgang G. Mayr and Karl W. Stumpf, HAER

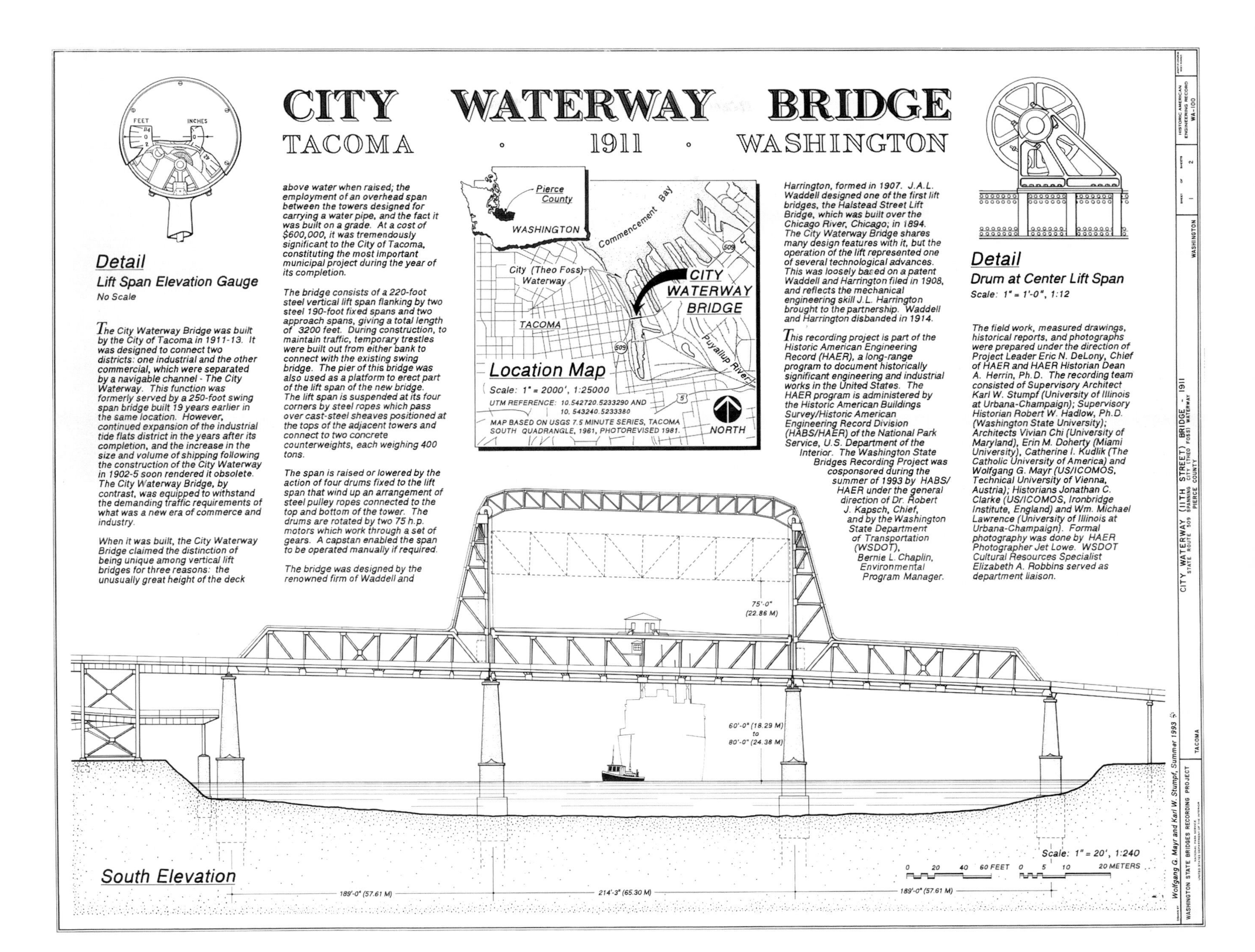
CITY WATERWAY BRIDGE
TACOMA • 1911 • WASHINGTON
FEET
INCHES
Detail
Lift Span Elevation Gauge
No Scale
The City Waterway Bridge was built by the City of Tacoma in 1911-13. It was designed to connect two districts: one industrial and the other commercial, which were separated by a navigable channel - The City Waterway. This function was formerly served by a 250-foot swing span bridge built 19 years earlier in the same location. However, continued expansion of the industrial tide flats district in the years after its completion, and the increase in the size and volume of shipping following the construction of the City Waterway in 1902-5 soon rendered it obsolete. The City Waterway Bridge, by contrast, was equipped to withstand the demanding traffic requirements of what was a new era of commerce and industry.
When it was built, the City Waterway Bridge claimed the distinction of being unique among vertical lift bridges for three reasons: the unusually great height of the deck above water when raised; the employment of an overhead span between the towers designed for carrying a water pipe, and the fact it was built on a grade. At a cost of $600,000, it was tremendously significant to the City of Tacoma, constituting the most important municipal project during the year of its completion.
The bridge consists of a 220-foot steel vertical lift span flanking by two steel 190-foot fixed spans and two approach spans, giving a total length of 3200 feet. During construction, to maintain traffic, temporary trestles were built out from either bank to connect with the existing swing bridge. The pier of this bridge was also used as a platform to erect part of the lift span of the new bridge. The lift span is suspended at its four corners by steel ropes which pass over cast-steel sheaves positioned at the tops of the adjacent towers and connect to two concrete counterweights, each weighing 400 tons.
The span is raised or lowered by the action of four drums fixed to the lift span that wind up an arrangement of steel pulley ropes connected to the top and bottom of the tower. The drums are rotated by two 75 h.p. motors which work through a set of gears. A capstan enabled the span to be operated manually if required.
The bridge was designed by the renowned firm of Waddell and Harrington, formed in 1907. J.A.L. Waddell designed one of the first lift bridges, the Halstead Street Lift Bridge, which was built over the Chicago River, Chicago; in 1894. The City Waterway Bridge shares many design features with it, but the operation of the lift represented one of several technological advances. This was loosely based on a patent Waddell and Harrington filed in 1908, and reflects the mechanical engineering skill J.L. Harrington brought to the partnership. Waddell and Harrington disbanded in 1914.
This recording project is part of the Historic American Engineering Record (HAER), a long-range program to document historically significant engineering and industrial works in the United States. The HAER program is administered by the Historic American Buildings Survey/Historic American Engineering Record Division (HABS/HAER) of the National Park Service, U.S. Department of the Interior. The Washington State Bridges Recording Project was cosponsored during the summer of 1993 by HABS/HAER under the general direction of Dr. Robert J. Kapsch, Chief, and by the Washington State Department of Transportation (WSDOT), Bernie L. Chaplin, Environmental Program Manager.
Pierce County
WASHINGTON
Commencement Bay
City (Theo Foss) Waterway
CITY WATERWAY BRIDGE
TACOMA
Puyallup River
509
5
Location Map
Scale: 1" = 2000', 1:25000
UTM REFERENCE: 10.542720.5233290 AND 10.543240.5233380
MAP BASED ON USGS 7.5 MINUTE SERIES, TACOMA SOUTH QUADRANGLE, 1961, PHOTOREVISED 1981.
NORTH
Detail
Drum at Center Lift Span
Scale: 1" = 1'-0", 1:12
The field work, measured drawings, historical reports, and photographs were prepared under the direction of Project Leader Eric N. DeLony, Chief of HAER and HAER Historian Dean A. Herrin, Ph.D. The recording team consisted of Supervisory Architect Karl W. Stumpf (University of Illinois at Urbana-Champaign); Supervisory Historian Robert W. Hadlow, Ph.D. (Washington State University); Architects Vivian Chi (University of Maryland), Erin M. Doherty (Miami University), Catherine I. Kudlik (The Catholic University of America) and Wolfgang G. Mayr (US/ICOMOS, Technical University of Vienna, Austria); Historians Jonathan C. Clarke (US/ICOMOS, Ironbridge Institute, England) and Wm. Michael Lawrence (University of Illinois at Urbana-Champaign). Formal photography was done by HAER Photographer Jet Lowe. WSDOT Cultural Resources Specialist Elizabeth A. Robbins served as department liaison.
75'-0" (22.86 M)
60'-0" (18.29 M) to 80'-0" (24.38 M)
South Elevation
189'-0" (57.61 M)
214'-3" (65.30 M)
189'-0" (57.61 M)
Scale: 1" = 20', 1:240
0 20 40 60 FEET 0 5 10 20 METERS
Wolfgang G. Mayr and Karl W. Stumpf, Summer 1993
WASHINGTON STATE BRIDGES RECORDING PROJECT
CITY WATERWAY (11TH STREET) BRIDGE - 1911
STATE ROUTE 509 SPANNING CITY (THEO FOSS) WATERWAY
PIERCE COUNTY
TACOMA
WASHINGTON
HISTORIC AMERICAN ENGINEERING RECORD
WA-100
SHEET 1 OF 2

Tacoma's City Waterway Bridge, named in honor of historian Murray Morgan.
Jet Lowe, HAER

The 3,200-foot bridge stretched across the Thea Foss Waterway to the muddy tide flats east of the city, creating a permanent link between Tacoma's downtown business center and the manufacturing district. Built by the International Contract Company of Seattle, the structure included a 214-foot Pratt vertical-lift span, flanked by two 190-foot fixed Pratt truss spans. The lift span itself is a simple truss design, suspended at its four corners by steel cables that connect to a 400-ton counterweight on either end. The bridge also featured one of the latest innovations for movable spans—electric motors to raise and lower the lift. The City Waterway Bridge offered shipping a 200-foot-wide passage, and a vertical clearance of 60 feet when the span was down, and 135 feet with the span up.

Originally, the deck's 55-foot-wide roadway carried two tracks for streetcars. The approach on the bridge's eastern end was a timber trestle more than 1,100 feet in length, which today has been replaced by a concrete-and-steel structure. From the west, the approach is a 100-foot concrete retaining wall and a 475-foot steel-and-concrete viaduct rising nearly 80 feet above the bay and buildings below.

The heavy silt and sand of Commencement Bay proved to be a complicating factor for the bridge's builders. To

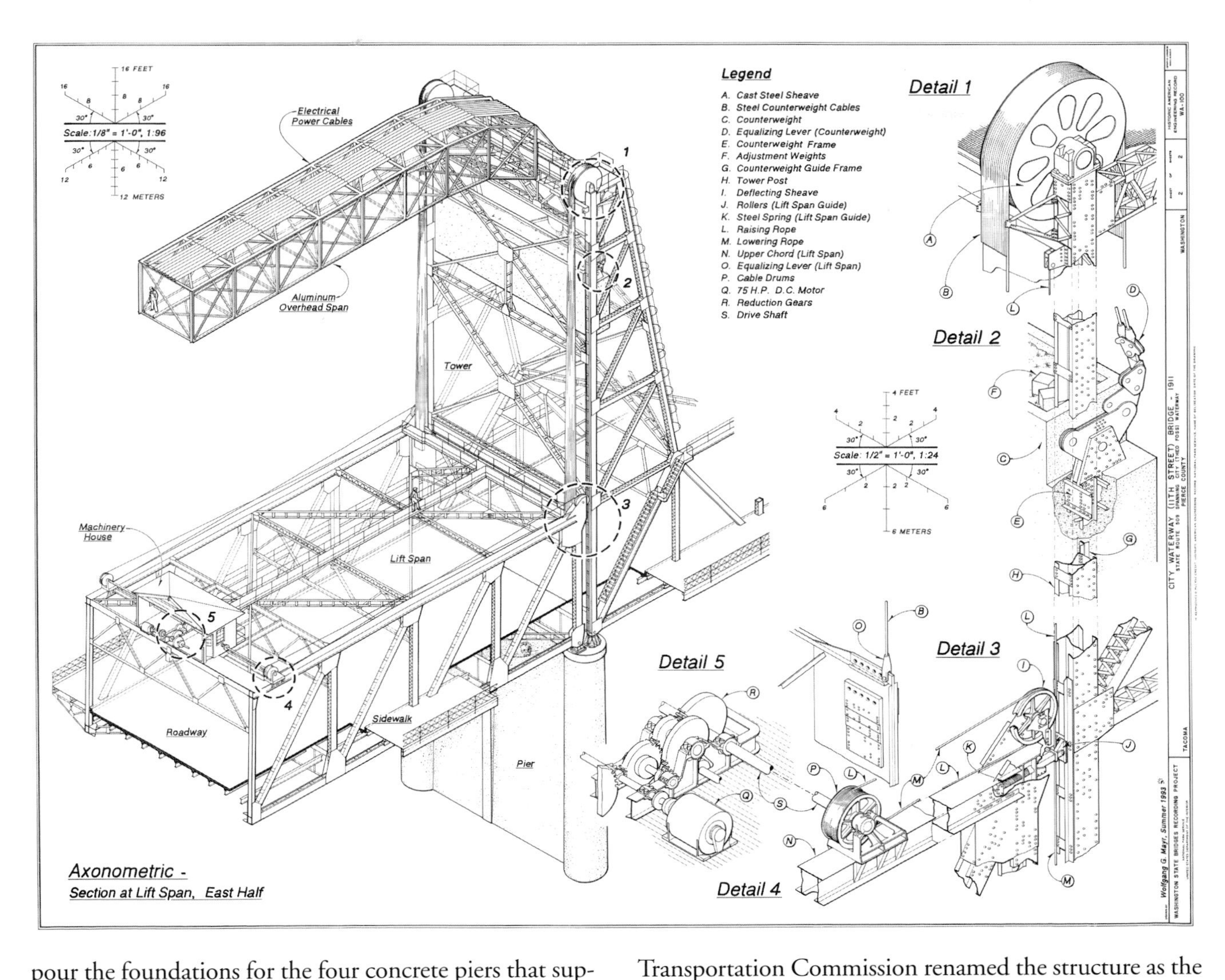

Details of the lift mechanism. Drawing 1993.

Wolfgang G. Mayr, HAER

pour the foundations for the four concrete piers that support the lift span and trusses, workmen built cribs 21 feet (to 30 feet at the bases) by 80 feet that were sunk to a depth of 54 feet below high water.

The City Waterway Bridge became a vital link in Tacoma's commercial expansion in the early decades of the twentieth century. Today, it is the sole surviving movable span of three built across the Thea Foss Waterway in the years before World War I. In 1997 the Washington State Transportation Commission renamed the structure as the "Murray Morgan Bridge." Morgan (1916–2000), a noted Puget Sound country historian, served as a bridge tender on the span while writing his classic popular history of Seattle, *Skid Road* (1951).

This unique landmark has a way of attracting attention. In 2003 a pair of peregrine falcons took up residence on the bridge, and in the spring four fuzzy hatchlings appeared in the nest. At the same time, an intense debate

Historic lighting has been retained on the East 34th Street Bridge (1948) between East B and East D streets, Tacoma.

Craig Holstine, WSDOT

over the bridge's future stirred controversy—advancing age and wear had made it a candidate for removal. Many interested parties are involved, from federal and state agencies to local citizens groups. Whether the bridge will be preserved or demolished remains undecided.[2]

East 34th Street Bridges, Tacoma, Pierce County

Two of the city's most elegant structures stand on the hill crest south of the Interstate 5 freeway overlooking Commencement Bay. Despite the usually intense freeway traffic, drivers inevitably find the open-spandrel concrete arches of the East 34th Street Bridges to be eye-catching.

In 1937 the MacRae Brothers of Seattle constructed the first of the two bridges, spanning from Pacific to A streets, under the supervision of City Engineer C. D. Forsbeck and City Bridge Engineer O. A. Anderson. Federal highway and Emergency Administration of Public Works funds made the replacement of an untreated timber trestle possible here.

The bridge's twin rib arches, set on 24-foot centers and spandrel columns standing atop the ribs, support the beam and girder deck. At the time of its construction, the main arch was the longest span (243 feet) with the greatest rise of any arch built in Washington before 1940. Total length, including the concrete-girder span approaches on both ends, is 485 feet. The main decorative features are found on its deck, where metal lampposts with urn-shaped lamps have been retained atop concrete balustrade railings.

When it came time to build another bridge east between B and D streets, the city wished to maintain design continuity. Tacoma officials signed a contract with Guy F. Atkinson of San Francisco to build a second open-spandrel concrete deck arch in 1948. Although less stylish on its deck, the second East 34th Street Bridge is the larger of the two. Its total length is 563 feet, and the twin arch ribs that mirror those of the older structure are a 273-foot span.

The open-spandrel concrete arch of Tacoma's East 34th Street Bridge (1937), between Pacific and A streets.

Office of Archaeology and Historic Preservation

The East 34th Street Bridge (1948) between East B and East D streets, Tacoma, mirrors the graceful arch of the earlier bridge.

Office of Archeaology and Historic Preservation

Together the graceful bridges reflected an advancement in concrete arch design, with a progressive reduction in the amount of materials required for structural stability.[3]

Tacoma Narrows Bridge, Pierce County

Today, motorists traveling between Puget Sound's eastern shore and the Kitsap Peninsula cross the second bridge built at the Tacoma Narrows. Like its ill-fated predecessor, "Galloping Gertie" (1940), the Tacoma Narrows Bridge became the world's third longest suspension span when completed in 1950. It holds the dubious distinction of being the first suspension bridge in the United States to be built after the failure of an earlier suspension structure at the same location.

Talk of a replacement started almost immediately after Galloping Gertie's collapse in November 1940. Yet, nearly eight years elapsed before work started on the second Tacoma Narrows Bridge. Charles E. Andrew, consulting engineer to the Washington State Toll Bridge Authority, chose an engineer from the Oregon State Highway Commission, Dexter R. Smith, to design the new structure. Construction began on the $11 million bridge in April 1948, and it opened in October 1950.

It included design elements aimed at preventing the aerodynamical-caused violent twisting and "galloping"

The second Tacoma Narrows Bridge (1950) revolutionized suspension bridge engineering.

WSDOT

that destroyed the first bridge. The new design featured deep, open Warren trusses (instead of shallow, solid plate girders), and a larger (four-lane) roadway with a greater width-to-span length ratio for better support stiffening. Engineers also installed mechanical devices (hydraulic jacks) and deck grating between the deck's traffic lanes to heighten resistance to twisting forces and enhance dampening effects.

The finished structure includes 5,000 feet of suspended spans (a 2,800-foot cable-suspended steel main span, and two 1,100-foot cable-suspended steel side spans), two towers 467 feet high (above the piers), the toll plaza, reinforced-concrete T-beam approach spans, a deck roadway 46 feet 8 inches wide with four traffic lanes and two sidewalks, and anchorages. The bridge's main suspension cables measure 20¼ inches in diameter.

After the bridge opened to motorists, toll collections mounted steadily. By 1965, the span brought in $19 million, and fares ended some thirteen years ahead of schedule. Heavy use has continued, providing ample testimony to the important link that the bridge provides between Tacoma and the Kitsap Peninsula. From more than 16,000 vehicles a day in 1967, the volume climbed to 90,000 per day in 2003.

In the summer of 2002, Washington signed a contract to build a long-delayed new parallel span at the Tacoma Narrows. The project, including conversion of the existing bridge to one-way traffic, has a price tag of $615 million, with financing costs increasing the total to $850 million. Completion is scheduled for April 2007.[4]

Fairfax Bridge, Pierce County

In Europe certain geographic conditions make the arch an economical and popular bridge form. In east central Pierce County along the scenic Carbon River, which winds through a deep, rocky gorge northwest of Mount Rainier, engineers found those same conditions. The graceful, three-hinged, steel-lattice arch constructed near the small town of Fairfax was one of only two such structures in the state.

Some twenty-five miles southeast of Tacoma, coal deposits and timber stands here brought enterprising miners and loggers to exploit the natural resources, and by the turn of the twentieth century, Fairfax was a thriving community. However, the only access to the remote area was by railroad, and trains stopped only twice a day. Residents of this isolated corner of Pierce County had few political friends and remained dependent upon the railroad for transportation to Tacoma.

In 1916, James O'Farrell, the son of a pioneer Puyallup family, was elected from this district to the Pierce County Board of Commissioners and dedicated himself to bringing a road and bridge to the area. It took five years, but O'Farrell was successful. Built in 1921 for $80,000, the Fairfax Bridge was the final link on the Carbon River Road. With a deck rising 250 feet above the rushing mountain waters, the span was reportedly the highest structure in the state for a time.

The three-hinged, steel-deck arch is flanked by two 14-foot towers and eight timber-trestle approach spans. The arch itself is comprised of a parabolic-curve steel bottom section, a spandrel-braced rib, and a horizontal Warren stiffening truss that supports the 17-foot-wide roadway. Latticed channels provide bracing for both the upper and lower arch chords.

Recognizing O'Farrell's singular contribution, the Fairfax Bridge was dedicated in his honor. Soon, the structure took on a significance beyond its utility for local residents. In the 1920s, the National Park Service and Pierce County worked jointly in a flurry of road building to connect the small towns lying near the base of Mount Rainier. By 1925, motorists from Seattle, Tacoma, and other cities were traveling to the picturesque upper Carbon River Gorge, crossing the Fairfax Bridge en route to Mount Rainier National Park, already the third most visited national park in the United States.

Over the next decade, with the onset of the Great Depression, the area's coal and timber industries declined. The once-thriving community grew smaller as businesses

The Fairfax Bridge stands 250 feet above the Carbon River near Mount Rainier National Park.

Jet Lowe, HAER

closed and families moved away. Today, most travelers crossing the Fairfax Bridge are Mount Rainier National Park tourists.[5]

McMillin Bridge, Pierce County

In 1934 a startling and innovative bridge was built at a quiet crossing of the Puyallup River in rural Pierce County, though it failed to make even a small splash in bridge engineering circles around the country. The McMillin Bridge's 170-foot main span was then the longest reinforced-concrete truss or beam span in the nation. Its major design feature, the remarkable hollow-box system, was suggested by innovative bridge engineer Homer M. Hadley of the Portland Cement Association.

When a flood swept through the Puyallup Valley in the winter of 1933, washing out roads and an old steel span at the McMillin site, county engineers reviewed proposed replacement designs. Cost considerations proved to be the deciding factor, due to the local impact of the Great Depression. A concrete structure was $826 cheaper than the lowest bid for a steel bridge, and, over the long term, it offered lower maintenance costs.

The cellular design, common in Europe, was not widely used in the United States at the time. While the through Pratt truss type was a common steel bridge form, rarely was reinforced-concrete used. Hadley's suggested design produced a unique and striking structure. The U-shaped chords and end posts, plus the H-shaped intermediate posts, contrast with rectangular walkway openings that

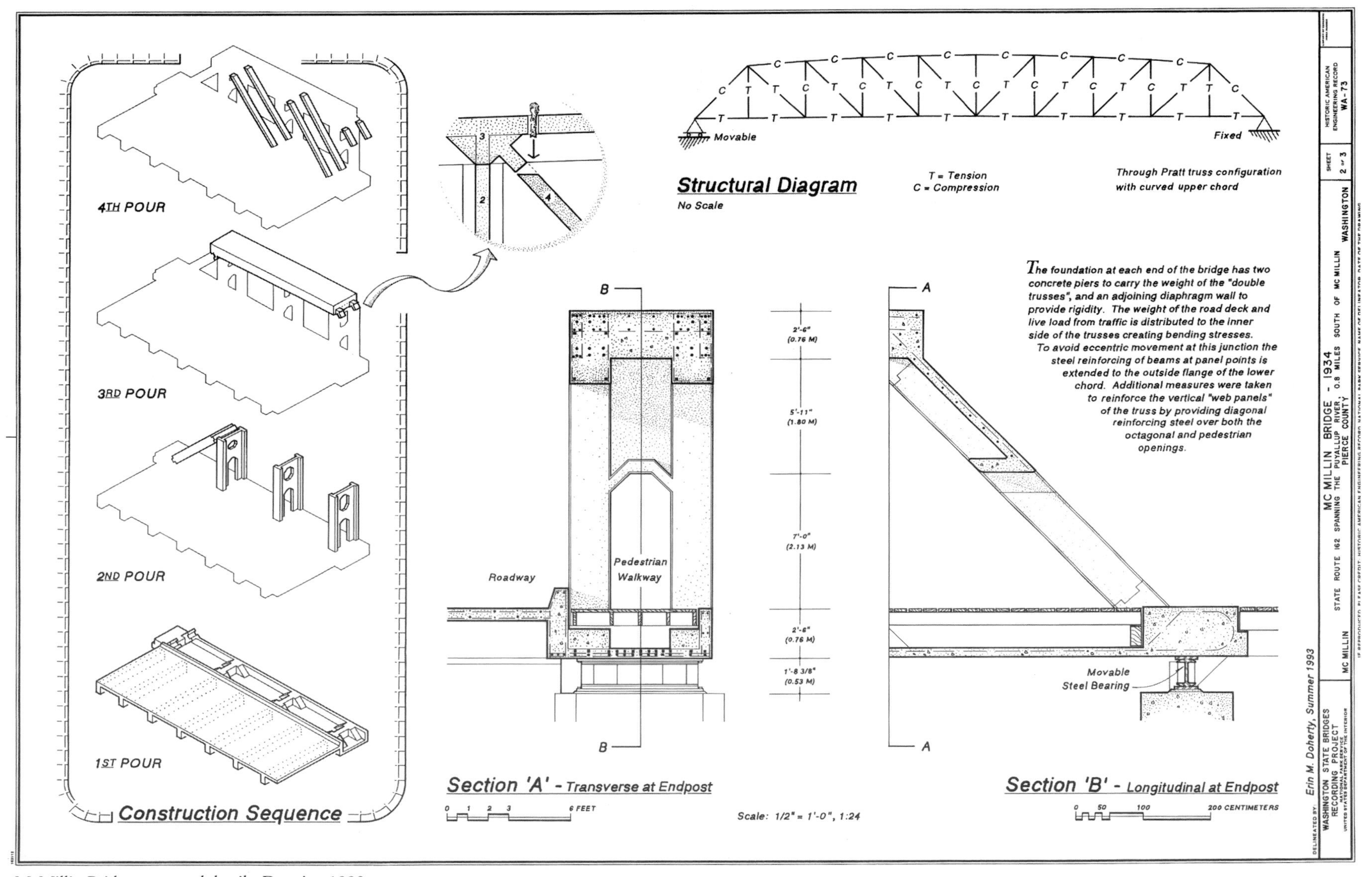

McMillin Bridge structural details. Drawing 1993.

Erin M. Doherty, HAER

pass through each of the seven-foot-wide trusses. Irregular, octagonal holes are located above each walkway portal. Though an unusual design, it had little influence on subsequent bridge design. Another bridge suggested by Hadley, the Purdy Bridge, garnered much wider notice in the engineering press.[6]

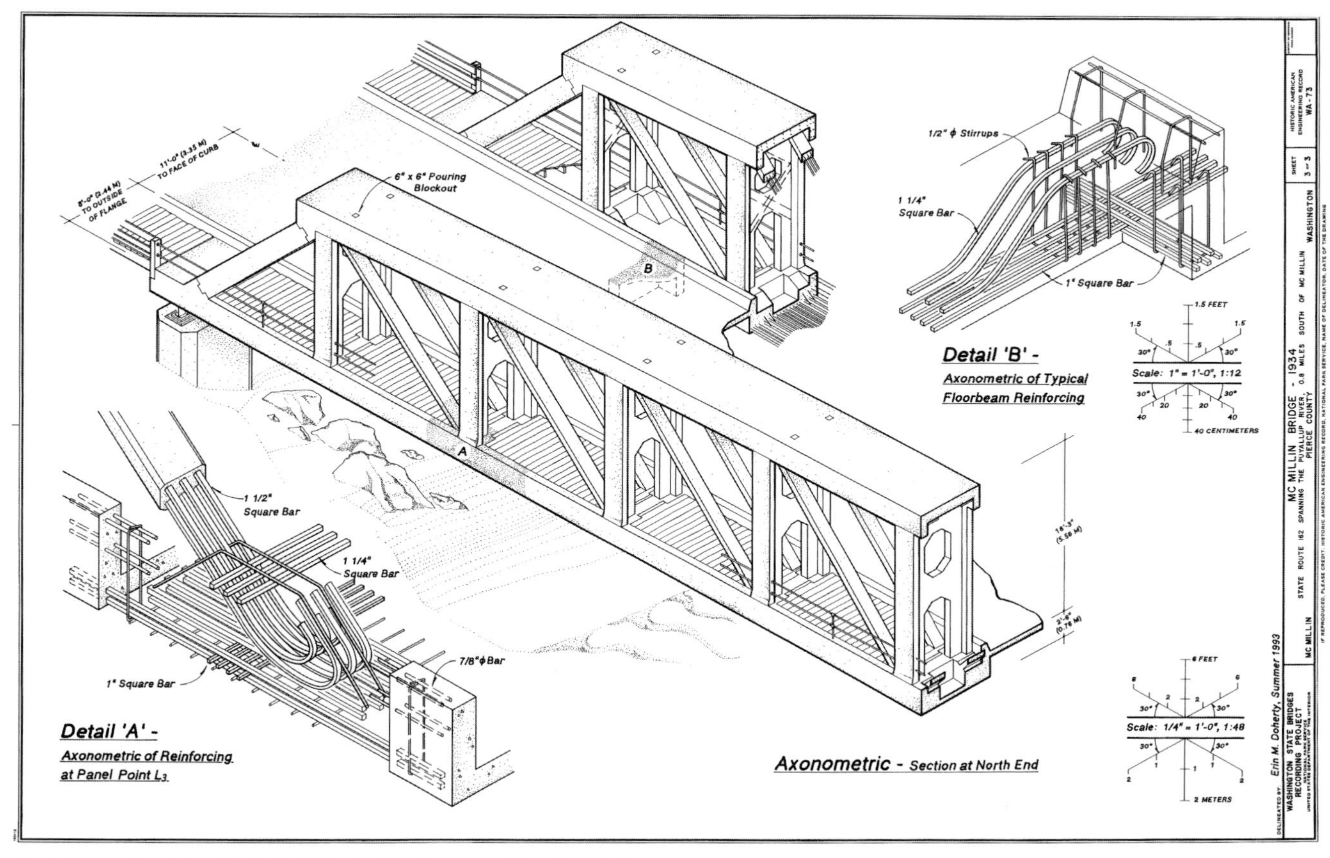

McMillin Bridge concrete truss details. Drawing 1993.
Erin M. Doherty, HAER

Homer M. Hadley suggested the hollow concrete box design for the innovative McMillin Bridge over the Puyallup River. Photo 1993.
Jet Lowe, HAER

Patton Bridge, King County

Yet another Homer Hadley bridge crosses the Green River near Auburn in south King County. Constructed in 1950, the Patton Bridge is an unusual combination of concrete and steel box girders. The 430-foot-long structure consists of two anchor arms, two cantilever arms, and a suspended section. Multiple-box, two-cell, reinforced-concrete girders comprise the anchor and cantilever arms. Welded-steel box girders with a reinforced-concrete deck make up

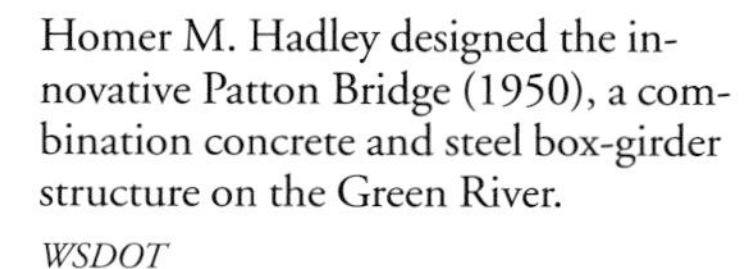

Homer M. Hadley designed the innovative Patton Bridge (1950), a combination concrete and steel box-girder structure on the Green River.
WSDOT

the 200-foot-long suspended span, which was the longest such span in Washington when built.[7]

Pe Ell Bridge, Lewis County

The Pe Ell Bridge is one of only three historic covered bridges remaining in Washington. It crosses the narrow, rocky channel of the Chehalis River two miles south of the small town of Pe Ell. The bridge bears the community's name because it carries Pe Ell's domestic water (as it has since 1934) in a ten-inch-diameter water pipe from a nearby reservoir. The February 2001 Nisqually Earthquake caused a slight movement of the pipe, and possibly some undetected structural damage to the bridge.

Measuring 63 feet long between approach span seats, the main span is a six-panel, covered Howe truss. Fifteen-foot-long approaches on both ends bring the bridge's total length to 93 feet. The approaches are constructed of 4 x 12-inch wood planks, resting atop 4 x 12-inch stringers on hewn log caps. The north approach is partially collapsed, reportedly due to recent all-terrain vehicular impact. Access to the bridge from the north by larger vehicles ended after the bank supporting the steep driveway off the main road eroded away, leaving only a narrow pedestrian path.

The Pe Ell Bridge is an intact representative of a significant bridge type. Covered bridges apparently were not as numerous in Washington, as compared to Oregon, where more than 400 covered bridges stood as late as the 1920s. However, similar historical data is not available for Washington. The loss of covered bridges to modern transportation needs has occurred regionally and nationwide.

The Pe Ell Bridge's Howe pony truss design reflects the predominant style used in covered bridges. In Oregon, for example, forty-one of the state's forty-nine covered bridges remaining in 1988 were Howe trusses. Developed by Massachusetts architect William Howe in 1840, the design relies upon heavy timber beams for bearing compression loads, and steel or iron rods to withstand tension loads. The Pe Ell Bridge varies from the typical Howe design only in its lack of diagonal cross braces in the center panel. Unlike many other covered bridges, the Pe Ell structure rather lacks in aesthetic appeal, being covered in corrugated metal purchased at a local hardware store in 1934.

The Pe Ell Bridge, erected across the Chehalis River in 1934, is one of Washington's three remaining historic covered bridges. Photo 2001.

Craig Holstine, WSDOT

The town of Pe Ell reportedly is named for Pierre Louis (P. L.) Charles, a former Hudson's Bay Company employee of French-Canadian descent. Pioneer farming on Pe Ell Prairie eventually gave way to logging and lumbering in economic importance, especially after 1900 when Frederick Weyerhaeuser purchased 900,000 acres of surrounding Northern Pacific Railway timberland. In 1914 the NP completed a line through the vicinity, and Pe Ell grew to be the largest settlement between Chehalis and South Bend.

By 1903, the Weyerhaeuser Timber Company owned the land south of Pe Ell where the town sought to tap a domestic water supply. In November of that year, the Washington Light and Water System Company acquired the right to install and maintain a water pipe from an intake on Weyerhaeuser land, and to "construct, operate and maintain over the Chehalis River… a water pipe line attached to a wire cable for a distance of about 200 feet… [and] about thirty-five (35) feet above the waters of the river." The legal successor to Washington Light and Water sold the water system to the Town of Pe Ell in 1923.

Ten years later, when the water system needed repairs, the town council adopted an ordinance allowing citizens to vote on a plan "for the making of certain additions, improvements, betterments, and repairs to the Municipal Water Works [including] the construction of a new crossing over the Chehalis River." In 1934 James Donahue, Pe Ell's water superintendent, reportedly built the bridge carrying the water pipe. Metal numbers ("1934") are nailed on the wooden truss, and the same date is written on the interior side of the corrugated metal covering.[8]

Modrow Bridge, Cowlitz County

One of the state's most unique structures, the Modrow Bridge, is nestled in the verdant countryside of south central Cowlitz County over the meandering Kalama River. In 1960 the American Institute of Steel Construction gave it a prestigious national "First Place Award for Steel Bridges with Spans under 400 Feet Costing Less than $500,000," and hailed the span as "the most imaginative and sensitive bridge in the competition."

Designed by a Seattle firm, Harry R. Powell and Associates, and built in 1958–59, the Modrow Bridge is mostly used by local residents and fishermen. The 200-foot-long structure consists of two 16-foot, reinforced-concrete slab spans at the northwest end; a 120-foot, open-spandrel, rib steel-deck arch main span; and three 16-foot, reinforced-concrete slab approach spans at the southeast end.

The arch rises more than 20 feet to its crown. The arch's centerpiece is its rib, consisting of two tapered, welded, steel-plate girder sections 17 feet apart. The rib sections are hinged at each end and at the span's centerline. Use of a three-hinged arch rib frees the structure from thermal stresses. Steel spandrel columns are supported along each of the plate girder rib sections at 15-foot centers, and support a steel crossbeam located under the reinforced-concrete roadway deck.

Harry R. Powell designed the Modrow Bridge, an award-winning structure on the Kalama River, built in 1958–59. Photo 2002.

Craig Holstine, WSDOT

The Modrow Bridge is the second steel, open-spandrel, rib deck-arch bridge built in Washington incorporating welded members. Earlier steel arches used riveted connections. This, combined with the addition of high tensile bolts, marked a major step in the evolution of steel bridge design and construction, leading to the demise of riveted connections in the mid-1960s.[9]

The Yale Bridge (1932) on the Lewis River is one of the state's few remaining historic suspension bridges. Photo 1993.

Jet Lowe, HAER

Yale Bridge, Clark and Cowlitz Counties

The Yale Bridge is a remarkable example of experimentation in economical bridge design. Situated on the Lewis River, which forms the boundary between Clark and Cowlitz counties, this structure is Washington's only steel, short-span, suspension bridge. Construction of the Ariel Dam in 1931 raised the river about ninety feet, requiring the demolition of the first Yale Bridge located near this site, twelve miles upstream from the dam. The new reservoir made it difficult to build the falsework required for a traditional, less expensive bridge. The unconventional solution, designed by State Department of Highways Engineer Harold Gilbert, was a 532-foot-long steel suspension span.

The short length of the span, and the cost-saving construction methods and materials, resulted in an economical bridge. The 300-foot steel truss span, with an 18-foot road deck 50 feet above the river at high water, is supported by four galvanized steel cables attached to two lattice steel towers. The Yale Bridge's innovative tower cable connection system simplified the bridge erection process and allowed the use of smaller ($2^{7}/_{8}$ inch) diameter suspension cables.

The Yale Bridge has continued to serve its local, rural population since completion in 1932. In 1957–58, the original timber approach spans were replaced with steel beam structures.[10]

Conrad Lundy Jr. Bridge, Skamania County

The deep, narrow Wind River Canyon snakes through the Gifford Pinchot National Forest in south Skamania County. Just northwest of the little Columbia River town of Carson, noted for its mineral hot springs resort, stands the Conrad Lundy Jr. Bridge, Washington's only symmetrical, two-span, steel-deck truss supported by a central steel tower.

The 1957 bridge, named for a Skamania County commissioner, replaced a suspension bridge that carried local traffic for more than three decades. The original Wind River bridge was a wood suspension structure erected in 1913. In 1925 it was rebuilt as a single-lane, steel suspension bridge, 554 feet long. Locals soon dubbed it the "swinging bridge" for a tendency to roll under the weight of loaded logging trucks.

By 1956, the bridge's useful life was at an end. The U.S. Bureau of Public Roads designed a new steel-deck truss bridge with two lanes to safely accommodate logging trucks and tourist traffic. The site required an innovative design; the canyon here is steep, and quite narrow at the bottom.

A central tower stands in the canyon on pedestals set amid boulders and stradling the river. Two identical Warren deck truss sections, 20 feet wide and 40 feet deep, extend from the tower to the canyon's sides. The 494-foot-long continuous truss stands some 192 feet above the river. Built by the C. M. Corkum Company of Portland for $507,000, the Conrad Lundy Jr. Bridge was one of only two steel deck-truss bridges constructed in the 1950s. (The Lake Washington Ship Canal Bridge in Seattle was the other.)

As work on the new span neared completion in 1957, local citizens tried to save the old suspension bridge. Some thought it might be preserved as a tourist attraction, but the cost of rehabilitating the structure proved too expensive, and it was demolished.[11]

Named for a Skamania County commissioner, the Conrad Lundy Jr. Bridge was designed by the U.S. Bureau of Public Roads.
WSDOT

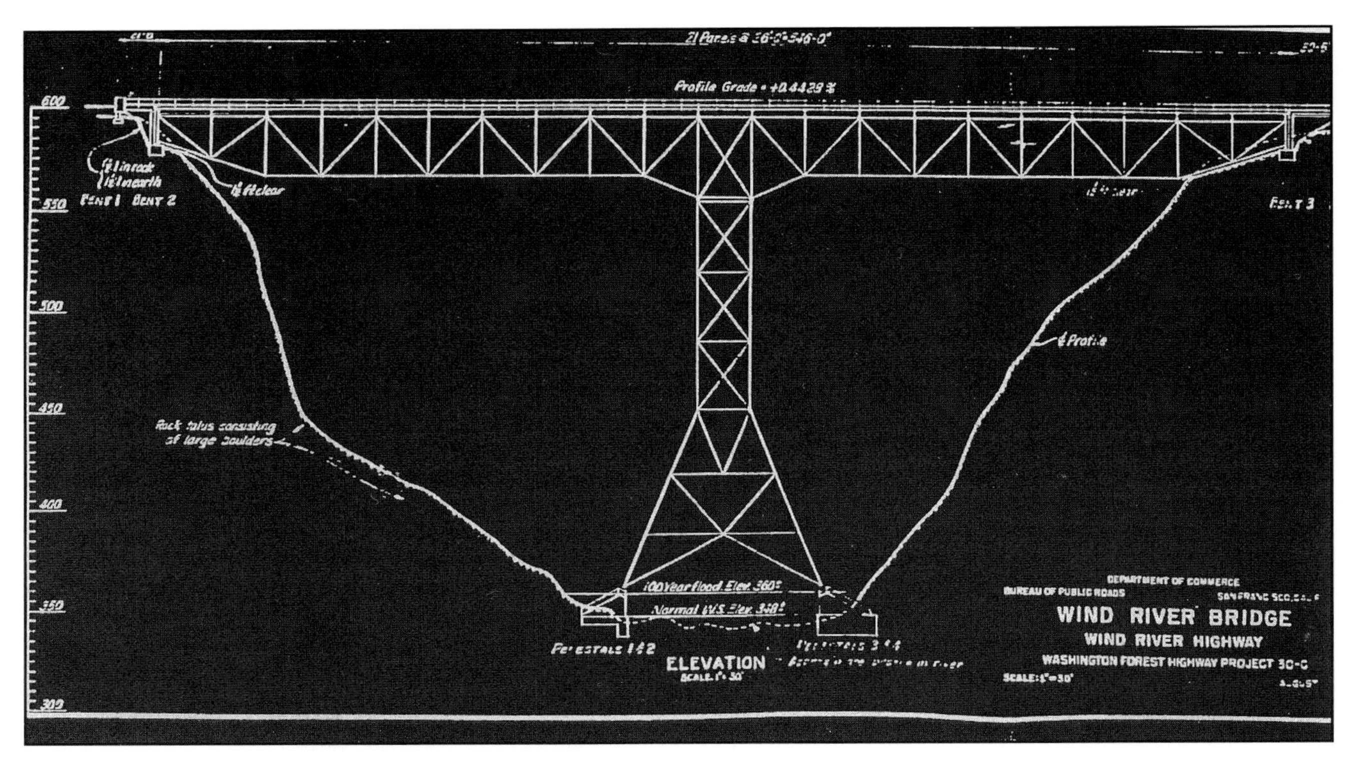

The Conrad Lundy Jr. Bridge. Drawing September 3, 1957.

Skamania County Public Works, Engineer's Office

NOTES

1. Lisa Soderberg, "Capitol Boulevard Bridge," HAER Inventory, 1979; Lisa Soderbery, "Lower Custer Way Bridge," HAER Inventory, 1979; and Oscar R. "Bob" George, "Upper Custer Way Bridge," National Register of Historic Places Nomination, 2001.
2. Jonathan Clarke, "City Waterway Bridge," HAER No. WA-100, 1993; Lisa Soderberg, "City Waterway Bridge," HAER Inventory, 1979.
3. Lisa Soderberg, "East Thirty-fourth Street Bridge, Pacific to A Street [1937]," HAER Inventory, 1979; Lisa Soderberg, "East Thirty-fourth Street Bridge, East B to East D Streets [1948],"HAER Inventory, 1979.
4. WSDOT, "Tacoma Narrows Bridge History," www.wsdot.wa.gov/TNBhistory (narrative by Richard Hobbs); Robert H. Krier and Craig Holstine, "Tacoma Narrows Bridge (Second)," NRHP Nomination, 1993; Charles Andrew, *Final Report on Tacoma Narrows Bridge* (Tacoma, WA: Washington Toll Bridge Authority, 1952); Charles Andrew, "Design of a Suspension Structure to Replace the Former Narrows Bridge—Part 1," *Pacific Builder and Engineer* 51, October 1945, 43–45; Charles Andrew, "Redesign of Tacoma Narrows Bridge," *Engineering News-Record* 135, November 29, 1945, 716–21; Charles Andrew, "Tacoma Narrows Bridge Number II ... The Nation's First Suspension Bridge Designed to Be Aerodynamically Stable," *Pacific Builder and Engineer* 56, October 1950, 54–57, 101; Charles Andrew, "Unusual Design Problems—Second Tacoma Narrows Bridge," *Proceedings of the American Society of Civil Engineers* 73, December 1947, 1483–97; Robert W. Hadlow, "Tacoma Narrows Bridge," HAER No. WA-99, 1993; Joe Gotchy, *Bridging the Narrows* (Gig Harbor, WA: Peninsula Historical Society, 1990).
5. Jonathan Clarke, "Fairfax Bridge," HAER No. WA-72, 1993.
6. Wm. Michael Lawrence, "McMillin Bridge," HAER No. WA-73, 1993.
7. Robert Krier, J. Byron Barber, Robin Bruce, and Craig Holstine, "Patton Bridge," NRHP Nomination, 1991.
8. Craig Holstine, "Pe Ell Bridge," NRHP Nomination, 2002; James Norman, *Oregon Covered Bridges: A Study for the 1989–90 Legislature* (Salem: Oregon State Highway Division, Environmental Section, 1988); Pe Ell and Doty Bridges Files, Washington State Office of Archaeology and Historic Preservation, Olympia; Records of the Town of Pe Ell, maintained in the Pe Ell Town Hall; Lisa Soderberg, "Weyerhaeuser/ Pe Ell Bridge," HAER Inventory, 1979.
9. Oscar R. "Bob" George, "Modrow Bridge." NRHP Nomination, 2001; Margot Vaughn, editor, "The Modrow Bridge," *Cowlitz County Quarterly* 26, Spring 1984, 35–36.
10. Robert W. Hadlow, "Yale Bridge/Lewis River Bridge," HAER No. WA-87, 1993.
11. Oscar R. "Bob" George, "Conrad Lundy, Jr. Bridge," NRHP Nomination, 2001.

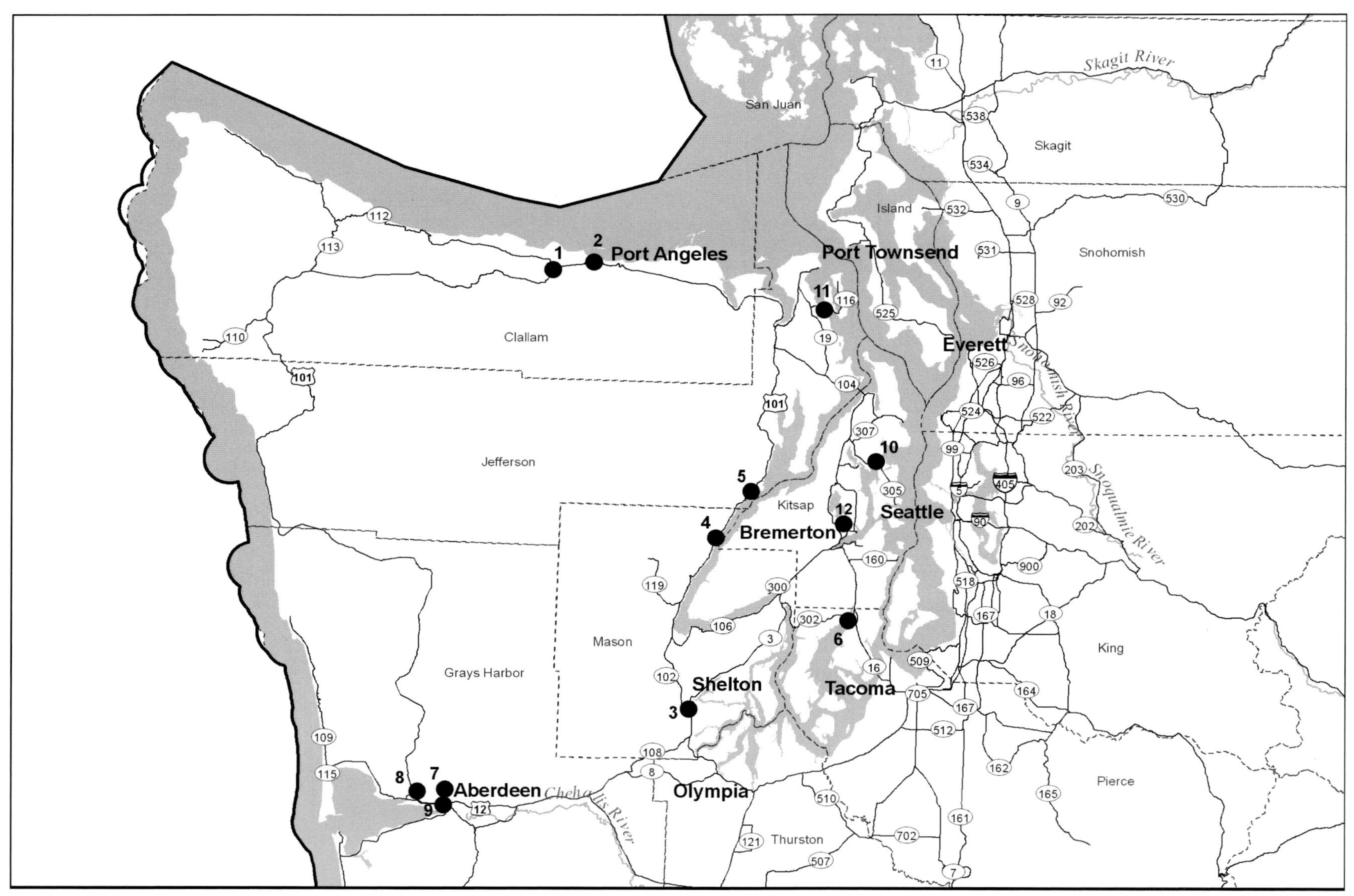

Premier Historic Bridges of the Kitsap and Olympic Peninsulas

● Bridge Locations — State Routes — Major Rivers — County Boundaries

Bridge No.	Bridge Name	Bridge No.	Bridge Name
1	Elwha River Bridge	7	Wishkah River Bridge
2	8th Street Timber Trestle Bridges	8	Hoquiam River Bridge
3	Goldsborough Creek Bridge	9	Chehalis River Bridge
4	Hamma Hamma River Bridges	10	Agate Pass Bridge
5	Duckabush River Bridge	11	Portage Canal Bridge
6	Purdy Bridge	12	Port Washington Narrows (Warren Avenue) Bridge

Bridges of the Kitsap and Olympic Peninsulas

Of the many significant bridges found on the Olympic and Kitsap peninsulas, fourteen are outstanding. For the state's earliest drivers, a new roadway circuiting the northern Olympic Peninsula crossed the Elwha River Bridge (1913), which was a vital link in the development of a long-sought highway for that remote locality.

Not far away, on 8th Street in south Port Angeles, two unique timber trestle bridges (erected in 1936) cross over Tumwater Creek and Valley Creek. Some of the Olympic Peninsula's unusual, large-dimensional timber was used in their construction.

On the east side of the Olympic Peninsula, along the western shore of Hood Canal, motorists cross four unusual reinforced-concrete, tied-arch spans at Goldsborough Creek (1923), the north and south forks of the Hamma Hamma River (both 1924), and the Duckabush River (1934). The only other bridge of this type in the state stands in far-off southeast Washington and dates from 1923.

Aberdeen and Hoquiam on Grays Harbor at the southwest base of the Olympic Peninsula have been wood-products shipping centers to the world for over a century. The three rivers emptying into eastern Grays Harbor have necessitated the erection of a number of movable bridges over the years. The three standing here today are among the most important and historically significant—the Wishkah River Bridge (1923), Hoquiam River Bridge (1928), and the Chehalis River Bridge (1955). All are bascule drawbridges.

On the Kitsap Peninsula, just five miles northwest of Gig Harbor, the rare Purdy Bridge remains as one of the earliest reinforced-concrete, box-girder spans ever built in the United States.

Also in western Puget Sound, the 1950 Agate Pass Bridge—a modern steel cantilever truss span, important in part for its role in significantly changing transportation patterns and community development in the area—connects the north end of Bainbridge Island to the Kitsap Peninsula.

Near Hadlock in east Jefferson County, the 1951 Portage Canal Bridge was the second steel box-girder suspended span built in the United States. It connects Indian and Marrowstone islands to the northeast side of the Olympic Peninsula.

Seven years later, the state's longest continuous plate-girder bridge was erected across Bremerton's Port Washington Narrows, a key location in linking communities on the Kitsap Peninsula.

Elwha River Bridge, Clallam County

In the early years of automobile transportation, engineers faced the daunting task of encircling the Olympic Peninsula with an all-season, all-weather highway. Constructing a roadway directly across the rugged Olympic Mountains was not feasible. Logging railroads had tapped the majestic rain forest's incredible wealth, but the need to provide roads for log trucks, as well as local residents and tourists in this remote region, propelled highway and bridge building efforts.

The bridge over the picturesque Elwha River served as a vital link in that system. It is believed to be the oldest highway deck truss, and the oldest Warren truss specifically constructed for highway use in the state. When erected in 1913 by the Portland Bridge Company, it consisted of a

Built in 1913, the Elwha River Bridge is probably the oldest highway deck truss in Washington.

Office of Archaeology and Historic Preservation

210-foot Warren deck truss. With its two timber approach spans, the structure's length totals 576 feet.

Steel floor beams were welded to the deck's substructure to increase load capacity, and the treated, laminated-timber deck appears to have replaced an earlier wooden deck. Despite improvements over the years, numerous deficiencies have been identified, including cracked stringer-floor beam welds; rusted steel chord members, rivet heads, and lateral wind braces; and worn roller bearings needing replacement.

A 1992 inspection concluded: "This bridge is in fairly advanced states of deterioration and will require partial rehabilitation within the next two years." Clallam County has classified the bridge as "functionally obsolete," and is seeking to have the structure replaced.[1]

8th Street Timber Trestle Bridges, Port Angeles, Clallam County

The Tumwater Creek and Valley Creek bridges on 8th Street in south Port Angeles are the longest timber trestle bridges in the state roadway system. Built in 1936 a few blocks apart, they are also among the highest and oldest. Each stands 100 feet high and measures 755 feet.

Trestle bridges normally do not represent sophistication or innovation in design and engineering. These bridges, however, were unconventional adaptations of designs drawn for trestles elsewhere, and the result of collaboration between two skilled engineers facing financial limitations. Conrad O. Mannes, of the Washington Emergency Recovery Administration, modified plans for two viaducts to be built in Portland, Oregon, for use at the Port Angeles sites. Mannes reduced the amount of timbers in the bents (supporting towers) by battering (tapering) the two-column bents and narrowing the roadway. His plan used 8,000 "split-ring" connectors instead of costlier steel gusset plates, and changed the specifications for treating the timbers (fifty percent creosote, fifty percent mineral oil).

In Port Angeles, the 8th Street Bridge over Tumwater Creek is one of the state's oldest, longest, and highest timber trestle vehicular bridges. 2005 photo.

Trent de Boer, WSDOT

District engineer A. M. Young of the Washington Emergency Recovery Administration suggested another cost-saving measure, using skilled laborers on Seattle's relief rolls to frame the timber bents and unskilled laborers in Port Angeles to erect the structures. Open-lattice framing and cross-bracing between the columns provided added strength and gave each structure "an attractive appearance," an unusual attribute for a timber trestle. The reinforced-concrete deck slab was attached to spikes projecting from timber stringers set on 29-foot spans, in "virtually T-beam construction."

The construction was done by an electric, "stiff-legged" derrick, with a 76-foot boom. As work progressed, the derrick advanced along the top of each of the bridges in turn. On the ground, a steam winch mounted on caterpillar treads assisted the derrick. H. E. Dodge, a Port Angeles city engineer, and T. J. Murphy, the Public Works Administration engineering inspector, oversaw construction by the Angeles Gravel and Supply Company. Mannes characterized the overall project as "a novel design in bridge building."[2]

Goldsborough Creek Bridge, Shelton, Mason County

Although diminutive in size, the Goldsborough Creek Bridge was designed to attract the attention of visitors arriving in south Shelton. The MacRae Brothers built the smallest of the state's five historic, reinforced-concrete tied-arch bridges in 1923. Today, State Route 101 bypasses the downtown, but State Route 3 enters Shelton across this 57-foot structure.

The relatively low height of the arches allows for unobstructed views into town. Its arches rest atop concrete abutments with timber pile bulkheads. From curb to curb, the two-lane deck is 24 feet wide. The bridge is pedestrian-friendly, with sidewalks on the outside of the arches.[3]

Goldsborough Creek Bridge in Shelton, the smallest of the state's concrete tied-arch bridges.

Office of Archaeology and Historic Preservation

Hamma Hamma River Bridges, Mason County

Motorists on State Route 101 along the west side of Hood Canal enjoy the illusion of a serene, sheltered boulevard where the concrete arches of the North and South Hamma Hamma River bridges frame a charming pathway in the trees. Near its mouth, the river forms north and south channels through a large marsh before emptying into Hood Canal.

The North Hamma Hamma River Bridge. 1993 photo.

Jet Lowe, HAER

The Hamma Hamma River bridges under construction, ca. 1924, on what would become State Route 101.

Washington State Archives, WSDOT Record

The Duckabush River Bridge.
Washington State Archives, WSDOT Records

When completed in 1936, the 550-foot Purdy Bridge was the longest reinforced-concrete, box-girder bridge in the nation. Asahel Curtis, photographer.
Washington State Archives, WSDOT Records

Here, the State Department of Highways located a pair of bridges for the shortest crossing of the marsh and river forks.

Built in 1924 by the Colonial Building Company, these are identical reinforced concrete, tied-arch spans, each with a 150-foot-long, 30-foot-high, three-hinged arch. The Hamma Hamma structures are fine examples of the concrete, ribbed, through-arch bridge commonly built in the United States in the 1920s and 1930s. They are similar to three other concrete tied-arch spans constructed by the State Department of Highways about the same time—the Goldsborough Creek and Duckabush River bridges on the Olympic Peninsula, and the Indian Timothy Memorial Bridge in southeast Washington. In 1924 the Hamma Hamma structures were the longest hinged through-arch bridges in the state.

The 22-foot wide concrete road deck is supported by hangers inside the arch and by spandrel columns outside. Originally, six struts connected the two arches of each span, but two were removed to provide higher vertical clearance (now 16 feet). Several design features enhance each bridge's attractiveness. The railing is balustrade shaped with arched openings. Both the overhead struts and the approach railing are scored to simulate inset panels. The sides of the arch ribs are hammered, leaving the aggregate exposed to highlight the contrasting textures and hues.[4]

Duckabush River Bridge, Jefferson County

Another of the state's concrete tied-arch spans crosses the Duckabush River on State Route 101 south of Brinnon. Built in 1934, it has the the greatest rise in its parallel arches of the five bridges of its type in Washington. Three struts connect the arches above the roadway, which is carried on a 24-foot-wide deck along the 168-foot length of the structure.

The bridge's abutments are relatively small due to the way in which the bridge was designed. Most arches require massive foundations to resist horizontal thrust. In this bridge, however, the deck slab—hung by suspenders from the arches—is subject to both tensile and compressive forces, and thus absorbs some of the force normally put on the foundations. Hence the need for large abutments was eliminated in this economical design.[5]

Purdy Bridge, Pierce County

Since 1892, a bridge has crossed Purdy Creek on Burley Lagoon—connecting the town of Purdy on the east to a narrow arm of land extending out from the opposite (or west) shore of Henderson Bay. When built in 1936, the bridge here was one of the few reinforced-concrete, box-girder spans in the nation, although the type was common in Europe. At the time, it was the longest of its kind in the United States. The Purdy Bridge received considerable attention, drawing notice in engineering articles, as well as in Carl Condit's seminal study, *American Building Art.* Condit termed it "the nearest American rival" to similar girder spans in Europe.

The Purdy Bridge is a 550-foot-long, continuous box girder. It is one of a handful designed and built in Pierce and King counties during the 1930s. Workmen poured

63151
ASAHEL CURTIS

concrete around a hollowed, steel box girder to form the roadway slab and top flange, 15 feet wide and 7 inches deep at the crown. The long, 190-foot middle section was designed as a simple suspended beam. The piers were also cellular and flush with the sides of the deck box girder.

The choice of this type of bridge for the Purdy site was both logical and economical. The span's design was suggested by the innovative Homer M. Hadley, then the regional structural engineer for the Portland Cement Association. Because Henderson Bay was a navigable waterway used by commercial fishing vessels (and thus under U.S. War Department jurisdiction), federal regulations dictated a vertical clearance of 18 feet at high tide. Additionally, strong tidal currents in the bay, varying up to 17 feet, meant that a pier depth of 20 feet was needed. Deep-water piers were expensive, but the hollow-box, concrete-girder design suggested by Hadley required only two piers, thus offering savings. The total cost was $62,000.

Significant as an early—and rare—example of the hollow-box concrete girder in America, the Purdy Bridge is just one of several types of unique, cellular, concrete bridges designed by Hadley during his long and distinguished career. However, the reinforced-concrete, box-girder form never was widely adopted in the United States. The Purdy Bridge, and a few others like it, are exceptional in the history of American bridge building.[6]

Aberdeen's Wishkah River Bridge is a single-leaf bascule. 1993 photo.

Jet Lowe, HAER

Wishkah River Bridge, Aberdeen, Grays Harbor County

The Wishkah River Bridge (1923) in Aberdeen was once described as having "all the glamour of a dockside crane." Its geometric pattern of a lattice steel truss, triangular tower, and concrete counterweight seems appropriate for this gritty setting in a hard-working lumber and fishing town. In the days when oceangoing vessels docked on the Wishkah River, a "fixed" bridge at this location, with a high enough clearance to allow ship passage, would have required the construction of very high approaches. To avoid such complications, the City of Aberdeen decided to build a movable lift span.

As a cost-saving measure, city officials chose a single-leaf span rather than a more conventional, and more expensive, double-leaf span. They selected M. M. Caldwell of Seattle as consulting engineer to supervise construction by the Puget Sound Bridge and Dredging Company of Seattle, which won the contract with a low bid of $183,635. The Strauss Bascule Bridge Company of Chicago designed the "heel trunnion" bascule, one of three basic designs that had made the company world famous.

The structure consists of a movable, single "leaf," 145-foot Warren through truss that, when tilted upward into the open position, rotates about the main trunnion at its east end. There, the counterweight tower is based upon a right triangle—a short approach span serves as the horizontal element; a diagonal brace serves as the hypotenuse; and the vertical legs serve as support posts for the counterweight trunnion, which sits at the apex of the triangular frame. The main advantage of the heel trunnion (a characteristically Art Deco design of its times) is that it does not require a counterweight pit or high approaches.

Following the bridge's completion in 1923, the state purchased the structure in 1935. In recent years, WSDOT has improved the electrical/mechanical lift system while maintaining the bridge's essential form and function.[7]

Hoquiam River Bridge, Hoquiam, Grays Harbor County

The town of Hoquiam straddles the Hoquiam River on flatlands at the head of Grays Harbor. Here, for decades, lumbermen, mill workers, and fishermen built a local economy based on natural resources from the land and sea. As early as 1910, city officials planned a bridge across the Hoquiam River near the railroad tracks and the Simpson Lumber Mill yard. It took eighteen years, but they succeeded.

Function over form, the Hoquiam River Bridge (foreground) includes a double-leaf bascule span. A railroad swing bridge stands open in the mid-distance. 1993 photo.

Jet Lowe, HAER

The Hoquiam River Bridge, completed in 1928, is a patented Strauss double-leaf, underneath-counterweight bascule span—the only one built in Washington before 1940. The designer, the J. B. Strauss Bascule Bridge Company, was one of the nation's most innovative and important design firms in the early twentieth century. The bridge, funded jointly by the City of Hoquiam, Grays Harbor County, and the state government (and purchased by Washington in 1935), was built for $350,000 by one of the Northwest's most prominent engineering and construction firms, the Puget Sound Bridge and Dredging Company.

Four different kinds of structures are combined in the bridge. Central is the 200-foot-long, two-leaf bascule span, providing a vertical clearance of 36 feet at high tide. The piers where the counterweights and machinery are housed are set 173 feet apart. The approaches stand mainly on land—these consist of concrete decks on timber bents, concrete girders on concrete piers, and steel trusses on concrete piers. The west approach is a timber trestle 969 feet long, and the east approach is 760 feet in length.

The Chehalis River Bridge in Aberdeen was the longest double-leaf bascule in the state when completed in 1955.
Craig Holstine, WSDOT

Visually, this combination lacks a harmony that would lend grace or beauty. Additionally, the counterweights for the spans hang openly out of the backs of the towers. The designers even left the counterweights' concrete rough and unfinished.

Unfortunately, this bridge on Simpson Avenue is less practical for today's traffic. The 20-foot-wide roadway is narrow by modern standards. Occasionally errant motorists damage the concrete railings or other exposed sections. The through trusses, with a vertical clearance at the portals of only 14 feet 6 inches, are notably vulnerable to vehicular damage.

The large and complex Hoquiam River Bridge is not pretty, but it is functional—an unadorned and hardworking bridge, appropriate for its locale.[8]

Chehalis River Bridge, Aberdeen, Grays Harbor County

A. J. Tetzlaff began his career as an Aberdeen bridge operator in 1921, tending a swing bridge located just upstream from the mouth of the Chehalis River. Before long, the city purchased another pivot-style structure to replace the aging swing span. When the West Bridge was installed in 1925, Tetzlaff blew the whistle to announce its opening. Thirty years later, his whistle signaled the bridge's last opening. By Tetzlaff's own estimation, he had opened the bridge at least 45,000 times (more than four times daily), mostly for ships passing between Grays Harbor and a log yard and lumber mill situated a short distance upstream. But his career as a bridge operator was not over. In 1955, Tetzlaff assumed duties as the day tender on the newest bridge to be located here.

Bridge engineer Willis B. Horn of the State Department of Highways designed the new movable structure for maximum navigational clearance. When completed, the double-leaf bascule span was the longest of its type in the state (at 271 feet). Today it is one of the eight longest bascule spans in North America. The bridge is particularly noteworthy for its pivoted counterweight system that maintains the bascule leaves' balance when open. The structure also has unusual counterweight pits. Because tidal ranges at this location are more than ten feet, the pits are designed to flood at high tide, thereby dampening the buoyancy effect that otherwise would lift up the piers.

The MacRae Brothers of Seattle built the bascule spans, main piers, and steel-girder approach spans, while the Manson Construction and Engineering Company, also of Seattle, erected the reinforced-concrete approach spans. Total cost of construction was $5 million.[9]

Agate Pass Bridge, Kitsap County

"A glorious mishmash of metal," said the *Bremerton Sun* of the Agate Pass Bridge in recognition of its fiftieth anniversary in October 2000. To engineers, it is an outstanding example of a modern steel cantilever truss bridge, and remarkable for its low-alloy steel known as "ASTM A 242."

However one might view this structure, the Agate Pass Bridge is the only link between Bainbridge Island and the Kitsap Peninsula. Its completion changed life there quickly and forever. Soon, residents witnessed the opening of the island's first supermarket and bank, and year-round real estate offices. Islanders accustomed to relative isolation suddenly found themselves living in "rural suburbia," although for Seattle commuters, access to Bainbridge Island still required a ferry ride to and from Elliott Bay. The bridge, of course, reduced travel time between Bainbridge Island and communities on the north Kitsap Peninsula, and eliminated ferry service at nearby tiny Suquamish and Indianola.

Built for about $1.6 million by the Manson Construction and Engineering Company of Seattle, under the

"A glorious mishmash of metal"—the Agate Pass Bridge, connecting Bainbridge Island and the Kitsap Peninsula.
Washington State Archives, WSDOT Records

supervision of Bridge Engineer George Stevens of the Highway Department, the 1,229-foot bridge opened in 1950. Its 540-foot main section is the second longest main span of any steel bridge built in Washington in the 1941–50 period (the Columbia River Bridge at Kettle Falls was longer, at 600 feet). The bridge's riveted steel portion totals 1,020 feet in length, and is flanked by two concrete T-beam approach spans. For navigation purposes, the bridge has a vertical clearance of 75 feet above high tide.

While some Bainbridge Island residents may have had misgivings about the new bridge, it represented the fulfillment of a quarter century of hopes and dreams for a span across Agate Passage. The island's once thriving lumber and shipbuilding industries had long passed, and berry farms were the mainstay of the local economy; thus residents looked to the bridge as a key to better jobs, goods, and services.

The Agate Pass Bridge was immediately popular. For its first year of service, tolls (thirty-five cents per car) were collected on the Kitsap Peninsula side. In the first four days of operation, more than 10,000 motorists paid $1,600 in revenue. A year later, in September 1951, the state legislature retired the outstanding debt and removed the toll plaza. The bridge that some islanders thought would cause "the end of the world" proved to be a pathway to the future.[10]

Portage Canal Bridge, Jefferson County

The Portage Canal Bridge is one of Homer Hadley's most unusual contributions in Washington's bridge building history. Designed in 1950, constructed in 1951, and opened to traffic in January 1952, this was the second bridge built in the United States with a steel box-girder suspended span, and one of the first of its type anywhere in the world. Soaring 70 feet above mean sea level, the bridge crosses Portage Canal, just east of Hadlock in eastern Jefferson County. Via State Route 116, the bridge connects the Quimper Peninsula to the west with Indian and Marrowstone islands to the east.

The 670-foot-long bridge consists of five spans. Its 250-foot main span is flanked by 170-foot spans. Beyond the latter secondary spans, 40-foot cantilever spans project to connect with the approaches. A pair of spread steel box girders support a reinforced-concrete roadway deck suspended on 55-foot sections cantilevered from the adjacent piers. The cross section of these cantilevers (and for the remainder of the bridge) is a two-cell, reinforced-concrete box girder.

In 1951, Homer Hadley pioneered the use of steel box girders in the Portage Canal Bridge.

Craig Holstine, WSDOT

The use of lighter steel box girders allowed Hadley to extend the span's length. This resulted in three main spans, rather than five spans, eliminating the need for, and cost of, two additional piers. Hadley took a pioneering step with this design, demonstrating the potential use of steel box girders for longer-span structures. Steel box-girder bridges now have wide acceptance throughout the world, due in part to Hadley's efforts. Their popularity derives from aesthetics, the efficient use of structural steel, a better resistance to corrosion, and suitability for curved alignment (due to their torsional stiffness).

In southeast Washington in 1979, WSDOT constructed the five-span Red Wolf steel box-girder bridge over the Snake River at Clarkston. That bridge, the current

Washington record holder, includes a maximum span of 420 feet. In other states, steel box-girder spans as long as 750 feet now carry traffic, due in large part to precedents set in the Pacific Northwest.[11]

Port Washington Narrows (Warren Avenue) Bridge, Bremerton, Kitsap County

With war against the Axis powers looming in the summer of 1941, officials realized that Bremerton, home of the bustling Puget Sound Naval Shipyard, was rapidly outgrowing its infrastructure. Indeed, between 1941 and 1942, the city's population would explode from 30,000 to more than 100,000. The aged and narrow Manette Bridge, built in 1930 across the city's Port Washington Narrows, was barely adequate for the increasing traffic between the Manette Peninsula to the north and downtown Bremerton to the south. The Kitsap County Defense Bridge Committee lobbied the government, but support for a new bridge failed to materialize—and would not, for another fifteen years.

Finally, after issuance of more than $5 million in bonds, the Washington Toll Bridge Authority contracted with the Tudor Engineering Company of San Francisco, California, to design a bridge for the Warren Avenue crossing. In late November 1958, the Peter Kiewit Sons Company of Vancouver, Washington, completed the structure, which operated as a toll bridge until October 1972.

By the mid-1950s, engineers had learned that steel plate girders could span distances previously requiring steel trusses. Unlike trusses, plate-girder bridges can be widened, enhancing their attractiveness to bridge builders anticipating future needs for increasing traffic capacities. The Port Washington Narrows Bridge represents the fulfillment of this design promise.

Bremerton's Port Washington Narrows Bridge on Warren Avenue.

Craig Holstine, WSDOT

Its main 250-foot span is exceeded in Washington only by the 1955 Wenatchee Avenue Bridge over the Wenatchee River for steel plate girder length. When built, Bremerton's three-span, 606-foot bridge was the longest continuous, riveted-steel plate girder bridge in the state. Its unusual design includes deeper girder depth near the main piers, which enables the girders to withstand greater stresses, and enhances the spans' aesthetic appearance.

Today, the Port Washington Narrows Bridge serves Warren Avenue on State Route 303. As the main connection between Bremerton's downtown area and the communities to the north, the bridge continues to play a vital role in the city's development.[12]

NOTES

1. Lisa Soderberg, "Elwha River Bridge," HAER Inventory, 1979; Gene H. Unger, P.E., Inspection Report, July 3, 1992, WSDOT Bridge Preservation Office, Olympia; Clallam County Engineer's response to WSDOT Historic Bridge Inventory correspondence from Craig Holstine, 1993; Robert H. Krier and Craig Holstine, "An Assessment of the Current Status and Condition of Bridges and Tunnels in Washington State Listed in the NRHP," Archaeological and Historical Services Short Report DOT93-10, Eastern Washington University, Cheney, August 1993, 10.
2. "Composite Trestle-Type Bridges on 100-Ft., Two-Column Bents," *Engineering News-Record,* August 6, 1936, 197–98.
3. Lisa Soderberg, "Goldsborough Creek Bridge," HAER Inventory, 1979.
4. Wm. Michael Lawrence, "South Hamma Hamma Bridge," HAER No. WA-96, 1993, and "North Hamma Hamma Bridge," HAER No. WA-97, 1993; Lisa Soderberg, "North Hamma Hamma River; South Hamma Hamma River Bridges," HAER Inventory, 1979.
5. Lisa Soderberg, "Duckabush River Bridge," HAER Inventory, 1979.
6. Wm. Michael Lawrence, "Purdy Bridge," HAER No. WA-101, 1993; Lisa Soderberg, "Purdy Bridge," HAER Inventory, 1979.
7. Wm. Michael Lawrence, "Wishkah River Bridge," HAER No. WA-92, 1993.
8. Lisa Soderberg, "Hoquiam River Bridge," HAER Inventory, 1979; Wm. Michael Lawrence, "Hoquiam River Bridge," HAER No. WA-93, 1993.
9. Oscar R. "Bob" George, "Chehalis River Bridge," NRHP Nomination, 2001.
10. Robert H. Krier, J. Byron Barber, Robin Bruce, and Craig Holstine, "Agate Pass Bridge," National Register of Historic Places Nomination, 1991; Todd Westbrook, "Spanning Half a Century; Agate Pass Bridge Celebrates its 50th Birthday Today," *Bremerton Sun,* October 16, 2000; Douglas Crist, "The Span of Time; How Bainbridge Got Its Bridge," in *Bainbridge Island Almanac, 2002* (Bainbridge Island: Sound Publishing, 2002), 11–17; Glenn Hartmann, Western Shore Heritage Services, Bainbridge Island, personal communication, 2005.
11. Oscar R. "Bob" George, "Portage Canal Bridge," NRHP Nomination, 2001.
12. Oscar R. "Bob" George, "Port Washington Narrows (Warren Avenue) Bridge," NRHP Nomination, 2001.

Washington's Unique Historic Bridge Legacy

Eric DeLony, Retired Chief, Historic American Engineering Record (HAER), National Park Service

In 1993, we were invited by Bernie Chaplin and Elizabeth Robbins of the Washington State Department of Transportation (WSDOT) to document a selection of the Evergreen State's historic bridges. This was part of a HAER initiative begun in 1975 to look at the historic bridges of the United States.[1]

The Cispus Valley Bridge under construction by the CCC, Lewis County, 1939. Donald Woods, photographer.

Gifford Pinchot National Forest

Portal and deck view of the Cispus Valley Bridge in 1993. Flooding destroyed the structure in January 2002.

Jet Lowe, HAER

HAER is the Historic American Engineering Record, a program the National Park Service created to compile a record of drawings, photographs, and histories of America's engineering, industrial, and technological heritage. Documentation consisting of measured and interpretive drawings, large-format photographs, and written histories is created primarily by student architects, landscape architects, engineers, and historians hired to work on field projects during the summer.

The documentation becomes part of the national collection at the Library of Congress. Along with the Park Service, which administers the program, and the Library of Congress, where the collection is maintained, the American Society of Civil Engineers participated in HAER's creation and continues as a partner advising the program through its national membership.[2]

Our concern for historic bridges stemmed from the collapse of the Silver Bridge into the Ohio River in 1967, one of the worst bridge disasters of the twentieth century, and the subsequent Congressional action to rid the nation of structurally deficient and functionally obsolete bridges.[3] The trouble with this well-intentioned legislation was that it threatened to remove all evidence of America's bridge-building art.

HAER began its historic bridge initiative by working with the states to complete statewide historic bridge inventories, cosponsoring such a project in Washington in 1979.[4] By the mid-1980s, we began working with state transportation departments that had completed inventories to document a selection of their more significant spans. A team was fielded in Washington during the summer of 1993 in cooperation with WSDOT and the state historic preservation office.[5]

Due in part to these efforts, historic bridges now are beginning to be appreciated by the public, modern highway engineers, and planners. No one can argue that a selection of our very best historic bridges, including representative examples of the more common types, should not be saved if possible. Out of this work has evolved not only the preservation of historic bridges, but also the beginning of contemporary bridge scholarship, of which this book is an excellent example.

The Wandipore Bridge built in the Himalayas, ca. 1643, was a timber cantilever bridge with a suspended span.

Thomas Pope, *Treatise on Bridge Architecture* (1811).

Notable Bridges

I would like to highlight some of the Evergreen State's historic bridge types, placing them in a national context, and conclude by saying something about bridge preservation. When we were selecting bridges to be recorded as part of the documentation project, I remember being struck by the number of cantilevers in Washington. Cantilever bridges are well suited for crossing the state's many deep canyons carved by the Columbia and other rivers. At least fifteen remain in Washington, and there may be more of them here than in any other state. We documented seven in 1993.[6]

If you hold your arm straight out from your shoulder, it is acting as a cantilever. The form originated in the Far East with the Shogun's Bridge, dating from the fourth century AD, that still spans eighty-four feet over the Daiya-gawa River in Nikko, Japan. Another ancient example is the Wandipore Bridge (ca. 1643) high in the Himalayan mountains in Bhutan, a cantilever of layered timbers projecting forty feet and carrying a simple timber platform—the suspended span. It was illustrated in Thomas Pope's *Treatise on Bridge Architecture*, the first American book on bridges, published in 1811.

The book was a summary of world bridge building and featured Pope's own "Flying Pendant Lever Bridge." Though never built, Pope proposed to span the Hudson River with a flying pendant of 3,000 feet and the East River with a span of 1,800 feet. This was the cantilever's first introduction in the United States.

The cantilever was not practical and did not achieve widespread use until the structural behavior of trusses was better understood half a century later. These mathematical issues were resolved by a German engineer, Heinrich Gerber, who built the Hassfurt Bridge over the River Main in Germany in 1867 with a central span of 124 feet—the first modern cantilever.

Pope's design for a "Flying Pendant Lever Bridge" over the Hudson River.

Thomas Pope, *Treatise on Bridge Architecture* (1811).

Ten years later, American engineer C. Shaler Smith built the world's longest cantilever for the Cincinnati Southern Railroad over the 275-foot-deep Kentucky River gorge. It had three spans of 375 feet each. The American Society of Civil Engineers selected the bridge for the 1878 Paris Exposition as one of the prime examples of American bridge ingenuity.

The equivalent engineering definition of the extended-arm analogy is that a cantilever is a continuous girder with hinges at the points of zero moments (the extended-arm theory is much easier to understand). The form was statically determinant, which meant that it was easy to calculate and the members did not have the inherent deficiency of the continuous beam or girder developing indiscernible internal stresses and possibly failing should one of the piers or abutments subside. Unstable soil conditions plagued foundation, pier, and abutment design, so the ability of a bridge's superstructure to adjust should one of the piers or abutments sink was a significant design breakthrough.

The form's greatest advantage, however, is its ability to be built without falsework, the temporary wood scaffolding needed to support a bridge while under construction. The cantilever was the ideal solution for deep, broad river gorges where it was difficult if not impossible to erect falsework. Washington proved to be the perfect place for cantilever bridges.

Another notable structure type we find is the concrete arch. Though every state has concrete arch bridges, the Monroe Street Bridge (1911) spanning the Spokane River on the eastern edge of the state is an unusual example. It is one of the few mass-poured or plain concrete arch bridges in America where no steel reinforcing was used in the main central arch span. Plain concrete arches are pure compression structures and enjoyed surprisingly wide application in the first decade of the twentieth century—the dawn of the age of concrete in America. Around a dozen were constructed exceeding one hundred feet.

In the early twentieth century, when concrete construction commenced, engineers were not aware of the possibilities of combining steel reinforcing with concrete. Instead, they viewed concrete as a substitute for stone, providing design and material economies and a particular solid, classic, stone-arch look. Consequently, early twentieth century American concrete bridges were designed in a Roman style—massive, cumbersome, stone-like structures, instead of the light, elegant, concrete arches pioneered by European engineers like Robert Maillart, whose bridges in the Swiss Alps are world renowned. The Monroe Street

Bridge is a good example of an early twentieth century American concrete type, and for a brief period following construction, it was the largest concrete arch in America.[7]

Today the Monroe Street Bridge is undergoing rehabilitation rather than replacement because, after nearly one hundred years, the concrete is still sound, and rehabilitation avoids the expense and difficulty of removing the massive arch ribs and building new piers and abutments in a precarious location. All of the ornate elements, such as the balustrades, and the covered wagon and cast-concrete bison skull motifs incorporated in the pavilions of this landmark structure, will be restored or replicated. The rehabilitation is designed to enable widening the bridge to six lanes should the city ever need them in the future.

Exposed-Timber Frame Bridges, Including Railroad Bridges

"Many of Washington's first successful highway bridge engineers began their careers in railroad construction" is the lead sentence in Chapter 4. This is significant because railways pioneered bridge engineering historically. One cannot get a comprehensive understanding of bridge building without looking at railroad bridges.

The Doty Bridge (1926) was a covered timber Howe truss crossing the Chehalis River in Lewis County. This Milwaukee Railroad bridge was listed in the NRHP until it burned in the 1980s.

Office of Archaeology and Historic Preservation

Though railway bridges are not a subject of this volume, they need to be mentioned because Washington was the first and, to my knowledge, remains the only state to have included railroad bridges in its first statewide historic bridge survey. The reason that so few states have included rail bridges is that the surveys are sponsored by highway departments, whose primary responsibility is vehicular spans. Most states have provided only a cursory review of their rail bridges.

Another reason railway bridges have not been included is that railroads, other than Amtrak, are privately owned, thus not falling under any government purview to survey or protect their bridges. Consequently, other than for Washington, we have no comprehensive listing of railroad bridges nationally. In my view, this is a significant shortcoming of our national bridge scholarship.

The CCC erected this timber Howe truss on the North Fork Sultan River, Snohomish County, in 1937. Listed in the NRHP, it was lost to flooding in the late 1980s.

Office of Archaeology and Historic Preservation

Washington has examples of nearly all railroad bridge types, but the most impressive was a collection of exposed-timber frame bridges—more than any other state. The Milwaukee Road (a.k.a. Chicago, Milwaukee, St. Paul, and Pacific Railroad), which traversed the state, was a big advocate of timber spans, and, given the abundant local supply, employed wood for a number of its bridges. According to Lisa Soderberg's 1979–80 survey and David Plowden's book, *Bridges: The Spans of North America* (1974), several notable exposed-timber trusses survived in

The Milwaukee Railway Bridge on the Dungeness River west of Sequim is one of the two exposed-timber bridges remaining in Washington.

Steve Hauff Collection, Port Angeles

INSET:
The same bridge under construction, ca. 1914.

Steve Hauf Collection, Port Angeles

the 1970s. (This included not just railroad bridges, but vehicular timber trusses as well).

Among them was the world's longest covered railroad bridge, the 450 foot-span Skykomish River Bridge (1932) at Sultan, east of Monroe. According to local sources, the bridge was replaced by a timber trestle in 1969. Today, only two exposed-timber truss bridges remain in Washington—the Dungeness River Bridge (1930) near Sequim, a classic Howe truss of timber compression members and metal tension rods and fittings that is part of a hiking trail; and the Goodyear Bridge over the Clearwater River, also on the Olympic Peninsula. Named for a State Commissioner of Public Lands, this Pratt truss was reportedly built by the Civilian Conservation Corps in the 1930s. It is now deteriorated and closed to traffic.[8]

Wooden bridge historian Joseph Conwill of Rangeley, Maine, pointed out to me that about a dozen exposed-timber trusses remain in the lower forty-eight states, while eighty-five are known to survive in Canada. Exposed-timber truss bridges are important not only for their rarity, but

Clearwater River (Goodyear) Bridge on the Olympic Peninsula, deck view. Built by the CCC in the 1930s or early 1940s, it may be the only wood truss vehicular bridge remaining in Washington.

Eric Carlsen, Olympic Region Department of Natural Resources, Forks

The Harlequin Bridge over the Stehekin River in Chelan County, a Baltimore-type wood truss unique in Washington. Built in 1948 by the U.S. Forest Service, it was removed ca. 2000 by the National Park Service. Photo 1989.

Jet Lowe, HAER

because they represent one of the earliest types in American bridge technology. Before Timothy Palmer enclosed the Permanent Bridge (1805) over the Schuylkill in Philadelphia—our first covered bridge—most bridges were exposed-wood frames. Exposed-timber bridges are extremely rare in the United States and the rest of the world.

We are more familiar with covered wood bridges. America has more covered bridges than any other country—about 750, according to Conwill. Washington has three, one of which is unique and worth highlighting.[9] This is the Harpole (Manning-Rye) Bridge (ca. 1922, 1928) in Whitman County, a massive, heavy timber Howe

The Howe trusses of the Harpole Bridge in Whitman County have been covered since the late 1920s.

Eric DeLony

Portal view of the Harpole Bridge showing the covered trusses.

Steve Hauff

truss like the exposed-timber bridges, but it is clad with vertical siding to protect the timbers. Only one other like the Harpole is known to exist: the Ahnola River Bridge over the Similkameen River west of Keremos, British Columbia. The Harpole Bridge is privately owned, and there is no law protecting privately owned historic properties. The timbers are so massive that it will be some time before they deteriorate to the point of failure, but the last time I saw the structure more than a decade ago, there was clear evidence of decay. This bridge in my opinion is so important that it warrants special attention by the community, state, and citizens.

Bridge Innovation

In Chapter 4, the authors have written short biographies of Washington's more prominent bridge engineers. From these vignettes, it is indisputable that Washington's bridge engineers have explored, experimented, and built some of the most innovative structures in the world, especially during the twentieth century. I doubt whether many bridge scholars will recognize their names, but when you look at their achievements, they are remarkable.

The bridges that first come to mind are the four concrete pontoon bridges on Lake Washington in King County and on Hood Canal west of Seattle. The concept of a concrete pontoon bridge belongs to Homer M. Hadley. It is counterintuitive to think of a buoyant concrete bridge, but due to Lake Washington's extreme depth (200 feet), a floating bridge was the only economical and practical solution. It would be difficult to sink piers to that depth. Though portions of two bridges sank in storms, Washington's floating concrete bridges have served admirably and advanced the art of floating bridge technology considerably.

The supervisory engineer for the Seattle Department of Transportation, Joanne McGovern, sent me information on Seattle's collection of movable spans, mostly bascules, and one that is a swing span, the Southwest Spokane Street Bridge (1991) over the Duwamish River, the world's only hydraulically operated double-leaf concrete swing bridge. Though not yet historic, this bridge illustrates the continuation of innovative, imaginative, and creative bridge design in Washington.

The swing span is ideal in the tight, precarious location. The bridge was designed to carry a major low-level road in an intensely busy and trafficked area, both vehicular and maritime. The juxtaposition of the low-level crossing, an elevated high-level through-bridge, and the navigable channel dictated a swing span. So as not to impede the shipping channel, the swing piers are located on either shore rather than the middle of the river, which was standard for swing spans. (Most swing spans have been replaced because they impair navigation.) This resulted in paired leafs of 413 feet, each weighing 7,200 tons, providing an eventual channel width of 250 feet. The existing channel is 150 feet wide with a vertical clearance of fifty-five feet, thus accommodating sixty percent of the river traffic without opening. When larger ships require passage, it takes twelve minutes to swing the leafs open.

The Southwest Spokane Street Swing Bridge.
Seattle Municipal Archives

The bridge has won many distinguished engineering awards, including a Design for Transportation Award from the National Endowment for the Arts, attesting to the bridge's aesthetics and innovative design. In addition to advocating the retention of historic bridges, "pontists" (historic bridge engineers, scholars, and enthusiasts) advocate recognition for outstanding, innovative, and aesthetic bridge designs.

"Galloping Gertie"

How can one discuss innovative bridge design without including the infamous "Galloping Gertie," the first Tacoma Narrows Bridge (1940)? This was the quintessential statement of world suspension bridge design at mid-twentieth century. Here was the climax of a new way of looking at suspension bridges that was introduced by Austrian engineer Josef Melan in 1888. He called his revolutionary bridge design "deflection theory," a new structural model that first was executed in the United States by Leon Moisseiff on the Manhattan Bridge (1909).

Deflection theory recognizes that the dead load of the cables, stiffening structure, and deck significantly moderates distortion under live load—wind, snow, pedestrians, and vehicles. The immediate consequence of this new way of visualizing the behavior of suspension bridges was the realization that the stiffening truss, common to most bridges, had little to do with reducing deflections on large suspension spans. Deflection theory suggested that the truss's bulk, depth, and weight could be substantially reduced without impairing the structure's stiffness. This allowed for drastic reductions of the deck-stiffening element, resulting in substantial cost savings. But equally important, a sleek, elegant, razor-thin look achieved its visual apogee in the first Tacoma Narrows Bridge.

Thus, the Tacoma bridge's design was derived from deflection theory, and Leon Moisseiff, the bridge's engineer, was the indisputable leading theoretician of suspension bridge computational analysis. It was not only the third-longest suspension bridge in the world, but also the ultimate fulfillment of thirty years of experimentation to make the decks of suspension bridges ultra-thin. This, along with sinuous cables and elegant, tapered towers, resulted in a modern look. The stiffening girders were only eight feet deep, incredibly shallow for a suspended span of 2,800 feet.

Only four months after opening, the girders of Moisseiff's Tacoma bridge lay at the bottom of The Narrows, brought down by a steady forty miles per hour wind, one-third less than the structure's design pressure of 120 miles per hour. A baffled Moisseiff was quoted to be "completely at a loss to explain the collapse." Though no lives were lost, this was the twentieth century's most spectacular bridge failure.

Filmed in motion pictures, the dramatic Tacoma Narrows Bridge failure is shown to civil engineering students throughout the world. The collapse took with it a generation of suspension bridge engineering theory, but out of this failure came a complete reevaluation of suspension bridge design and the new science of bridge aerodynamics.

During the ensuing investigation, engineers drew from theoretical concepts such as lift, eddy formation, vortex shedding, negative dampening, and flutter, which were developed by aeronautical engineers in designing airplanes, but were applicable to the design of suspension bridges.[10] The paired suspension bridge presently under construction that will double the capacity of this crossing benefits from the demise of its predecessor.

Devine Intervention

Let us move ahead nearly four decades to the late 1970s and discuss another important event resulting from completion of another suspension structure. This was the Ed Hendler Bridge between Pasco and Kennewick, the first major cable-stayed suspension bridge in the lower forty-eight states.[11] As the new bridge was being completed, a local resident, Virginia Devine, brought suit against Drew Lewis, President Ronald Reagan's Secretary of Transportation, and the cities of Pasco and Kennewick. She posted $75,000 of her own money to secure a bond that extended a court injunction prohibiting demolition of the older Pasco-Kennewick Bridge (1922) that the new cable-stayed bridge was replacing. Also known as the Golden Rivet Bridge—commemorating its status as the last link in one of the nation's first transcontinental motor routes, the Yellowstone Trail linking Plymouth Rock and Puget Sound—it was the second cantilever highway bridge built over the Columbia River in Washington.

The issue was whether to keep the old bridge as a complementary crossing for nonmotorized traffic after the new Ed Hendler Bridge opened. Hendler was the Pasco mayor who led the fight to have the 1922 cantilever removed. The Pasco-Kennewick Bridge became nationally notorious as the subject of controversial legal attempts by the two cities to demolish a historic bridge, and a citizen effort to preserve the older bridge as a pedestrian crossing. Thus, Pasco-Kennewick was the center of one of the classic preservation battles of the early 1980s.

The suit, brought forth by the Benton Franklin Riverfront Trailway and Bridge Committee, under Devine's leadership, eventually ended up in the Ninth Circuit Court of Appeals in San Francisco. On March 15, 1983, the court ruled in the plaintiff's favor, finding that the Secretary of Transportation did not in good faith carry out his duty to seriously consider alternatives to demolishing this historic span.[12]

The case put the Federal Highway Administration on notice that it must seriously address alternatives to demolishing historic bridges. Though the bridge eventually was torn down, it was an important test case of Section 4(f) of the National Transportation Act of 1966, and one of the first cases where the National Trust for Historic Preservation intervened to help defend a historic bridge. The Pasco-Kennewick controversy held significant implications for future bridge preservation efforts nationwide. This is particularly poignant now, since Section 4(f) is scheduled to be debated by the 109th Congress.[13]

The Murray Morgan Bridge over the Thea Foss Waterway in Tacoma, still in service, is the focus of ongoing preservation efforts.

Eric DeLony

City Waterway (Murray Morgan, Eleventh Street) Bridge (1911)

I wish to end this Afterword with a discussion of Tacoma's City Waterway Bridge, another structure HAER recorded in 1993. Though at one time scheduled for demolition in 2006, there has been strong local support to save the structure. "We would like the state to explore every approach" to rehabilitating the bridge, Tacoma Mayor Bill Baarsma told WSDOT project engineer Ron Landon several years ago.

Since then a coalition of local leaders, preservationists, and the business community under the leadership of Jim Hoard, president of Save Our Bridge, and State Representtative Dennis Flannigan, who legislated an approximately $25 million package through the 2004 legislature, has been working to save the bridge. Supplementing the efforts of Hoard and Flannigan is David Allen, who is credited with helping revitalize the downtown, including such projects as the University of Washington-Tacoma campus, Washington State History Museum, Tacoma Actors Guild theater district, and the Thea Foss Waterway. As an advisor to the City Manager and City Council, Allen has been working to find potential sources of supplementary fundings for the restoration of the bridge should state monies prove inadequate. State officials have identified funds that could be applied towards the estimated $80 million cost of rehabilitation (or $130 million for replacement). David Evans and Associates Inc., of Salem, Oregon, arrived at these figures. The firm's experience includes engineering services in support of the recently completed restoration of the Hawthorne Bridge in Portland, Oregon.

When visiting with Dave Nicandri, Washington State's historical society director, several years ago, I remember thinking how well the bridge fit into the Union Station and Pacific Avenue redevelopment. He was showing me the recently completed state museum near the renovated Union Station railroad depot. The museum would serve as an important anchor to this redeveloped historic district. I remember saying to Nicandri, "Imagine the trusses, towers, and overhead structure painted aluminum silver and lit up at night." Night lighting the bridge has proved to be spectacular.

In addition to its potential landmark values, the bridge is historically significant. It is not the oldest vertical lift in the West. That accolade belongs to the Hawthorne Bridge in Portland, Oregon, built in 1910 and designed by J .A. L. Waddell, the same engineer who designed the City Waterway Bridge. Waddell was one of the foremost consulting engineers practicing in the United States at the turn of the century. In addition to his numerous patents for bridges, and his extensive bridge engineering practice, he was recognized abroad as one of America's most prominent bridge engineers, authoring a classic engineering text, *Bridge Engineering* (1916), that is the bible of contemporary bridge scholars.

Where Have All the Bridges Gone?

This is a good question, considering that Washington has as rich and varied a collection of bridges as any state, as

Reportedly built in the early 1950s by a logging company, this bridge on the upper Cowlitz River in Lewis County collapsed under the weight of a logging vehicle, ca. 1981. Photo by David Harvey, 1979.

Office of Archaeology and Historic Preservation

one will know after reading this volume. I do not intend to belabor the losses, but "lost" bridges are significant, as pointed out in the previous paragraphs. Recent statistics suggest that half, if not more, of our nation's historic bridges were lost in the last twenty years—two decades in which transportation and preservation consciousness reached unprecedented levels. This is an alarming and sobering statistic. While we are not quite at the threshold of saving the few surviving examples, we are fast approaching that point.

Though we have a national historic bridge program, Title 23, Section 144(o), enacted by Congress in 1987, it is not working well enough to stem the loss of historic bridges. To address this shortcoming, the Federal Highway Administration, the American Association of State Highway and Transportation Officials (called "AASHTO"), the SRI Foundation, the National Trust for Historic Preservation, and HAER (within the National Park Service) organized an invitation-only workshop of experts to articulate and define the issues confronting historic bridges. The workshop was held in Washington, D.C., December 3–4, 2003.

In the spirit of stewardship, streamlining, and sound environmental and historic bridge management, the workshop's goal was to provide federal and state transportation agencies, Congress, and the interested public with recommendations and solutions on how to preserve this heritage at risk. Specifically, the workshop articulated and defined efficient and economical strategies for historic bridge preservation and management.[14]

Federal surface transportation legislation enacted in 1991 inspired an important transformation within the transportation community, broadening its mission from the traditional task of providing a safe and efficient highway system, to acknowledging that these activities play a critical role in preserving our nation's natural and cultural heritage.[15] This volume on the Evergreen State's historic bridges is directly related to that visionary legislation. Hopefully, it will inspire Washingtonians to realize the value of historic bridges, get out there to appreciate them, and to advocate for their retention when threatened.

The North Fourth and Dock Street Bridge in Tacoma was listed in the NRHP until its removal.

Office of Archaeology and Historic Preservation

NOTES

1. The HAER documentation project resulted in an exhibit, entitled "Trunnions, Bascules and Cantilevers," curated by Susan J. Torntore and Robert Hadlow at the Washington State Capitol Museum, Olympia, that was organized the year following the 1993 HAER recording project. Though I was not able to attend the opening, it was scheduled when the state legislature was in session so that members from communities in which bridges were recorded could attend.
2. Support of the program was expanded in 1987 through a protocol signed by the other "founding" engineering societies—the American Society of Mechanical Engineers, the Institute of Electrical and Electronics Engineers, the American Institute of Chemical Engineers, and the American Institute of Mining, Metallurgical, and Petroleum Engineers. Today, HAER has compiled information on more than 7,000 sites, structures, and objects, including a thousand bridges: www.cr.nps.gov/habshaer.
3. For information on initial efforts to recognize the value of America's historic bridges, see "The 19th Century Iron Bridge: 32,000 Bridges Determined Structurally Deficient or Functionally Obsolete," *Society for Industrial Archeology Newsletter* 4, no. 6 (November 1975).

See also, "HAER's Historic Bridge Program," *IA: The Journal of the Society for Industrial Archeology* 15, no. 2, 1989, 57–71.

4. The same year that HAER cosponsored Lisa Soderberg's Washington bridge survey, we fielded a team in Tacoma that looked at the adaptive reuse and rehabilitation potential of the Burlington Northern's Union Passenger Station and the surrounding warehouse and industrial district along Pacific Avenue. In 1987, HAER teams, cosponsored by the National Park Service's Pacific Northwest Regional Office, documented hydroelectric facilities on the Nooksack River and sites in the North Cascades National Park. Next, 1989 saw the beginning of a two-year project documenting Seattle City Light's dams and power plants on the Skagit River and Puget Power's White River Hydroelectric Plant. In 1992, we fielded a team in Mount Rainier National Park documenting park roads and bridges, while 1993 was the year of the Washington Historic Bridges Recording Project. In 1995, we documented the dams and hydroelectric plants on the Elwha River and at Glines Canyon in Olympic National Park, and in 1997, we studied the Grand Coulee Dam and Power Plant on the Columbia River. Washington's engineering and industrial heritage is well represented in the national collection at the Library of Congress. The HAER collection is one of the first to be digitized for the National Digital Library. HAER documentation on Washington sites can be found online: memory.loc.gov/ammem/collections/habs_haer/.
5. In 1986, Ohio was the first state to document its most outstanding bridges, followed by New York and Wisconsin in 1987, Arkansas in 1988, Massachusetts and Oregon in 1990, Washington in 1993, Iowa in 1995, Texas and Iowa again in 1996, Pennsylvania in 1997 and 1998, Portland's Willamette River bridges and the Chicago River bascules in 1999, and Merrill Butler's City Beautiful Los Angeles River bridging system and Texas again in 2000.
6. These cantilever bridges are located at Aurora Avenue (Seattle), Deception Pass, Grand Coulee, Kettle Falls, Longview, Lyon's Ferry, and Fort Spokane.
7. The plain concrete bridges of more than one hundred feet that I am aware of include the Illinois Central's Big Muddy River Bridge (1902) at Grand Tower, Illinois; the Anthony Kill Bridge (ca. 1903) in Mechanicville, New York; the Union Pacific's Santa Ana River Bridge (1904) in Riverside, California; the Vermilion River Bridge (1905) in Danville, Illinois; the Rocky River Bridge (1906) in Cleveland, Ohio; the 16th Street Bridge (1906) over Piney Creek, and the Connecticut Avenue (Taft Memorial) Bridge (1907), both in Washington, D.C.; the Walnut Lane Bridge (1908) over Wissahickon Creek in Philadelphia; the DL&W's Delaware River Bridge (1908) in Portland, Pennsylvania; the Edmundson Avenue Bridge (1909) over Gwynns Falls in Baltimore; the Larimer Avenue Bridge (1909) in Pittsburgh; and the Monroe Street Bridge (1911), Spokane.
8. Besides the Skykomish River Bridge, exposed-timber truss bridges lost since the 1970s in Washington include the Little Sheep Creek/Red Mountain Railroad Bridge (1896) near Northport, Stevens County; the Winslow Railroad Bridge (1917) near Orin, Stevens County; the Nooksack River Bridge (1931) near Everson, Whatcom County; the McClure Bridge (1912, 1953), Whitman County; the North Fourth and Dock Street Bridge (date unknown), Tacoma, Pierce County; the North Fork Sultan River Bridge (1937), Snohomish County; the Cispus Valley Bridge (1939) over the Cispus River, and the Upper Cowlitz River Bridge (ca. 1950s), both Lewis County; the Company Creek (Harlequin) Bridge (1948) over the Stehekin River, Chelan County; and the Norman Bridge (1924, 1950) spanning the Snoqualmie River, King County.
9. Covered bridges have benefited from a modest federal funding program in recent years to rehabilitate selected examples, including the Pe Ell Bridge in Washington.
10. Thomas R. Winpenny, *Manhattan Bridge: The Troubled Story of a New York Monument*, foreword by Eric DeLony (Easton, PA: Canal History and Technology Press, National Canal Museum, 2004).
11. Though cable-stayed design dates from the Renaissance when Faustus Verantius published a prototype in *Machinae Novae* (1595), the form's modern popularity dates from post-World War II reconstruction in Germany. More than forty were built worldwide before one was completed in Alaska in 1973. Five years later, the Pasco-Kennewick bridge was the first built in the lower forty-eight states. Cable stays have become popular—every city seems to want one of these signature bridges to grace its threshold. The aesthetics of a light deck, slender tower, and uncluttered appearance achieved by closely spaced cables is the main reason for the cable-stay's appeal.
12. Nancy C. Shanahan, brief of *Amicus Curiae* in support of plaintiff-appellant, the Benton Franklin Riverfront Trailway and Bridge Committee, U.S. Court of Appeals for the Ninth Circuit, National Trust for Historic Preservation, Western Regional Office, San Francisco, 1981, 36.
13. Section 4(f) is the provision in the Department of Transportation Act that prohibits federal approval or funding of transportation projects that require "use" of any historic site unless, (1) there is "no feasible and prudent alternative to the project," and (2) the project includes "all possible planning to minimize harm."
14. One of the results of the workshop is a report published in June 2004 that can be accessed on the SRI Foundation Web site: www.srifoundation.org.
15. Intermodal Surface Transportation Efficiency Act of 1991 (ISTEA).

Washington State Historic Highway Bridges
National Register of Historic Places/ Historic American Engineering Record

Bridges Listed in (NR), Determined Eligible (NR/DE), or Nominated (NR/Nominated) for the National Register of Historic Places in Washington State. Bridges documented by, or for, the Historic American Engineering Record (HAER) are also identified. (As of June 2005.)[1]

Asotin County—

Grande Ronde River Bridge (1941) (#129/2)
Location: State Route 129, 13 miles southwest of Anatone
Type: riveted steel girder
Length: 283 feet
Builder: Henry Hagman
Designer: Washington State Department of Highways, R. W. Finke, Bridge Engineer
Ownership: WSDOT
Inventory: NR

Indian Timothy Memorial Bridge (1923) (#12/903)
Location: Alpowa Creek, off U.S. Route 12
Type: reinforced concrete through-arch
Length: 213 feet
Builder: Colonial Building Company
Designer: Washington State Department of Highways
Ownership: WSDOT
Inventory: NR/HAER

Benton County—

Benton City-Kiona Bridge (1957) (#225/1)
Location: State Route 225 crossing the Yakima River
Type: concrete tied-cantilever (precursor of cable-stayed)
Length: 400 feet
Builder: Everett McKellar, Chelan County
Designer: Homer M. Hadley
Ownership: WSDOT
Inventory: NR/Nominated

Benton County and Franklin County—

Pioneer Memorial "Blue" Bridge (1954) (#395/40)
Location: Route 397 crossing the Columbia River between Kennewick and Pasco
Type: steel tied-arch span with flanking Warren trusses
Length: 2,520 feet
Builder: P. J. Jarvis, Inc., Cascade Construction Company, Robert W. Austin Company, and United States Steel Corporation
Designer: Washington State Department of Highways, George Stevens
Ownership: WSDOT
Inventory: NR/Nominated

Benton County and Umatilla County, OR—

Columbia River Bridge at Umatilla (1955) (#82/280S)
Location: Southbound Interstate 82 crossing the Columbia River
Type: steel cantilevered Warren through truss
Length: 3,308 feet
Builder: Austin Construction Company, Cascade Construction Company, and American Bridge Company
Designer: Tudor Engineering Company
Ownership: WSDOT

Chelan County—

Wenatchee Avenue Northbound Bridge (1933) (#285/20E)
Location: State Route 285 crossing the Wenatchee River
Type: steel cantilever deck-arch
Length: 640 feet
Builder: Puget Sound Bridge and Dredging Company
Designer: Washington State Department of Highways
Ownership: WSDOT

Wenatchee Avenue Southbound Bridge (1955) (#285/20W)
Location: State Route 285 crossing the Wenatchee River
Type: steel plate girder
Length: 608 feet
Builder: Paul Jarvis
Designer: Washington State Department of Highways, William A. Bulley, Lead Designer
Owner: WSDOT
Inventory: NR/Nominated

West Monitor Bridge (1907) (#306)
Location: crossing the Wenatchee River one mile north of Monitor
Type: steel pin-connected Pratt truss
Length: 320 feet
Builder: Puget Sound Bridge and Dredging Company
Designer: unknown
Ownership: County
Inventory: NR

Chelan County and Douglas County—

Columbia River Bridge at Wenatchee (1908)
Location: Wenatchee and East Wenatchee on pedestrian/bicycle trail
Type: steel pin-connected cantilever
Length: 1,060 feet
Builder: Washington Bridge Company
Designer: unknown
Ownership: Local
Inventory: NR

Columbia River Bridge at Wenatchee (1950) (#285/10)
Location: Wenatchee and East Wenatchee on State Route 285
Type: steel cantilever tied arch
Length: 1208 feet
Builder: General Construction Company of Seattle, General Contractor
Designer: Washington State Department of Highways, R. W. Finke, Bridge Engineer
Ownership: WSDOT
Inventory: NR

Clallam County—

Elwha River Bridge (1913) (#30000 BR 1)
Location: Old Highway 12
Type: Warren deck truss
Length: 576 feet
Builder: Portland Bridge Company
Designer: unknown
Ownership: County
Inventory: NR

Tumwater Creek Bridge (1936) (#PT ANGL 2)
Location: 8th Street, Port Angeles
Type: wood trestle on two-column timber bents
Length: 755 feet
Height: 100 feet
Builder: Angeles Gravel and Supply Company
Designers: Conrad O. Mannes and A. M. Young, W.E.R.A.
Ownership: Local
Inventory: NR/DE

Valley Creek Bridge (1936) (#PT ANGL 1)
Location: 8th Street, Port Angeles
Type: wood trestle on two-column timber bents
Length: 755 feet
Height: 100 feet
Builder: Angeles Gravel and Supply Company
Designers: Conrad O. Mannes and A. M. Young, W.E.R.A.
Ownership: Local
Inventory: NR/DE

Clark County and Cowlitz County—

Yale (Lewis River) Bridge (1932) (#503/26)
Location: crossing the Lewis River
Type: steel suspension
Length: 532 feet
Builder: Gilpin Construction Company
Designer: Harold H. Gilbert
Ownership: WSDOT
Inventory: NR/HAER

Clark County and Multnomah County, OR—

Vancouver/Portland Columbia River Interstate 5 Bridge East (1916) (#5/1E)
Location: Interstate 5 crossing the Columbia River
Type: vertical lift
Length: 3,531-foot main channel span
Builder: Porter Brothers and the Pacific Bridge Company
Designer: Waddell and Harrington
Ownership: ODOT
Inventory: NR/HAER

Vancouver/Portland Columbia River Interstate 5 Bridge West (1958) (#5/1W)
Location: Interstate 5 crossing the Columbia River
Type: vertical lift
Length: 3,538 feet
Builder: Guy F. Atkinson Company of Portland
Designer: Oregon State Highway Department
Ownership: ODOT

Columbia County and Franklin County—

Snake River/Lyons Ferry Bridge (1927) (#261/125)
Location: originally crossing the Columbia River at Vantage; presently crossing the Snake River at Lyons Ferry on State Route 261
Type: steel cantilever truss
Length: 1,636 feet originally; now 2,040 feet
Builder: 1927 at original location; 1968 at present location, by J. W. Hardison Company, Munson Construction and Engineering Company, and the United States Steel Corporation's American Bridge Division
Designer: Washington State Department of Highways, George Stevens, lead engineer for rebuild
Ownership: WSDOT
Inventory: NR/HAER

Cowlitz County—

Jim Creek Bridge (1945) (#503/112)
Location: State Route 503 crossing Jim Creek
Type: open-spandrel concrete arch
Length: 308 feet
Builder: Macri Brothers and Company
Designer: Washington State Department of Highways, R. W. Finke, Bridge Engineer
Ownership: WSDOT
Inventory: NR

Modrow Bridge (1958) (#3535001)
Location: Modrow Road crossing the Kalama River
Type: steel open-spandrel rib deck-arch
Length: 200 feet
Builder: Hart Construction Company of Tacoma
Designer: Harry R. Powell
Ownership: County
Inventory: NR/Nominated

Cowlitz County and Columbia County, OR—

Longview (Lewis and Clark) Columbia River Bridge (1929) (#433/1)
Location: State Route 433 crossing the Columbia River
Type: steel cantilever span with 5 warren trusses and a suspended span
Length: 3,892 feet
Builder: J. H. Pomeroy and Company of Seattle
Designer: Strauss Engineering Corporation, Joseph B. Strauss, Engineer
Ownership: WSDOT
Inventory: NR/HAER

Douglas County—

Chief Joseph Dam Bridge (1958) (#26.5ENE)
Location: Pearl Hill Road crossing Foster Creek
Type: Howe wooden deck truss
Length: 297 feet
Builder: Cherf Brothers and Sandy Kay Builders
Designer: U.S. Army Corps of Engineers, Seattle District
Ownership: County
Inventory: NR/Nominated

Douglas County and Okanogan County—

Columbia River Bridge at Bridgeport (1952) (#17/401)
Location: State Route 17 crossing the Columbia River
Type: continuous riveted steel deck truss
Length: 1,150 feet
Builder: U.S. Army Corps of Engineers, Seattle District
Designer: U.S. Army Corps of Engineers, Seattle District
Ownership: WSDOT
Inventory: NR/HAER

Ferry County—

Curlew Bridge (1908) (#2)
Location: crossing the Kettle River
Type: steel Parker truss
Length: 213 feet
Builder: William Oliver of Spokane
Designer: unknown
Ownership: County
Inventory: NR

Ferry County and Stevens County—

Barstow Bridge (1947) (#224)
Location: crossing the Kettle River
Type: steel through Pratt truss
Length: 180 feet
Builder: purchased from the U.S. War Assets Administration, erected 1947
Designers: U.S. War Department, bridge design; W. T. Batcheller of Seattle, erection plan and piers design
Ownership: Stevens County
Inventory: NR

Columbia River Bridge at Kettle Falls (1941) (#395/545)
Location: U.S. Route 395 crossing the Columbia River
Type: steel cantilever through truss with a suspended span
Length: 1,266 feet
Builder: S. S. Mullen Inc. and the L. Romano Engineering Company
Designer: Washington State Department of Highways, R. W. Finke, Bridge Engineer
Ownership: WSDOT
Inventory: NR/HAER

Grays Harbor County—

Chehalis River Bridge (1955) (#101/115)
Location: U.S. Route 101 crossing the Chehalis River, Aberdeen
Type: simple trunnion bascule with double-leaf, Pratt deck trusses
Length: 2,638 feet
Builder: MacRae Brothers of Seattle and Manson Construction and Engineering Company of Seattle
Designers: Bridge Division of the Washington State Department of Highways: Willis B. Horn, lead designer for the bascule span; George Stevens, Chief Bridge Engineer
Ownership: WSDOT
Inventory: NR/Nominated

Hoquiam River Bridge (1928) (#101/125W)
Location: U.S. Route 101 crossing the Hoquiam River, Hoquiam
Type: double-leaf Strauss trunnion bascule
Length: 1,980 feet
Builder: Wallace Bridge and Structural Steel Company of Seattle
Designer: Washington State Department of Highways
Ownership: WSDOT
Inventory: NR/HAER

Wishkah River Bridge (1924) (#12/12N)
Location: crossing the Wishkah River, Aberdeen
Type: heel-trunnion bascule with single-leaf, Warren through truss
Length: 202 feet
Builder: Puget Sound Bridge and Dredging Company of Seattle
Designer: Strauss Bascule Bridge Company of Chicago
Ownership: WSDOT
Inventory: NR/DE/HAER

Jefferson County—

Duckabush River Bridge (1934) (#101/266)
Location: U.S. Route 101 crossing the Duckabush River
Type: concrete tied arch
Length: 168 feet
Builder: West Coast Construction Company
Designer: unknown
Ownership: WSDOT
Inventory: NR

Portage Canal Bridge (1951) (#116/5)
Location: State Route 116 Portage Canal
Type: steel box-girder suspended span
Length: 670 feet
Builder: Manson Construction and Engineering Company of Seattle
Designer: Homer M. Hadley
Ownership: WSDOT
Inventory: NR/Nominated

Jefferson County and Kitsap County—

Hood Canal (William A. Bugge) Bridge (1961/1982) (#104/5)
Location: State Route 104 crossing Hood Canal
Type: floating concrete pontoons with two steel Warren-truss transition spans and two retractable draw spans
Length: 7,967 feet
Built: 1961; west half replaced, 1982
Designer: WSDOT
Ownership: WSDOT
Inventory: NR/DE/HAER

King County—

12th Avenue South Bridge (1911) (#14S/14N)
Location: crossing Dearborn Street, Seattle
Type: steel deck-arch
Length: 361 feet
Builder: Seattle Engineering Department
Designer: unknown
Ownership: Local
Inventory: NR

14th Avenue South (South Park) Bridge (1931) (#3179)
Location: crossing the Duwamish River
Type: Scherzer rolling-lift bascule bridge
Length: 1,285 feet
Builder: Wallace Bridge and Structural Steel Company and the King County Engineering Department
Designer: Scherzer Rolling Lift Bridge Company of Chicago
Ownership: County
Inventory: NR

Alaskan Way Viaduct (1950–53) (#99/540)
Location: State Route 99 through Seattle and under Battery Street
Type: double-deck concrete girder and T-beam
Length: 10,546 feet southbound; 11,156 feet northbound
Builder: MacRae Brothers and the Morrison-Knudsen Company, both of Seattle
Designers: Ralph W. Finke, City of Seattle; Ray M. Murray and George Stevens, Washington State Department of Highways
Ownership: WSDOT

Aurora Avenue (George Washington Memorial) Bridge (1931) (#99/560)
Location: Aurora Avenue North crossing the Lake Washington Ship Canal on State Route 99, Seattle
Type: steel cantilever Warren deck truss
Length: 2,955 feet
Builder: United States Steel Products Company of Seattle, Pacific Bridge Company of Portland, and N. Nygren of Seattle
Designer: Jacobs and Ober of Seattle
Ownership: WSDOT
Inventory: NR/HAER

Ballard Bridge (1917) (#20)
Location: crossing the Lake Washington Ship Canal at 15th Avenue Northwest, Seattle
Type: double-leaf trunnion bascule
Length: 295 feet
Builder: Hans Pederson and the Dyer Brothers of San Francisco
Designer: City of Seattle, D. R. Huntington, City Architect
Ownership: Local
Inventory: NR

Baring Bridge (1899) (#509A)
Location: Northeast Index Creek Road crossing the South Fork of the Skykomish River
Type: timber suspension
Length: 334 feet
Designer: Unknown
Ownership: County
Inventory: NR/DE

Cowen Park Bridge (1936) (#15)
Location: Fifteenth Avenue North crossing Cowen Park, Seattle
Type: open-spandrel reinforced-concrete arch
Length: 358 feet
Builder: A. W. Quist
Designers: Washington State Department of Highways, Clark H. Eldridge and A. J. Mahoney
Ownership: Local
Inventory: NR

Duwamish (First Avenue South) River Bridge (1956) (#99/530E)
Location: First Avenue South crossing the Duwamish River, Seattle
Type: Pratt deck-truss, double-leaf, trunnion bascule
Length: 3,010 feet
Builder: General Construction Company of Seattle
Designers: Bruce V. Christy and Winfred T. Robertson
Ownership: WSDOT

Foss River Bridge (1951) (#2605A)
Location: crossing the Foss River near Skykomish
Type: pony truss
Length: 80 feet
Designer: unknown
Ownership: County
Inventory: NR/Nominated

Fremont Bridge (1917)
Location: crossing the Lake Washington Ship Canal, Seattle
Type: double-leaf trunnion bascule
Length: 502 feet
Builder: United States Steel Products Company and the Pacific States Construction Company
Designer: unknown
Ownership: Local
Inventory: NR

Green River Gorge Bridge (1914) (#3022)
Location: Green River Gorge Road crossing the Green River
Type: Baltimore Petit deck truss
Length: 428 feet originally; now 437 feet
Builder: Charles G. Huber

Designer: Charles D. Calley, Bridge Engineer
Ownership: County
Inventory: NR/DE

Judd Creek Bridge (1953) (#3184)
Location: Vashon Highway Southwest, Vashon Island
Type: concrete box-girder
Length: 370 feet
Designer: Homer M. Hadley
Ownership: County
Inventory: NR/DE

Lake Washington Ship Canal Bridge (1962) (#5/570)
Location: Interstate 5 crossing the Lake Washington Ship Canal, Seattle
Type: steel double deck, Warren through-truss (for lower level) and deck-truss (for upper level)
Length: 4,429 feet
Builder: Scheumann and Johnson of Seattle, Allied Structural Steel Company of Chicago, MacRae Brothers of Seattle, and S. S. Mullen of Seattle
Designers: Washington State Department of Highways, George H. Andrews and Ed Wilkerson, Engineers
Ownership: WSDOT

Meadowbrook Bridge (1921) (#1726A)
Location: 394th Place Southeast crossing the Snoqualmie River
Type: Pratt/Parker, riveted-steel through-truss
Length: 373 feet
Builder: Ward and Ward Inc.
Designer: Donald H. Evans, King County Bridge Engineer
Ownership: County
Inventory: NR/DE/HAER

Miller River Bridge (1922) (#999W)
Location: Northeast Old Cascade Highway crossing the Miller River near Skykomish
Type: steel Pratt truss
Length: 228 feet
Builder: David, Bigelow, and Stratton
Designer: Donald H. Evans, King County Bridge Engineer
Ownership: County
Inventory: NR/DE

Montlake Bridge (1924) (#513/12)
Location: crossing the Lake Union Ship Canal, Seattle
Type: double-leaf trunnion bascule
Length: 345 feet
Builder: C. L. Creelman, the Wallace Equipment Company, and the Westinghouse Electrical and Manufacturing Company
Designer: Seattle City Bridge Engineering Department
Ownership: Local
Inventory: NR/HAER

Mount Si Bridge (1904/1955) (#2550A)
Location: Mount Si Road crossing the Middle Fork of the Snoqualmie River
Type: pin-connected steel Pratt/Parker through-truss
Length: 292 feet
Built: 1904, moved to its present location, 1955
Designer: unknown
Ownership: Local
Inventory: NR/Nominated

North Queen Anne Drive Bridge (1936) (#92)
Location: Wolf Creek Canyon, Queen Anne Hill, Seattle
Type: steel deck-arch
Length: 327 feet
Builder: General Construction Company
Designer: Clark Eldridge
Ownership: Local
Inventory: NR

Patton Bridge (1950) (#3015)
Location: Green Valley Road crossing the Green River near Auburn
Type: concrete cantilever and steel-box girder
Length: 430 feet
Designer: Homer M. Hadley
Ownership: Local
Inventory: NR

Raging River Bridge (1915) (#1008E)
Location: SE 68th Street, a spur off the Preston-Fall City Road at milepost 10.0 crossing the Raging River
Type: concrete Luten arch
Length: 70 feet
Designer: Daniel Benjamin Luten
Ownership: County
Inventory: NR/DE

Ravenna Park Bridge (1913) (#58)
Location: Tenth Avenue Northeast crossing the Ravenna Park Ravine, Seattle

Type: steel deck-arch
Length: 354 feet
Builder: J. R. Wood and Company of Seattle
Designer: Frank M. Johnson
Ownership: Local
Inventory: NR

Saltwater State Park Bridge (1914) (#3139)
Location: Marine View Drive crossing a ravine, City of Des Moines
Type: concrete girder
Length: 570 feet
Builder: Nelse Mortensen and Company
Designer: T. P. Blum, King County Bridge Engineer
Ownership: Local
Inventory: NR/DE

Schmitz Park Bridge (1935–36) (#13)
Location: crossing the Schmitz Park Ravine, Seattle
Type: concrete beam
Length: 175 feet
Builder: Schuchle Brothers
Designer: Washington State Department of Highways, Clark Eldridge
Ownership: Local
Inventory: NR

South Twin Bridge (1951) (#3143)
Location: 16th Avenue South, City of Des Moines
Type: concrete box-girder
Length: 375 feet
Designer: unknown
Ownership: Local
Inventory: NR/DE

Stossel Bridge (1951) (#1023A)
Location: Northeast Carnation Farm Road crossing the Snoqualmie River
Type: steel Warren through-truss
Length: 330 feet
Designers: Cecil C. Arnold and Raymond G. Smith, Cecil C. Arnold and Associates
Ownership: County
Inventory: NR/Nominated

Tolt Bridge (1922) (#1834A)
Location: crossing the Snoqualmie River about one-half mile southwest of Carnation
Type: riveted-steel Pratt/Parker through-truss
Length: 200 feet
Builder: J. A. McEachern
Manufacturer: Kansas City Structural Steel Company
Ownership: County
Inventory: NR/DE

University Bridge (1915–19) (#3)
Location: crossing the Lake Washington Ship Canal, Seattle
Type: bascule
Length: 218 feet
Builder: Booker, Kiehl, and Whipple and the United States Steel Products Company
Designer: unknown
Ownership: Local
Inventory: NR

Yesler Way Bridge over 4th Avenue (1909) (#36)
Location: on Yesler Way, Seattle
Type: continuous steel stringer
Length: 78 feet
Designer: unknown
Ownership: Local
Inventory: NR/DE

Kitsap County—

Agate Pass Bridge (1950) (#305/10)
Location: State Route 305 crossing Agate Passage
Type: steel cantilever truss
Length: 1,229 feet
Builder: Manson Construction and Engineering Company of Seattle
Designer: unknown
Ownership: WSDOT
Inventory: NR

Port Washington Narrows (Warren Street) Bridge (1958) (#303/12)
Location: State Route 303 crossing Washington Narrows, Bremerton
Type: steel-plate girder
Length: 1,717 feet
Builder: Peter Kiewit Sons Company of Vancouver
Designer: Tudor Engineering Company of San Francisco
Ownership: WSDOT
Inventory: NR/Nominated

Kittitas County—

Lake Keechelus Snowshed Bridge (1951) (#90/110)
Location: Interstate 90 east of Snoqualmie Pass

Type: concrete snowshed
Length: 500 feet
Designer: Washington State Department of Highways, George Stevens, Engineer
Ownership: WSDOT
Inventory: NR/HAER

Klickitat County—

B-Z Corner Bridge (1957) (#110)
Location: crossing the White Salmon River
Type: open-spandrel rib deck-arch
Length: 182 feet
Builder: West Coast Steel Works of Portland of Oregon
Designer: Harry R. Powell and Associates
Ownership: County
Inventory: NR/Nominated

Klickitat River Bridge (1954) (#142/9)
Location: State Route 142 crossing the Klickitat River
Type: pre-stressed concrete
Length: 141 feet
Builder: Louis Elterich Company and Guy J. Norris of Port Angeles
Designers: Harry R. Powell and associate, Leonard K. Narod
Ownership: WSDOT
Inventory: NR/Nominated

Klickitat County and Hood River County, OR—

Hood River-White Salmon Bridge (1924)
Location: Columbia River toll bridge between White Salmon, WA, and Hood River, OR
Length: 4,418 feet
Builder: 1924 by the Gilpin Construction Company; alterations, 1930s–1940s
Ownership: Port of Hood River
Inventory: NR/DE

Klickitat County and Wasco County, OR—

Columbia River Bridge at The Dalles (1954) (#197/1)
Location: U.S. Route 197 crossing the Columbia River
Type: cantilevered Warren through-trusses with steel-girder spans
Length: 3,339 feet
Builder: Guy F. Atkinson Company of Portland and the Ostrander Company of San Francisco
Designer: Ralph A. Tudor
Ownership: ODOT

Lewis County—

Pe Ell Covered Bridge (1934)
Location: crossing the Chehalis River approximately two miles south of Pe Ell
Type: corrugated-metal covered wooden Howe through-truss
Length: 93 feet
Builder: James Donahue
Designer: probably James Donahue
Ownership: Local
Inventory: NR/DE

Lincoln County and Stevens County—

Spokane River Bridge at Fort Spokane (1941) (#25/6)
Location: State Route 25 crossing the Spokane River near confluence with the Columbia River
Type: steel cantilever truss
Length: 953 feet
Builder: Angeles Gravel and Supply Company, and C and F Teaming and Trucking Company
Designer: Washington State Department of Highways, R. W. Finke, Engineer
Ownership: WSDOT
Inventory: NR/HAER

Spokane River Bridge at Long Lake Dam (1949) (#231/101)
Location: State Route 231 crossing the Spokane River below Long Lake Dam
Type: concrete arch
Length: 481 feet
Builder: Henry Hagman
Designer: Washington State Department of Highways, George Stevens, Engineer
Ownership: WSDOT
Inventory: NR/HAER

Mason County—

Goldsborough Creek Bridge (1923) (#3/3)
Location: State Route 3 crossing Goldsborough Creek in Shelton
Type: reinforced-concrete tied arch
Length: 57 feet
Builder: MacRae Brothers
Designer: unknown
Ownership: WSDOT
Inventory: NR

North Hamma Hamma River Bridge (1924) (#101/403)
Location: U.S. Route 101 crossing forks of the Hamma Hamma River
Type: concrete tied arch
Length: 154 feet
Builder: Colonial Building Company
Designer: unknown
Ownership: WSDOT
Inventory: NR/HAER

South Hamma Hamma River Bridge (1924) (#101/404)
Location: U.S. Route 101 crossing forks of the Hamma Hamma River
Type: concrete tied arch
Length: 154 feet
Builder: Colonial Building Company
Designer: unknown
Ownership: WSDOT
Inventory: NR/HAER

Okanogan County and Douglas County—

Grand Coulee Bridge (1935) (#155/101)
Location: State Route 155 crossing the Columbia River below Grand Coulee Dam
Type: cantilever steel Warren through truss
Length: 1,066 feet
Builder: U.S. Bureau of Reclamation
Designer: Washington State Department of Highways
Ownership: WSDOT
Inventory: NR/HAER

Pierce County—

City Waterway (Murray Morgan) Bridge (1911) (#509/5)
Location: Eleventh Street crossing Thea Foss Waterway, Tacoma
Type: steel vertical lift
Length: 3,200 feet
Builder: International Contract Company of Seattle
Designer: Waddell and Harrington
Ownership: WSDOT
Inventory: NR/HAER

East 34th Street Bridge, Pacific to A Street (1937) (#E-7)
Location: Pacific to A Street, Tacoma
Type: open-spandrel concrete arch
Length: 478 feet
Builder: MacRae Brothers of Seattle
Designers: C. D. Forsbeck, City Engineer of Tacoma, and Bridge Engineer O. A. Anderson
Ownership: Local
Inventory: NR

Fairfax (James R. O'Farrell) Bridge (1921) (#165/10)
Location: State Route 165 crossing the Carbon River, 11.6 miles northwest of Mount Rainier National Park
Type: three-hinged steel deck-arch
Length: 494 feet
Builder: Pierce County and the Washington State Department of Highways
Designer: E. A. White, Pierce County Engineer
Ownership: WSDOT
Inventory: NR/HAER

McMillin (Puyallup River) Bridge (1934) (#162/6)
Location: crossing the Puyallup River
Type: concrete through truss
Length: 170 feet
Builder: Dolph Jones
Designers: suggested by Homer M. Hadley, designed by the W. H. Witt Company of Seattle
Ownership: WSDOT
Inventory: NR/HAER

North 21st Street Bridge (1911) (#N2)
Location: crossing Buckley Gulch between Fife and Oakes
Type: continuous concrete beam
Builder: Creelman, Putnam and Healy of Tacoma
Designer: Waddell and Harrington
Ownership: Local
Inventory: NR/HAER

North 23rd Street Bridge (1910) (#N3)
Location: crossing Buckley Gulch between North Fife and Oakes
Type: concrete beam
Length: 312 feet
Designer: unknown
Ownership: Local
Inventory: NR

Purdy Creek Bridge (1936) (#302/105)
Location: crossing Henderson Bay
Type: continuous box girder
Length: 550 feet
Builder: Pierce County
Designers: suggested by Homer M. Hadley, Designed by W. H. Witt Company of Seattle

Ownership: WSDOT
Inventory: NR/HAER

Tacoma Narrows ("Galloping Gertie") Bridge Ruins (1940)
Location: lying on the bottom of the Tacoma Narrows
Type: suspension
Length: 5,939 feet
Builder: 1940 by Pacific Bridge Company of San Francisco, General Construction Company of Seattle, and the Columbia Construction Company; the bridge collapsed, 1940
Designer: Leon S. Moisseff
Ownership: WSDOT
Inventory: NR

Tacoma Narrows Bridge (1950) (#16/110)
Location: on State Route 16, 7.3 miles north of junction with Interstate 5
Type: steel suspension
Length: 5,979 feet
Builder: Bethlehem Pacific Coast Steel Corporation and the John A. Roebling Sons Company
Designer: Dexter R. Smith
Ownership: WSDOT
Inventory: NR/DE/HAER

Winnifred Street Bridge (1941) (#1130)
Location: Winnifred Street crossing the Burlington Northern Railroad tracks, Town of Ruston
Type: concrete box girder
Length: 215 feet
Builder: S. R. Gray
Designer: W. H. Witt Company of Seattle
Ownership: Local
Inventory: NR

Skagit County—

Baker River Bridge (1917) (#20/259)
Location: City of Concrete
Type: open-spandrel concrete arch
Length: 186 feet
Builder: J. R. Wood Superior Company and Washington Portland Cement Company of the City of Concrete
Designers: Bowerman and McCloy of Seattle
Ownership: WSDOT
Inventory: NR/HAER

Dalles Bridge (1952) (#40090)
Location: Concrete-Sauk Valley Road crossing the Skagit River
Type: steel through-arch
Length: 506 feet
Builder: General Construction Company of Seattle
Designers: Cecil C. Arnold and Raymond G. Smith of Seattle
Ownership: County
Inventory: NR/Nominated

Rainbow Bridge (1957) (#40039)
Location: Pioneer Parkway crossing the Swinomish Channel, La Conner
Type: fixed steel through-arch
Length: 797 feet
Builder: Neukirch Brothers of Seattle
Designer: Harry R. Powell
Ownership: County
Inventory: NR/Nominated

Skagit County and Island County—

Deception Pass Bridge (1935) (#20/204)
Location: State Route 20 crossing Deception Pass
Type: steel cantilever deck truss
Length: 976 feet
Builder: Puget Construction Company of Seattle
Designers: L. V. Murrow, Director of Highways, and O. R. Elwell, Bridge Engineer
Ownership: WSDOT
Inventory: NR/HAER

Skamania County—

Conrad Lundy Jr. Bridge (1960) (#207)
Location: Wind River Road crossing the Wind River Canyon
Type: steel Warren deck-truss
Length: 598 feet
Builder: C. M. Corkum Company of Portland
Designer: U.S. Bureau of Public Roads, San Francisco Office
Ownership: County
Inventory: NR/Nominated

Skamania County and Hood River County, OR—

Bridge of the Gods (1926)
Location: crossing the Columbia River south of Stevenson and northeast of Bonneville Dam
Type: steel cantilever through truss
Length: 1,858 feet
Builder: Puget Sound Bridge and Dredging Company

Designer: Te Wanna Toll Bridge Company
Ownership: Port of Cascade Locks
Inventory: NR/DE

Snohomish County—

Red Bridge (1954) (#537)
Location: Mountain Loop Highway crossing the Stillaguamish River
Type: Pratt/Parker through-truss
Length: 209 feet
Builder: Peter Kiewit Sons Company
Designer: U.S. Bureau of Public Roads, San Francisco Office
Ownership: County
Inventory: NR/Nominated

Snohomish River Bridge (1926) (#529/10E)
Location: State Route 529 crossing the Snohomish River, near Everett
Type: steel lift span
Length: 2,680 feet
Builder: J. A. McEachern Company of Seattle
Designers: Washington State Department of Highways, Charles E. Andrew, Bridge Engineer; J. A. L. Waddell, lift span designer
Ownership: WSDOT
Inventory: NR/DE

Snohomish River Bridge (1954) (#529/10W)
Location: State Route 529 crossing the Snohomish River, near Everett
Type: steel lift span
Length: 2,465 feet
Builder: Manson Construction and Engineering Company
Designers: Larry Robertson and Willis B. Horn, lead designers for structural portions of the lift spans; Derby Livesay and Carl West designed the operating system machinery
Ownership: WSDOT
Inventory: NR/Nominated

Steamboat Slough Bridge (1954) (#529/20E)
Location: State Route 529 crossing Steamboat Slough, near Everett
Type: swing span with flanking Warren through-trusses
Length: 910 feet
Builder: M. P. Butler of Seattle
Designer: Washington State Department of Highways, Willis B. Horn, Engineer
Ownership: WSDOT
Inventory: NR/Nominated

Spokane County—

Greene Street Bridge (1955) (#02)
Location: Greene Street crossing the Spokane River, Spokane
Type: open-spandrel rib concrete deck-arch
Length: 434 feet
Builder: Henry Hagman Construction Company of Cashmere
Designer: B. J. Garnett, City Engineer
Ownership: Local
Inventory: NR/Nominated

Maple Street Bridge (1958) (#16)
Location: Maple Street crossing the Spokane River, Spokane
Type: steel-plate girder spans
Length: 1,713 feet
Builder: Morrison-Knudsen Company of Seattle, substructure; superstructure by United States Steel Corporation's American Bridge Division of Portland
Designer: Tudor Engineering Company
Ownership: Local

Marshall Bridge (1949) (#2404)
Location: Community of Marshall
Type: concrete T-beam
Length: 547 feet
Builder: Clifton and Applegate of Spokane
Designer: W. L. "Pat" Malony
Ownership: County
Inventory: NR

Monroe Street Bridge (1911) (#371001001)
Location: crossing the Spokane River, Spokane
Type: concrete arch
Length: 965 feet
Builder: City of Spokane
Designers: J. C. Ralston, City Engineer, designed the bridge; Cutter and Malmgren of Spokane designed the ornamentation
Ownership: Local
Inventory: NR/HAER

Sunset Boulevard/Latah Creek Bridge (1914)
Location: crossing Latah Creek, Spokane
Type: concrete arch
Length: 1,070 feet
Builder: J. E. Cunningham of Spokane
Designer: W. S. Maloney
Ownership: Local
Inventory: NR

Stevens County—

Columbia River Bridge at Northport (1949) (#25/130)
Location: State Route 25 crossing the Columbia River
Type: steel cantilever through-truss
Length: 1,542 feet
Builder: MacRea Brothers of Seattle
Designer: unknown
Ownership: WSDOT
Inventory: NR

Thurston County—

Capitol Boulevard Bridge (1937)
Location: crossing the Deschutes River, Tumwater
Type: concrete beam
Length: 1,100 feet
Designer: Washington State Department of Highways, Clark H. Eldridge, Bridge Engineer
Ownership: Local
Inventory: NR

Lower Custer Way Crossing Bridge (1915)
Location: crossing the Deschutes River, Tumwater
Type: three-span concrete Luten arch
Length: 190 feet
Builder: Charles G. Huber
Designer: unknown
Ownership: Local
Inventory: NR

Upper Custer Way I-5 Overcrossing Bridge (1956) (#5/316)
Location: crossing the Deschutes River and Interstate 5, Tumwater
Type: concrete open-spandrel rib arch and prestressed concrete box-girders
Length: 530 feet
Builder: A. J. Cheff Construction Company (arch span)
Designer: Washington State Department of Highways, George Stevens, Bridge Engineer
Ownership: WSDOT

Wahkiakum County—

Grays River Covered Bridge (1905/1989) (#10)
Location: Worrel Road crossing Grays River
Type: covered timber Howe truss
Length: 188 feet
Builder: 1905; rebuilt 1989 by the Dulin Construction Company of Centralia
Designers: 1905 by Ferguson and Houston, Astoria, Oregon; Sargent Engineers of Olympia designed the 1989 reconstruction
Ownership: County
Inventory: NR/HAER

Walla Walla County—

Johnson Bridge (1929) (#933700012)
Location: crossing the Touchet River north of the Town of Touchet
Type: three-span concrete tee-beam
Length: 126 feet
Builder: George Harding
Designer: E. R. Smith
Ownership: County
Inventory: NR

Waitsburg Bridge (1925) (#760136001)
Location: crossing the Touchet River, City of Waitsburg
Type: concrete Luten arch
Length: 92 feet
Designer: unknown
Ownership: Local
Inventory: NR

Whatcom County—

Gorge Creek Bridge (1955) (#20/323)
Location: State Route 20 crossing Gorge Creek
Type: steel truss deck-arch
Length: 250 feet
Designers: Cecil C. Arnold and Raymond G. Smith
Ownership: WSDOT
Inventory: NR/Nominated

Middle Fork Nooksack River Bridge (1915) (#140)
Location: crossing the Middle Fork of the Nooksack River
Type: steel Pennsylvania Petit through-truss
Length: 416 feet
Builder: Toledo Bridge and Crane Company and the Weymouth Construction Company at original location on Guide Meridian Road; moved to present location, 1951
Designer: unknown
Ownership: County
Inventory: NR

Whitman County—

Harpole (Manning-Rye) Bridge (1922)
Location: crossing the Palouse River northwest of Colfax
Type: covered wooden Howe truss
Length: 150 feet
Builder: 1922 by the Spokane and Inland Empire Railroad; trusses covered 1928 by the Great Northern Railway
Designer: unknown
Ownership: Private
Inventory: NR/HAER

Yakima County—

Toppenish-Zillah Bridge (1947) (#485)
Location: crossing the Yakima River between Toppenish and Zillah
Type: concrete box girder
Length: 541 feet
Builder: Ramsey and Company
Designer: Homer M. Hadley
Ownership: Local
Inventory: NR

NOTE

1. This table does not include demolished bridges, railroad and pedestrian bridges, bridges built exclusively for water or sewer pipes, or tunnels.

Abutment. A substructure element supporting each end of a single span or the extreme ends of a multi-span superstructure and, in general, retaining or supporting the approach embankment.

Anchor Span. The span that counterbalances and holds in equilibrium the cantilevered portion of an adjacent span during construction.

Approach Span. The span or spans connecting the abutment with the main span or spans.

Arch Rib. The main support element used in open-spandrel arch construction; also known as arch ring.

Balustrade. A rail, and the row of posts that support it, along a stairway or on a bridge deck.

Bascule Bridge. A bridge over a waterway with one or two leaves that rotate from a horizontal to a near-vertical position, providing unlimited vertical clearance for marine traffic.

Beam. A linear structural member designed to span from one support to another.

Bearing. A support element transferring loads from superstructure to substructure, capable of permitting rotation or longitudinal movement.

Bent. A substructure unit supporting each end of a bridge span; made up of two or more columns or column-like members connected at their top-most ends by a cap, strut, or other member holding them in their correct positions; also called pier.

Box Beam/Girder. A hollow structural beam or girder with a square, rectangular, or trapezoidal cross section.

Cable-Stayed Bridge. A bridge in which the superstructure is directly supported by cables or stays, passing over or attached to a tower or towers located at the main pier(s).

Cantilever. A structural member that has a free end projecting beyond its support; length of span overhanging the support.

Cast-in-Place. Concrete poured within formwork on-site to create a structural element in its final position.

Centerlock. In bascule bridge decks, a device used to secure movable leaves (spans) in position and to transfer live (traffic) loads.

Chord. A horizontal truss member.

Compression Members. Structural members that withstand compression forces of applied loads.

Continuous Beam. A general term applied to a beam that spans uninterrupted over one or more intermediate supports.

Continuous Spans. Spans designed to extend without joints over one or more intermediate supports.

Counters. Cross members in steel truss bridges used to balance live (traffic) loads.

Counterweight. A weight used to balance, and to minimize resistance of, movable members or spans.

Cross Brace. Transverse brace between two main longitudinal members.

Dead Load. A static load due to the weight of the structure itself.

Deck. The portion of a bridge that provides direct support for vehicular and pedestrian traffic.

Deck Bridge. A bridge in which the supporting members are all beneath the roadway.

Deck Truss. A bridge whose roadway is supported from beneath by a truss.

Diagonal. A sloping structural member of a truss or bracing system.

Expansion Joint. A joint designed to provide means for expansion and contraction movements produced by temperature changes, load, or other forces.

Falsework. A temporary wooden or metal framework built to support a structure's weight during its construction and until it becomes self-supporting.

Fatigue. Cause of structural deficiencies, usually due to repetitive loading over time.

Fixed Span. A superstructure span having its position practically immovable, as compared to a movable span.

Flange. The horizontal parts of a rolled I-beam or built-up girder extending transversely across the web's top and bottom.

Footing. The enlarged, lower portion of a substructure that distributes the structural load either to the earth or to supporting piles; the most common footing is the concrete slab; "footer" is a colloquial term for footing.

Girder. A flexural member that is the main or primary support for the structure and usually receives loads from floor beams and stringers; any large beam, especially if built up.

Girder Bridge. A bridge whose superstructure consists of two or more girders supporting a separate floor system, as differentiated from a multi-beam bridge or a slab bridge.

Hanger. A tension member serving to suspend an attached member.

Haunch. The enlarged part of a beam near its supported ends, visible as the curved or angled bottom edge, that adds strength to the beam.

H-Beam. A rolled steel member having an H-shaped cross section and commonly used for piling.

Hinge. A point in a structure at which a member is free to rotate.

Howe Truss. Named for William Howe, who patented the design in 1840. A type of bridge truss having parallel chords, vertical (tension) rods at the panel points, and diagonals forming an X pattern.

I-Beam. A structural member with a cross-sectional shape similar to the capital letter "I."

Joint. In stone masonry, the space between individual stones; in concrete, a division in continuity of the concrete; in a truss, the point at which members of a truss frame are joined.

Laminated Timber. Small timber planks glued together to form a larger member.

Lattice Truss. A truss having two or more web systems composed entirely of diagonal members at any interval and crossing each other without reference to vertical members.

Leaf. The movable portion of a bascule bridge that forms the structure's span.

Live Load. Vehicular traffic.

Lower Chord. The bottom horizontal member of a truss.

Luten Arch. Named for Daniel B. Luten, usually refers to the earth- or concrete-filled, spandrel barrel arch, probably the most popular of his more than thirty bridge patents.

Main Beam. A beam supporting the spans and bearing directly onto a column or wall.

Member. An individual angle, beam, plate, or built piece intended to become an integral part of an assembled frame or structure.

Movable Span. A general term applied to a superstructure span designed to be swung, lifted, or otherwise moved longitudinally, horizontally, or vertically.

Open-Spandrel Arch. A bridge having open spaces between the deck and the arch members allowing "open" visibility through the bridge.

Panel. The portion of a truss span between adjacent intersection points of web and chord members.

Parker Truss. A Pratt truss with a polygonal top chord.

Pennsylvania Petit Truss. A Parker truss with sub-struts and/or sub-ties; derived its name in part due to extensive use by the Pennsylvania Railroad; patented in 1875.

Pier. A substructure unit that supports the spans of a multi-span superstructure at an intermediate location between its abutments.

Pile. A shaft-like member that carries loads through weak layers of soil to those capable of supporting such loads.

Pile Bent. A row of driven or placed piles with a pile cap to hold them in their correct positions; see Bent.

Pin. A cylindrical bar used to connect chords and web members.

Pin-Connected Truss. A general term applied to a truss of any type with its chord and web members connected at the panel points by pins.

Plate Girder. A large I-beam composed of a solid web plate with flange plates attached to the web plate by flange angles or fillet welds.

Pony Truss. A bridge consisting of a low through truss that has no overhead truss members.

Portal. The clear, unobstructed space of a through-truss bridge forming the structure's entrance.

Post. A member resisting compressive stresses located vertical to the bottom chord of a truss and common to two-truss panels; sometimes used synonymously with vertical column.

Posttensioning. A method of externally prestressing concrete in which the tendons are stressed after the concrete has been cast.

Pratt Truss. A truss with parallel chords and a web system composed of vertical posts with diagonal ties inclined outward and upward from the bottom-chord panel points toward the ends of the truss; also known as N-truss; named for Thomas and Caleb Pratt, who patented the design in 1844.

Precast Concrete. Concrete that has been cast and cured before being placed into position in a bridge or other structure.

Prestressed Concrete. Concrete in which cracking and tensile forces are greatly reduced by compressing it with tensioned cables or bars.

Pretensioning. A method of prestressing concrete in which the cables are held in a stretched condition until the concrete has hardened, then the pull on the cable is released, inducing internal compression into the concrete.

Reinforced Concrete. Concrete with steel reinforcing bars bonded within it to supply increased tensile strength and durability.

Rib. Curved structural member supporting a curved shape or panel.

Rigid-Frame Bridge. A bridge with moment-resistant connections between the superstructure and the substructure to produce an integral, elastic structure.

Riveted Connection. A rigid connection of metal bridge members that is assembled with rivets; riveted connections increase the structure's strength.

Roller. A steel cylinder intended to provide longitudinal movements by rolling contact.

Safety Hangers. Backup for original connections to provide redundancy; often added for seismic retrofit.

Simple Trunnion. Type of movable bridge consisting of a forward cantilever arm out over a channel and a rear counterweight arm, allowing for the leaf to rotate about the trunnion.

Slab Bridge. A bridge having a superstructure composed of a glue-laminated timber slab or a reinforced-concrete slab constructed either as a single unit, or as a series of narrow slabs, placed parallel to the roadway and spanning the space between the supporting abutments.

Span. The distance between piers, towers, or abutments.

Spandrel. The space bounded by the extrados (exterior curves) of arches and the horizontal member above.

Spandrel Column. A column constructed on an arch span's rib and serving as a support for the deck of an open-spandrel arch.

Stay. Diagonal brace installed to minimize structural movement.

Suspended Span. A simple span supported from the free ends of cantilevers.

Stirrup. A U-shaped bar providing a stirrup-like support for a member in timber and metal bridges; U-shaped bar placed in concrete construction to resist diagonal tension (shear) stresses.

Stringer. A longitudinal beam supporting the bridge deck.

Strut. A piece or member acting to resist compressive stress.

Suspension Bridge. A bridge in which the floor system is supported by catenary cables that are supported upon towers and are anchored at their extreme ends.

Sway Bracing. Diagonal bracing located at the top of a through truss perpendicular to the truss itself, usually in a vertical plane, and designed to resist horizontal forces.

Swing Span Bridge. A movable bridge in which the span rotates in a horizontal plane on a pivot pier to permit passage of marine traffic.

T-Beam. A rolled steel section shaped like a T; part of a reinforced-concrete floor in which the beam projects below the slab.

Three-Hinged Arch. An arch that is hinged at each support and at the crown.

Through Bridge. A bridge where the floor elevation is nearly at the bottom and traffic travels between the supporting parts.

Tie. A member carrying tension.

Tied Arch. A through-arch bridge in which the deck is tied to (suspended from) the arch.

Trestle. A bridge structure consisting of spans supported upon frame bents.

Trunnion. A heavy pin around which leaves and/or counterweights rotate in a movable bridge.

Truss. A jointed structure made up of individual members arranged and connected, usually in a triangular pattern, so as to support longer spans.

Truss Bridge. A bridge having a pair of trusses for the superstructure.

Tunnel. An underground passage open to daylight at both ends.

Two-Hinged Arch. A rigid frame that may be arch shaped or rectangular, but is hinged at both supports.

Upper Chord. The top longitudinal member of a truss.

Viaduct. A series of spans carried on piers at short intervals.

Vertical Lift Bridge. A movable bridge that can be raised vertically, with the movable span remaining in a horizontal position, by weights and pulleys operating in towers at each end of the structure.

Warren Truss. A triangular truss consisting of sloping members between the top and bottom chords, sometimes no vertical members; members form the capital letter "W"; named for James Warren, one of two British engineers who patented the design in 1848.

Web. The portion of a beam located between and connected to the flanges.

Web Members. Usually vertically or inclined in orientation, the intermediate members of a truss, not including the end posts.

Welded Joint. A joint in which the assembled elements and members are united through fusion of metal.

Unpublished Sources

Primary Records—

Eldridge, Clark. Manuscript autobiography (1896–1982), loaned by his son, C. W. Eldridge.
Evergreen Point Bridge Scrapbook, Special Collections, University of Washington Libraries, Seattle.
Gloyd, C. S. "History of Floating Bridges in the State of Washington." Unpublished ms., WSDOT Library, Olympia, n.d.
Hadley, Homer M. "Report on Proposed Pontoon Bridge at Seattle, Wa., Oct. 1, 1921." Copies in various locations, including WSDOT Bridge Preservation Office, Washington State Archives, and Office of Archaeology and Historic Preservation.
Hitch, Stephen J. "The Evergreen Point Floating Bridge in Crisis: An Approach to Evaluating Alternative Solutions," Unpublished M.S. Thesis, University of Washington, 1999.
Hood River County Historical Museum, Hood River, Oregon, Bridge Files.
Johnson, Frank Melvin. "The Ravenna Park Steel Arch Bridge." Unpublished M.S. Thesis, University of Washington, 1916.
Lake Washington Floating Bridge Scrapbook, Special Collections, University of Washington Libraries.
Port of Hood River, Hood River, Oregon, Photograph File.
"South End Mercer Island Improvement Club," typed history, 1920–54, folder 1, Gertrude Pool Papers, Mss 061, Washington State Library, Olympia.
University of Washington Libraries, Special Collections, Biography File.
Washington State Bureau of Vital Statistics, Olympia.
Washington State Department of Licenses, Licenses of Professional Engineers, Olympia.
Washington State Department of Transportation (WSDOT and State Department of Highways) Records, Washington State Archives, Olympia.
________. Bridge Preservation Office files, Tumwater.
________. Engineering Records, Plans Vault, Olympia.
________. Environmental Services Office, Cultural Resources Program files, Tumwater.
Washington State Office of Archaeology and Historic Preservation, Bridge files, Olympia.

Historic American Engineering Record (HAER) Reports, Library of Congress and Office of Archaeology and Historic Preservation, Olympia—

Berkley, Brian T. "Harpole Bridge," HAER No. WA-133, 1995.
Bruce, Robin. "'F' Street Bridge," HAER No. WA-31, 1990.
________. "Grays River Covered Bridge," HAER No. WA-28, 1991.
________. "McClure Bridge," HAER No. WA-25, 1989.
________. "Orient Bridge," HAER No. WA-32, 1992.
Clarke, Jonathan. "City Waterway Bridge," HAER No. WA-100, 1993.
________. "Dosewallips River Bridge," HAER No. WA-94, 1993.
________. "Fairfax (James R. O'Farrell) Bridge," HAER No. WA-72, 1993.
________. "Lake Keechelus Snowshed Bridge," HAER No. WA-110, 1993.
________. "Mount Baker Ridge Tunnel," HAER No. WA-109, 1993.
________. "South Fork Newaukum River Bridge," HAER No. WA-112, 1993.
________. "Vancouver-Portland (Columbia River) Interstate Bridge," HAER No. WA-86, 1993.
Croteau, Todd A. "Company Creek (Harlequin) Bridge # 2," HAER No. WA-115, 1993.
Donovan, Sally, and Gifford Pinchot National Forest personnel. "Cispus Valley Bridge (Forest Service Bridge No. 2306-3.6)," HAER No. WA-65, 1998 (Draft).
Hadlow, Robert W. "Aurora Avenue (George Washington Memorial) Bridge," HAER No. WA-107, 1993.
________. "Canoe Pass Bridge," HAER No. WA-104, 1993.
________. "Columbia River Bridge at Bridgeport," HAER No. WA-90, 1993.
________. "Columbia River Bridge at Grand Coulee Dam," HAER No. WA-102, 1993.
________. "Columbia River Bridge at Kettle Falls," HAER No. WA-91, 1993.
________. "Deception Pass Bridge," HAER No. WA-103, 1993.
________. "Longview (Lewis and Clark) Bridge," HAER No. WA-89, 1993.
________. "Snake River Bridge at Lyons Ferry," HAER No. WA-88, 1993.
________. "Spokane River Bridge at Fort Spokane," HAER No. WA-113, 1993.
________. "Tacoma Narrows Bridge," HAER No. WA-99, 1993.
________. "Washington State Cantilever Bridges, 1927–1941," HAER No. WA-106, 1993.
________. "Yale (Lewis River) Bridge," HAER No. WA-87, 1993.
Hagglund, Daniel E. "Lacey Murrow Floating Bridge (Lake Washington Floating Bridge) (Mercer Island Floating Bridge)," HAER No. WA-2, 1989.
Hobbs, Richard. "Hood Canal Floating (William E. Bugge Memorial) Bridge," HAER No. WA-173, 2004.
Holstine, Craig, and Darcy Fellin. "Outlet Creek Bridge," HAER No. WA-117, 1994.
Kopperl, Robert. "Meadowbrook Bridge (KC Bridge #1726A)," HAER No. KC HRI 0832, 2004.
Krier, Robert H., and Craig Holstine. "Latah Creek Bridge No. 4102," HAER No. WA-163, 1998.
Lawrence, Rhoda A. R. "North Twenty-First Street Bridge (Buckley Gulch Bridge, North Twenty-First Street Viaduct)," HAER No. WA-83, 1993.
Lawrence, Wm. Michael. "Baker River (Henry Thompson) Bridge," HAER No. WA-105, 1993.

________. "Hoquiam River Bridge," HAER No. WA-93, 1993.
________. "Indian Timothy Memorial Bridge," HAER No. WA-85, 1993.
________. "Lacey Murrow Memorial Bridge," HAER No. WA-2, 1993.
________. "McMillin Bridge," HAER No. WA-73, 1993.
________. "Montlake Bridge," HAER No. WA-108, 1993.
________. "North Hamma Hamma Bridge," HAER No. WA-97, 1993.
________. "Purdy Bridge," HAER No. WA-101, 1993.
________. "South Hamma Hamma Bridge," HAER No. WA-96, 1993.
________. "Spokane River at Long Lake Dam," HAER No. WA-95, 1993.
________. "Wishkah River Bridge [1924]," HAER No. WA-92, 1993.
Mighetto, Lisa, and Marcia Babcock Montgomery. "Raging River Bridge 234A," HAER No. WA-141, 1997.
Montgomery, Marcia Babcock, and Lisa Mighetto. "Novelty Bridge 404B," HAER No. OAHP 110598-02-KI, 1999.
Rickert, Wayne L., Jr. "Grays River Covered Bridge," HAER Report, 1988.
Sheldon, John C. E. "West Wishkah Bridge," HAER No. WA-22, 1988.
Soderberg, Lisa. "Chow Chow Bridge," HAER Report, 1983.
________. "Pasco-Kennewick Bridge," HAER No. WA-8, 1980.
Yearby, Jean P. "Grant Avenue Bridge (Prosser Steel Bridge)," HAER No. WA-4, 1985.

HAER Inventories—

Soderberg, Lisa. "12th Avenue South Over Dearborn Street Bridge," HAER Inventory, 1980.
________. "14th Avenue South [South Park] Bridge," HAER Inventory, 1980.
________. "Aurora Avenue Bridge," HAER Inventory, 1980.
________. "Capitol Boulevard Bridge," HAER Inventory, 1979.
________. "Chow Chow Bridge," HAER Inventory, 1979.
________. "City Waterway Bridge," HAER Inventory, 1979.
________. "Cowen Park Bridge," HAER Inventory, 1980.
________. "Curlew Bridge," HAER Inventory, 1979.
________. "Deception Pass, Canoe Pass Bridges," HAER Inventory, 1979.
________. "Doty Bridge," HAER Inventory, 1979.
________. "Duckabush River Bridge," HAER Inventory, 1979.
________. "East Thirty-fourth Street Bridge, East B to East D Streets [1948]," HAER Inventory, 1979.
________. "East Thirty-fourth Street Bridge, Pacific to A Street [1937]," HAER Inventory, 1979.
________. "Elwha River Bridge," HAER Inventory, 1979.
________. "'F' Street Bridge," HAER Inventory, 1979.
________. "Goldsborough Creek Bridge," HAER Inventory, 1979.
________. "Grays River Covered Bridge," HAER Inventory, 1980.
________. "Hoquiam River Bridge," HAER Inventory, 1979.
________. "Indian Timothy Memorial Bridge," HAER Inventory, 1979.
________. "Jack Knife Bridge," HAER Inventory, 1979.
________. "Johnson Bridge," HAER Inventory, 1979.
________. "Lacey Murrow/Lake Washington Floating Bridge," HAER Inventory, 1981.
________. "Latah Creek (Sunset Boulevard) Bridge," HAER Inventory, 1979.
________. "Longview Bridge," HAER Inventory, 1980.
________. "Lower Custer Way Bridge," HAER Inventory, 1979.
________. "Manning-Rye [Harpole] Covered Bridge," HAER Inventory, 1979.
________. "McClure Bridge," HAER Inventory, 1979.
________. "Middle Fork Nooksack River Bridge," HAER Inventory, 1979.
________. "Monroe Street Bridge," HAER Inventory, 1979.
________. "Montlank Bridge," HAER Inventory, 1980.
________. "Mount Baker Ridge Tunnel," HAER Inventory, 1980.
________. "North and South Hamma Hamma River Bridges," HAER Inventory, 1979.
________. "North Queen Anne Drive Bridge," HAER Inventory, 1980.
________. "Orient Bridge," HAER Inventory, 1979.
________. "Pickett Bridge," HAER Inventory, 1979.
________. "Prosser Steel Bridge," HAER Inventory, 1979.
________. "Purdy Bridge," HAER Inventory, 1979.
________. "Ravenna Park Bridge," HAER Inventory, 1981.
________. "Schmitz Park Bridge," HAER Inventory, 1980.
________. "University Bridge, Fremont Bridge, Ballard Bridge," HAER Inventory, 1980.
________. "Vancouver Portland Bridge, Columbia River Interstate," HAER Inventory, 1980.
________. "Waitsburg Bridge," HAER Inventory, 1979.
________. "West Monitor Bridge," HAER Inventory, 1979.
________. "Weyerhaeuser/Pe Ell Bridge," HAER Inventory, 1979.
________. "Wishkah River Bridge [1915]," HAER Inventory, 1979.

National Register of Historic Places (NRHP) Nominations and Reports—

Belshaw, William T. "Jack Knife Bridge," NRHP Nomination, 1973.
Bruce, Robin, Craig Holstine, Robert H. Krier, and J. Byron Barber. "Amendment to 'Historic Bridges and Tunnels in Washington State.'" NRHP Multiple Property Documentation, 1991.
Chapman, Judith A., and Elizabeth O'Brien. "Hood River-White Salmon NRHP Determination of Eligibility," Archaeological Investigations Northwest, Inc., Portland, OR, June 21, 2004.
Davidson, R. B. "NRHP Determination of Eligibility for the Grant Avenue Bridge, Prosser," n d.
Garret, Patsy. "Monroe Street Bridge," NRHP Nomination, 1975.
George, Oscar R. "Bob." "Alaskan Way Viaduct, Battery Street Tunnel," NRHP Nomination, 2001.
________. "Benton City-Kiona Bridge," NRHP Nomination, 2001.
________. "B-Z Corner Bridge," NRHP Nomination, 2001.
________. "Chehalis River Bridge," NRHP Nomination, 2001.
________. "Chief Joseph Dam Bridge," NRHP Nomination, 2001.
________. "Columbia River Bridge at The Dalles," NRHP Nomination, 2001.
________. "Columbia River Bridge at Umatilla," NRHP Nomination, 2001.
________. "Conrad Lundy, Jr. Bridge," NRHP Nomination, 2001.
________. "Dalles Bridge," NRHP Nomination, 2001.
________. "Duwamish River Bridge at First Avenue South," NRHP Nomination, 2001.

________. "Foss River Bridge," NRHP Nomination, 2001.
________. "Gorge Creek Bridge," NRHP Nomination, 2001.
________. "Greene Street Bridge, Spokane," NRHP Nomination, June 2001.
________. "Lake Washington Ship Canal Bridge," NRHP Nomination, June 2001.
________. "Maple Street Bridge," NRHP Nomination, 2001.
________. "Modrow Bridge," NRHP Nomination, 2001.
________. "Mount Si Bridge," NRHP Nomination, 2001.
________. "North 102nd Street Pedestrian Bridge," NRHP Nomination, 2001.
________. "Pioneer Memorial 'Blue' Bridge," NRHP Nomination, 2001.
________. "Portage Canal Bridge," NRHP Nomination, 2001.
________. "Port Washington Narrows Bridge," NRHP Nomination, 2001.
________. "Rainbow Bridge," NRHP Nomination, 2001.
________. "Red Bridge," NRHP Nomination, 2001.
________. "Snohomish River Bridge," NRHP Nomination, 2001.
________. "Steamboat Slough Bridge No. 529/20E," NRHP Nomination, 2001.
________. "Stossel Bridge," NRHP Nomination, 2001.
________. "Upper Custer Way Bridge," NRHP Nomination, 2001.
________. "Vancouver-Portland Southbound Interstate 5 Columbia River Bridge," NRHP Nomination, 2001.
________. "Wenatchee Avenue Southbound Bridge No. 285/20W," NRHP Nomination, 2001.
George, Oscar R. "Bob," and Craig Holstine. "Bridges and Tunnels Built in Washington State, 1951 to 1960," NRHP Multiple Property Documentation, 2001.
Hansen, David M. "Centralia Massacre (Armistice Day Riot) Site," NRHP Nomination, 1974.
Holstine, Craig. "Hood Canal Floating Bridge," NRHP Determination of Eligibility, 2002.
________. "Pe Ell Bridge," NRHP Determination of Eligibility, 2002.
________. "Snohomish River Bridge [1926]," NRHP Determination of Eligibility, 2005.
________. "U.S. Navy Railroad Bridge at Gorst, Kitsap County," NRHP Determination of Eligibility, 1994.
Krier, Robert H., and Craig Holstine. "Tacoma Narrows Bridge [1950]," NRHP Nomination, 1993.
________. "Wenatchee River Bridge No. 285/20E, NRHP Eligibility Evaluation," Short Report DOT98-20, Archaeological and Historical Services, Eastern Washington University, Cheney, 1998.
Krier, Robert H., J. Byron Barber, Robin Bruce, and Craig Holstine. "Agate Pass Bridge," NRHP Nomination, 1991.
________. "Barstow Bridge," NRHP Nomination, 1991.
________. "Columbia River Bridge at Bridgeport," NRHP Nomination, 1991.
________. "Columbia River Bridge at Kettle Falls," NRHP Nomination, 1991.
________. "Columbia River Bridge at Northport," NRHP Nomination, 1991.
________. "Columbia River Bridge at Wenatchee," NRHP Nomination, 1991.
________. "Donald-Wapato Bridge," NRHP Nomination, 1991.
________. "Grande Ronde River Bridge," NRHP Nomination, 1991.
________. "Jim Creek Bridge," NRHP Nomination, 1991.
________. "Lake Keechelus Snowshed Bridge," NRHP Nomination, 1991.
________. "Marshall Bridge," NRHP Nomination, 1991.
________. "Patton Bridge," NRHP Nomination, 1991.
________. "Spokane River at Long Lake Dam," NRHP Nomination, 1991.
________. "Spokane River Bridge at Fort Spokane," NRHP Nomination, 1991.
________. "Toppenish-Zillah Bridge," NRHP Nomination, 1991.
________. "Winnifred Street Bridge," NRHP Nomination, 1991.
Lentz, Flo, and Leonard Garfield. "Norman Bridge," NRHP Nomination, 1994.
Sivinski, Valerie, Penny Chatfield Sodhi, and John M. Simpson. "Tacoma Narrows Bridge Ruins," NRHP Nomination, 1991.
Soderberg, Lisa. "Historic Bridges and Tunnels in Washington State," NRHP Thematic Nomination, 1980.
Weyeneth, Robert R. "Properties Associated with Centralia Armistice Day, 1919," NRHP Multiple Property Documentation, 1991.

Miscellaneous Reports—

Andreas, A. D. "SR 104 Hood Canal Bridge: Task Force Preliminary Report, July 1979, and Summary Report, October 1979," WSDOT, 1979.
Axline, Jon. "Monuments above the Water: Montana's Historic Highway Bridges, 1860–1956," Montana Department of Transportation, Helena, 1993.
Axton, Susan, and Stephen Emerson. "Historic Documentation of the Monroe Street Bridge, Spokane County, Washington." Short Report 717. Archaeological and Historical Services, Eastern Washington University, Cheney, 2002.
"Bi-State Bridge Inventory for Columbia River Crossings," Washington State Department of Transportation, SW Region Planning, May 1998.
Boswell, Sharon A. "Documentation of the Historic Donald-Wapato Bridge No. 396, Yakima County, Washington," Northwest Archaeological Associates Report No. WA 02-25, Seattle, November 8, 2002.
ENTRIX, Inc. "Port Angeles 8th Street Bridge Replacements," Cultural Resources Technical Report, Seattle, July 2004.
Fulton, Ann. "Historic Features Study: SR 14 in the Columbia River Gorge National Scenic Area," Vancouver, WSDOT, 1997.
George, Oscar R. "Bob," and Craig Holstine. "Evaluation and Nomination to the NRHP of Washington State Bridges Built 1951–1960," WSDOT Environmental Affairs Office, Olympia, 2001.
Holstine, Craig. "Monroe Street Bridge," Spokane Register of Historic Places Nomination, 1990.
Krier, Robert H. "King County Historic Bridge Inventory Phase 3: Final Evaluation and Documentation," Short Report 485, Archaeological and Historical Services, Eastern Washington University, Cheney, 1995.
Krier, Robert H., and Craig Holstine. "An Assessment of the Current Status and Condition of Bridges and Tunnels in Washington State Listed in the NRHP," Short Report DOT93-10, Archaeological and Historical Services, Eastern Washington University, Cheney, 1993.

________. "Mora Road Bridge No. 110/25: An Evaluation of Significance," Short Report DOT98-03, Archaeological and Historical Services, Eastern Washington University, Cheney, 1998.
Krier, Robert H., Craig Holstine, Robin Bruce, and J. Byron Barber. "Inventory, Evaluation, and NRHP Nomination of Bridges in Washington State, 1941–1950: A Project Summary," Short Report DOT92-9, Archaeological and Historical Services, Eastern Washington University, Cheney, 1992.
Oregon-Washington Bridge Company, *Annual Reports [Hood River-White Salmon Bridge],* 1934–42. Hood River County Museum, Hood River, Oregon.
Palmer, Kevin A., and Christine Savage Palmer, "14th Avenue South Bridge," King County Landmark Registration Form, 1996.
Phelps, Myra L. *Public Works in Seattle: A Narrative History: The Engineering Department, 1875–1975*. Seattle: Seattle Engineering Department, 1978.
Sheridan, Mimi. "Baring Bridge," King County Landmark Registration Form, King County Office of Cultural Resources, Seattle, 1999.
Tudor Engineering Company. *Lake Washington Bridge Crossings*. Legislative Reconnaissance and Feasibility Report, 1968.
Washington State Department of Transportation (WSDOT; formerly State Department of Highways). *Biennial Reports*, various issues, beginning in 1908.
Washington State Department of Highways. "History of Roads and Highways in the State of Washington," circa 1938.

Published Sources

Books—

Advisory Board on the Investigation of Suspension Bridges. *The Failure of the Tacoma Narrows Bridge*; a reprint of original reports. A contribution to the work of the Advisory Board on the Investigation of Suspension Bridges by the United States Public Roads Administration and the Agricultural and Mechanical College of Texas. College Station, TX: School of Engineering, Texas Engineering Experiment Station, 1944.
America's Young Men: The Official Who's Who among the Young Men of the Nation. Los Angeles: Richard Blank Publishing Co., 1934.
Andrew, Charles. *Final Report on Tacoma Narrows Bridge*. Tacoma, WA: Washington Toll Bridge Authority, 1952.
Bagley, Clarence B. *History of King County, Washington*, 3. Chicago: S. J. Clark Publishing Co., 1931.
________. *History of Seattle from the Earliest Settlement to the Present Time*, 3. Chicago: S. J. Clarke Publishing Co., 1916.
Brown, David J. *Bridges: Three Thousand Years of Defying Nature*. St. Paul, MN: MBI Publishing Co., 1988.
Condit, Carl W. *American Building Art: The Twentieth Century*. New York: Oxford University Press, 1961.
Copeland, Tom. *The Centralia Tragedy of 1919: Elmer Smith and the Wobblies*. Seattle: University of Washington Press, 1993.
Corsilles, D. V., ed. *Rizal Park: Symbol of Filipino Identity*. Seattle: Magiting Corp., 1983.
DeLony, Eric. *Landmark American Bridges*. New York: American Society of Civil Engineers, 1993.
Dorpat, Paul, and Genevieve McCoy. *Building Washington: A History of Washington State Public Works*. Seattle: Tartu Publications, 1998.
Edson, Lelah Jackson. *The Fourth Corner: Highlights from the Early Northwest*. Bellingham, WA: Cox Brothers, 1951.
Gotchy, Joe. *Bridging the Narrows*. Gig Harbor, WA: Peninsula Historical Society, 1990.
Hitchman, Robert. *Place Names of Washington*. Tacoma: Washington State Historical Society, 1985.
Hadlow, Robert W. *Elegant Arches, Soaring Spans: C. B. McCullough, Oregon's Master Bridge Builder*. Corvallis: Oregon State University Press, 2001.
Harshbarger, Patrick, and Mary McCahon. *Delaware's Historic Bridges: Survey and Evaluation of Historic Bridges, with Historic Contexts for Highways and Railroads*, 2nd edition. Paramus, NJ: Lichtenstein Consulting Engineers, Inc., 2000.
Herbst, Rebecca, and Vicki Rottman, eds. *Historic Bridges of Colorado*. Denver: Colorado Department of Highways, 1986.
Historic Highway Bridges of California. Sacramento: California Department of Transportation, 1990.
Hyde, Charles K. *Historic Highway Bridges of Michigan*. Detroit: Wayne State University Press, 1993.
Jackson, Roy A. *Historic Highway Bridges of Florida*. Tallahassee: Florida Department of Transportation, 2004.
Kuykendall, Elgin *Historic Glimpses of Asotin County*. Clarkston, WA: *Clarkston Herald*, 1954.
Legler, Dixie, and Carol M. Highsmith. *Historic Bridges of Maryland*. Baltimore: Maryland Historical Trust Press, 2002.
Lewty, Peter J. *Across the Columbia Plain*. Pullman: Washington State University Press, 1995.
________. *To the Columbia Gateway*. Pullman: Washington State University Press, 1987.
Litvak, Dianna. *Spanning Generations: The Historic Bridges of Colorado*. Denver: Colorado Department of Highways, 2004.
McClelland, John, Jr. *Wobbly War: The Centralia Story*. Tacoma: Washington State Historical Society, 1987.
McCormick, Taylor, and Associates, Inc. *Historic Highway Bridges in Pennsylvania*. Philadelphia: Pennsylvania Department of Transportation, 1986.
Miles, Charles, and O. B. Sperlin, eds. *Building a State: Washington, 1889–1939*. Washington State Historical Society Publications, III. Tacoma: Pioneer, Inc., 1940.
Neil, Dorothy. *A Bridge over Troubled Water: The Legend of Deception Pass*. Langley, WA: South Whidbey Historical Society, 2002.
Norman, James. *Oregon Covered Bridges: A Study for the 1989–90 Legislature*. Salem: Oregon State Highway Division, Environmental Section, 1988.
Petroski, Henry. *Design Paradigms: Case Histories of Error and Judgement in Engineering*. New York: Cambridge University Press, 1994.
________. *Engineers of Dreams: Great Bridge Builders and the Spanning of America*. New York: Alfred A. Knopf, 1995.

________. *To Engineer Is Human: The Role of Failure in Successful Design.* New York: St. Martin's Press, 1985.
Plowden, David. *Bridges: The Spans of North America.* New York: W. W. Norton, 1974, 1984.
Potter, James E., and L. Robert Puschendorf, eds. *Spans in Time: A History of Nebraska Bridges.* Lincoln: Nebraska State Historical Society and Nebraska Department of Roads, 1999.
Quivik, Fredric L. *Historic Bridges in Montana.* Washington, D.C.: Historic American Engineering Record, National Park Service, 1982.
Scott, Richard. *In the Wake of Tacoma: Suspension Bridges and the Quest for Aerodynamic Stability.* Reston, VA: ASCE Press, 2001.
Smith, Dwight A., James B. Norman, and Pieter T. Dykman. *Historic Highway Bridges of Oregon.* Salem: Oregon State Highway Division, Environmental Section, 1985.
Summers, Camilla G. *About Kelso: An Historical Gem.* 1982.
Thomson, Reginald Heber. *That Man Thomson.* Seattle: University of Washington Press, 1950.
Who's Who in Engineering: A Biographical Dictionary of the Engineering Profession. New York: Lewis Historical Publishing Co., 1937, 1941, 1948, 1954.
Who's Who in the State of Washington, 1939–1940. Seattle: 1940.
Willis, Margaret, ed., *Skagit Settlers.* Mount Vernon, WA: Skagit County Historical Society, 1975.
Wood, Sharon. *The Portland Bridge Book.* Portland: Oregon Historical Society, 1989.

Articles and Miscellaneous Publications—

"A Pontoon Bridge with Submerged Pontoon." *Engineering News* 68 (1912): 148.
Andrew, Charles E. "Building the World's Largest Floating Bridge." *Civil Engineering* 10 (January 1940): 17–20.
________. "Design of a Suspension Structure to Replace the Former Narrows Bridge—Part 1." *Pacific Builder and Engineer* 51 (October 1945): 43–45.
________. "The Lake Washington Floating Bridge." *Pacific Builder and Engineer* 46 (July 6, 1940): 29–33.
________. "Redesign of Tacoma Narrows Bridge." *Engineering News-Record* 135 (November 29, 1945): 716–21.
________. "Tacoma Narrows Bridge Number II…The Nation's First Suspension Bridge Designed to be Aerodynamically Stable." *Pacific Builder and Engineer* 56 (October 1950): 54–57, 101.
________. "Unusual Design Problems—Second Tacoma Narrows Bridge." *Proceedings of the American Society of Civil Engineers* 73 (December 1947): 1483–97.
Andrews, George. "Steel Bridge—May to January." *Pacific Builder and Engineer* 67 (June 1961): 86.
"Approve Hangman Plans." *Spokesman-Review*, August 23, 1911.
Arnold, Cecil C. "Field Problems and Construction Methods." *Pacific Builder and Engineer* 46 (November 2, 1940): 28–30.
Averill, Walter A. "Collapse of the Tacoma Narrows Bridge." *Pacific Builder and Engineer* 46 (December 1940): 20–27.
Barber, J. Byron. "The Golden Era of Bridge Building." *The Pacific Northwesterner* 29, no. 1 (Winter 1984): 1–13.
Barber, Richard. "Massive Concrete Pontoons." *Pacific Builder and Engineer* 46 (August 3, 1940): 41–43.
"Beebe Bridge." *Highway News* 10, no. 3 (November-December, 1962): 20.
"Beebe Bridge Construction Fulfills Age-Old Need," *Highways News* 9, no. 1 (July-August 1960): 13–14, 20.
"Biggs Rapids Bridge," *Highway News* 10, nos. 1–2 (September-October, 1962): 16.
"Bridge Builder [Morton Macartney] Amazed by City." *Spokesman-Review,* August 22, 1951.
"Bridge Dedication Number, Lake Washington Floating Bridge-July 2, Tacoma Narrows Suspension Bridge-July 1." *Pacific Builder and Engineer,* July 6, 1940.
"Bridge Designer [J. F. Greene] Tests Spans He Built in City Long Ago." *Spokesman-Review,* August 25, 1953.
"Bridge Span Progresses." *Spokesman-Review*, August 17, 1912.
"Bridge that Floats." *Scientific American* 162 (February 1940): 75–77.
"Building Lake Washington Pontoons." *Engineering News-Record* (August 3, 1939): 50.
"Capture and Imprisonment by Japs." *Pacific Builder and Engineer* 51 (December 1945): 44–49.
"Clark Eldridge Didn't Let Age Dull Desire to Work." *Seattle Times,* November 9, 1990.
Clark, John H. "West Seattle Swing Bridge, Seattle, Washington." *Structural Engineering International* (January 1995): 23–25.
Clarke, Jonathan. "Material Concerns in the Pacific Northwest: Steel Versus Reinforced Concrete in Highway Bridge Design in Washington State, 1910–1930." *Construction History* 16 (2000): 33–61.
"Clarkston-Lewiston Bridge." *Western Construction News* (August 1939): 267–699.
Coleman, F. C. "New Pontoon Bridge over the Golden Horn at Constantinople." *Engineering News* 70 (November 20, 1913): 1018–20.
"Collapse of Spokane Bridge." *Pacific Builder and Engineer* 23 (February 23): 1917.
"Columbia River Toll Bridge at Hood River, Oregon, Now in Service." *Engineering News* 94, no. 1 (January 5, 1925): 130.
"Commissioners Accept New Steel [Mellen Street] Bridge." *Centralia News Examiner*, July 18, 1911.
Comp, T. Allan, and Donald Jackson. "Bridge Truss Types: A Guide to Dating and Identifying." Technical Leaflet 95, American Association for State and Local History, *History News* 5, no. 5 (May 1977).
"Composite Trestle-Type Bridges on 100-Ft., Two-Column Bents." *Engineering News-Record* (August 6, 1936): 197–98.
Conger, Kay. "Seattle Freeway Bridge to Be Sixth Ship Canal Crossing." *Highway News* (January-February 1958): 10–13.
Corser, Champ E. "The Transition Section." *Pacific Builder and Engineer* 46 (August 3, 1940): 44.
Crist, Douglas. "The Span of Time; How Bainbridge Got Its Bridge." In *Bainbridge Island Almanac, 2002.* Bainbridge Island: Sound Publishing, 2002, 11–17.
Cronk, Leon. "History of Old Beebe Bridge." *Chelan Valley Mirror,* July 11, 1963.

DeLony, Eric. "HAER's Historic Bridge Program." *The Journal of the Society for Industrial Archeology* 15, no. 2 (1989): 57–71.

DeLony, Eric, and Michael J. Auer. "Historic Bridges: Preservation Challenges." *Cultural Resources Management* 14, no. 1 (1991): 1–8.

Dimock, A. H. "Bridging of Inter-City Waterways at Seattle." *Pacific Builder and Engineer* (March 8, 1913): 153–55.

Dorpat, Paul. "Tampering with Lake Union." In *Seattle: Now and Then,* 2nd ed. Seattle: Tartu Publications, 1984.

Dugovich, William. "Seattle's Superfreeway." *Washington Highways* 14, no. 2 (May 1967): 2–5.

Dunford, J. A. "Record Rigid-Frame Bridge." *Engineering News-Record* (June 24, 1937): 939–42.

________. "Seattle Remodels Its University Bridge." *Engineering News-Record* (October 12, 1933): 439.

Durkee, L. R. "Floating Bridge More Stable than Fixed Spans." *Pacific Builder and Engineer* 46 (August 3, 1940): 38–41.

Dusenberry, Donald. "What Sank the Lacey Murrow?" *Civil Engineering* 63, no. 11 (November 1993): 54–59.

Eldridge, Clark H. "The Tacoma Narrows Bridge." *Civil Engineering* 10 (May 1940): 299–302.

________. "The Tacoma Narrows Suspension Bridge." *Pacific Builder and Engineer* 46 (July 6, 1940): 35–40.

"Engineer [Clark Eldridge] Has Long Career as Bridge Builder." *Seattle Times,* June 1, 1940.

Eriksson, Merv, C. Milo McLeod, and Dan Gard. *Identifying and Preserving Historic Bridges.* Technology and Development Program, Publication 8E82L47, U.S.D.A., Forest Service, Region One, Missoula, Montana, 2000.

"The Evergreen Point Bridge, 1960–1963." *Washington Highway News* (July-August 1963): 1.

"Experts Approve Pontoon Bridge." *Engineering News-Record* (July 14, 1938): 38.

Finke, R. W. "Seattle's Long-Planned Alaskan Way Viaduct Scheduled for Construction." *Pacific Builder and Engineer* (August 1949): 52–55.

Garber, Andrew. "Finding Indian Remains May Sink Bridge Project." *Seattle Times*, September 12, 2003.

Gloyd, C. S. "Concrete Floating Bridges." *Concrete International* (May 1988): 17–24.

Gordon, Charles M. "World's Largest Stack Drift Tunnel." *Pacific Builder and Engineer* 91 (July 22, 1985): 8–9.

Greene, J. F. "The Latah Creek Bridge, Spokane, Wash.," *Engineering News* 69, no. 13, 27 (March 1913): 614–17.

"Greene Bridge Plans Readied." *Spokesman-Review*, November 28, 1952.

Griffin, Walter R. "George W. Goethals, Explorer of the Pacific Northwest, 1882–1884." *Pacific Northwest Quarterly* 62 (October 1971): 133–37.

"Hadley, Floating Bridge Proposer, Dies in Lake." *Seattle Times,* July 6, 1967.

Hadley, Homer M. "Concrete Pontoon Bridge—A Review of the Problems of Design and Construction." *Western Construction News* (September 1939): 293–98.

Hadlow, Robert W. "Bridges over Washington: A Historic Engineering Record." *Columbia Magazine* 8, no. 1 (Spring 1994): 6–11.

Handley, Scott. "Home of Floating Bridges Soon to Open New Span…First Response to Idea Was Far from Positive." *Construction Data and News, Supplement to Engineering News-Record* 230, no. 22 (May 31, 1993): 3–19.

Hanrahan, Brenda, and Jim Manders. "Graving Work at Least Week from Restart." *Peninsula Daily News*, September 4, 2003.

Hensel, Doris. "Longtime Bridge Engineer [Clark Eldridge] Recalls 'Galloping Gertie' with Heartache." *Daily Olympian,* September 3, 1986.

Higgins, Mark. "Floating Bridge to Bear Name of Innovator Hadley." *Seattle Post-Intelligencer,* July 17, 1993.

"History of Division Street Bridge Recalled." *Spokesman-Review*, May 10, 1964.

"Historic Spokane Bridge Rebuilt." *Pacific Builder and Engineer* (September 6, 2004): 10–11.

"The Hood Canal Floating Bridge." *Highway News* (May-June 1961): 18–19.

Hunt, Thomas D. "Fourteenth Avenue South Bridge, Seattle." *Western Construction News* (June 10, 1931): 287.

Johnson, Edgar B., Lee J. Holloway, and Georg Kjerbol. "Unearthing Mt. Baker Tunnel." *Civil Engineering* 55, no. 12 (December 1985): 36–39.

Judd, Harold "Twin Bore Tunnels: Construction Methods Used to Bore 30 Tunnels through the Glacial Blue Clay of Mount Baker Ridge." *Pacific Builder and Engineer* 46 (October 5, 1940): 46–48.

Kemp, Emory L. "The Fabric of Historic Bridges." *The Journal of the Society for Industrial Archeology* 15, no. 2 (1989): 3–22.

Kuesel, Thomas R. "Floating Bridge for 100 Year Storm." *Civil Engineering* (June 1985): 60–65.

"Lacey Murrow, Former Director, Honored at WSC." *Highway News* 8 (May-June 1959): 19.

"Lake Union Bridge Provides New Traffic Link in Seattle." *Engineering News-Record* (March 8, 1932): 313.

"Lake Union Bridge, Seattle, Washington." *Western Construction News* (May 10, 1930): 226–29.

"Last Rites for Admiral Gregory Set." *Seattle Times,* September 14, 1960.

"L. R. Durkee Named Engineer of the Year." *Pacific Builder and Engineer* (March 1960): 3.

Lwin, M. Myint, and Donald O. Dusenberry. "Responding to a Floating Bridge Failure." *Public Works* (January 1994): 39–43.

Mapes, Lynda "Tribe's Letter Deepens Dilemma over Project." *Seattle Times/Seattle Post-Intelligencer*, December 12, 2004.

"Maple Street Bridge Plans and History Reviewed." *Highway News* (July-August 1958): 26–27.

"Maple Street Toll Bridge Opened July 1 for Spokane Traffic." *Highway News* (July-August 1958): 26–27.

McKay, Louise. "Drawbridges across the Okanogan." *Okanogan County Heritage* 17, no. 3 (Summer 1979): 3–12.

"Megler-Astoria: A Bridge for the Columbia's Mouth." *Highway News* 11, no. 1 (July-August 1963): 7–8.

Merry, Bill. "Public Pulse of Highway Progress." *Highway News* 11, no. 2 (September-October 1963): 2–7.

"Metalline Falls Bridge." *Highway News* 2, no. 6 (December 1952): 23.

"The Monroe St. Bridge, Spokane, Washington; A Concrete Bridge Containing a 281-ft. Arch." *Engineering News* (September 2, 1909): 241–43.

"Moorings for Lake Washington Bridge." *Engineering News-Record* (July 18, 1940): 46–48.

Moses, Paul. "The Bridges of Okanogan, 1910–2003." *Okanogan County Heritage* (June 2003): 2–10.

"Munster, Andreas W." [Obituary]. *Transactions, American Society of Civil Engineers* 95. New York: ASCE, 1931: 1565–66.

Murrow, Lacey "A Concrete Pontoon Bridge to Solve Washington Highway Location Problem." *Western Construction News* (July 1938): 249–52.

________. "Early History of the Lake Washington Floating Bridge." *Pacific Builder and Engineer* (July 1940).

"New $225,000 Bridge at Greene Street Planned by City." *Spokesman-Review*, February 15, 1949.

"New Seattle Bridge." *Washington Highway News* (November-December 1960): 16–19.

"Panel Finds Holes Helped Sink I-90 Bridge." *Seattle Times*, May 2, 1991, A1.

Parker, Harvey W., and Robert A. Robinson. "The World's Largest-Diameter Soil Tunnel." *Underground Space* 7 (November 1982/January 1983): 175–81.

PCI Design Handbook, 4th edition. Precast/Prestressed Concrete Institute, Chicago, 1999.

"Point of Weakness in Low-truss Highway Bridges." *Engineering News* 69, no. 11 (March 13, 1913): 523–24.

"Pontoon Bridge Proposed to Relieve Seattle Ferries." *Engineering News-Record* 87 (September 22, 1921): 511.

"Pontoon or Floating Drawbridges." *Engineering News* 62 (April 30, 1908): 474–79.

"Pontoons Redesigned for Seattle Crossing." *Engineering News-Record* 228, no. 3 (January 1992): 40–41.

"Port Body Buys Bridge of Gods." *Marine Digest* (December 30, 1961): 24.

Rapp, F. A. "Three Double-Leaf Bascule Bridges at Seattle, Wash." *Engineering News-Record* 84 (April 8, 1920): 718–22.

Reflections on the Beginnings of Prestressed Concrete in America. Precast/Prestressed Concrete Institute, Chicago, 1981.

"Reginald Heber Thomson." In Bagley, Clarence, *History of King County* 2. Chicago: S. J. Clarke Publishing Co., 1929, 92–100.

Robinson, Rita. "The Stacked-Drift Tunnel." *Civil Engineering* 60, no. 7 (July 1990): 40–42.

"Safeguarding Lone Highway Bridge from Fire and other Damage." *Engineering News* 94, no. 18 (April 30, 1925): 722–23.

Schmidt, Jesse. "Methow Valley Fords and Bridges." *Okanogan County Heritage* (December 1968): 7–10.

"Second Floating Bridge Seattle Need in 1955." *Engineering News-Record* 144 (February 16, 1950), 28.

"Service Performance of Grid Deck on University Bridge, Seattle." *Engineering News-Record* (September 20, 1934): 376.

"Sign Big Bridge Contract." *Spokesman-Review*, October 15, 1911.

Sines, C. E. "Five Million Dollar Vantage Bridge Progressing Smoothly." *Highway News* 9, no. 4 (January-February 1961): 32–33.

"Sliding Span for Proposed Pontoon Bridge." *Western Construction News* (February 17, 1938): 248.

Smith, Darrell, and Wesley Karney. "Salmon Creek Bridge." *Highway News* 5 (March 1956): 8–9, 28.

"Spokane May Inspect City Bridges." *Pacific Builder and Engineer* 21 (January 1916): 34.

"State Says Yes on Bridge Plan." *Spokesman-Review*, November 3, 1952.

"Temporary Crossings First to Be Arranged." *Wenatchee Daily World*, July 6, 1917.

"Twin Tunnels Driven through Clay for Lake Washington Bridge Project." *Western Construction News* 15 (July 1940): 246.

Vaughn, Margot, ed. "The Modrow Bridge." *Cowlitz County Quarterly* 26 (Spring 1984): 35–36.

Washington Toll Bridge Authority. *Bridges of 1940 Built by the Washington Toll Bridge Authority*. American Association of State Highway Officials, 26th Annual Convention Committee, 1940.

Wassam, Homer. "Movable Bridges of the Snohomish." *Everett Herald*, March 23, 1974, 6–7.

"Wenatchee Bridge Burned." *Wenatchee Daily World*, July 4, 1917.

Westbrook, Todd. "Spanning Half a Century; Agate Pass Bridge Celebrates its 50th Birthday Today." *Bremerton Sun*, October 16, 2000.

"Wind River's Bridge Crews Off for Winter." *Skamania County Pioneer*, December 19, 1958.

Woodin, Mark S. "Bridges, Now and Then." *Highway News* 3, no. 3 (September 1953): 61–64.

Wooliscroft, B. "World's First Draw Pontoon." *Pacific Builder and Engineer* 46 (August 3, 1940): 46.

Wright, Andrew G. "Buoyed by a Water-tight Design; Despite Two Failures, State Engineers Retain a Pioneering 1940s Concept." *Engineering News-Record* 230, no. 22 (May 31, 1993): 22.

Newspapers—

Bellevue American
Centralia Daily Chronicle
Centralia News Examiner
Chelan Valley Mirror
Daily Okanogan
Daily Olympian
Everett Herald
Goldendale Sentinel
Kettle River Journal
Mercer Island Reporter
Metalline Falls News
The Olympian
Oregonian
Peninsula Daily News
Seattle Post-Intelligencer
Seattle Times
Seattle Weekly
Skamania County Pioneer
Spokesman-Review

Tri-City Herald
University Herald
Wahkiakum County Eagle
Wenatchee Daily World

Selected Web Sites—

WSDOT—Tacoma Narrows Bridge History: *www.wsdot.wa.gov/TNBhistory*
WSDOT—Environmental Services Office: *www.wsdot.wa.gov/environment/culres/bridges*

Designers, Contractors, and Directors

A

James Allen, 38, 143
American Bridge Company (New York), 114
O. A. Anderson, 202
Charles E. Andrew, x, 38, 82–84, 88, 170, 189, 203
Angeles Gravel and Supply Company (Port Angeles), 133, 217
Arch Rib Construction Company, 101
C. Adrian Arnold, 84
Cecil C. Arnold, 82–83, 84, 192, 193
Guy F. Atkinson Company (Portland), 202

B

Beebe Orchard Company, 98, 114
Beers Building Company (Seattle), 79
Benton-Franklin Intercounty Bridge Company, 109
Duane Berentson, 178
Bethlehem Steel, 102
D. Boyington Company, 79
Barry Brecto, 76
J. C. Broad Company, 40, 71
William A. Bugge, 84, 100, 112, 135, 140, 181
Bureau of Public Roads (San Francisco), 193
Troy T. Burnham Company (Seattle), 116
M. P. Butler Company (Seattle), 190

C

C. and F. Teaming and Trucking Company (Butte), 133
M. M. Caldwell (Seattle), 221
Clinton Bridge Works (Iowa), 145
Colonial Building Company (Spokane), 40, 143, 218
C. M. Corkum Company (Portland), 211
Creech Brothers Contracting Company (Aberdeen), 71
J. E. Cunningham Company (Spokane), 128
Kirtland Cutter and Carl Malmgren (Spokane), 77, 125–26, 128

D

Charles Davis, 131
DeLong Corporation (New York), 100
Arthur H. Dimock, 85, 88, 159
J. H. Dirkes, 169
Professor F. Dischinger, 23–24
H. E. Dodge, 217
James Donahue, 209
Dulin Construction Company (Centralia), 75
"L. R." Durkee, 82–83, 85, 170

E

Clark Eldridge, 61, 85–86, 158–59, 170, 197
David Evans and Associates Inc. (Salem), 77, 236
F. T. Evans, 82–83
O. R. Elwell, 191

F

F. B. Farquharson, x, 61, 63
Ferguson and Houston (Astoria), 74
Ralph W. Finke, 100, 145
R. S. Fluent, 82–83
C. D. Forsbeck, 202

G

General Construction Company (Seattle), 112, 192
Ben C. Gerwick Company (San Francisco), 100
Harold Gilbert, 210
Gilpin Construction Company, 106
Jean Glebov, 82–83
Charles Stewart Gloyd, 176
Lieutenant George W. Goethals, 132
Arvid Grant, 98, 111
J. F. Greene, 128
Luther E. Gregory, 86–87, 88, 170
Gunderson Brothers Engineering (Portland), 190

H

Homer M. Hadley, 8, 47, 73, 87–88, 90, 142, 145, 169–70, 172, 177, 197, 205–7, 220, 224, 234
Henry Hagman Construction Company (Cashmere), 131, 134, 135, 145
J. W. Hardison Company, 112
John Lyle Harrington, 128
Hart Construction Company, 55
Willis B. Horn, 189, 222
O. H. Horton Company (Colfax), 68
William Howe, 6, 7, 32–33, 208
Charles G. Huber Company (Seattle), 15, 37
Samuel J. Humes, 88

I

International Contract Company (Seattle), 69, 200
Interstate Construction Company, 105

J

Jacobs and Ober (Seattle), 90, 157
Paul Jarvis, 146, 192
Paul Jarvis Inc. (Seattle), 107
Frank M. Johnson, 159

K

Peter Kiewit Sons Company (Vancouver), 193, 225
Elgin V. Kuykendall, 143

L

Derby Livesay, 189
Long-Bell Company (Longview), 101
David Luten, 5, 15, 30, 40

M

Morton Macartney, 77, 128, 129, 130
MacRea Brothers (Seattle), 114, 202, 217, 223
D. C. Maloney Company, 68
W. L. "Pat" Malony, 134–35
W. S. Malony, 128
Conrad O. Mannes, 216–17
Manson Construction and Engineering Company (Seattle), 189, 223
James Marsh, 143
Thomas McCrory, 157
J. A. McEachern Company (Seattle), 189
Joanne McGovern, 234
Ray B. McMinn, 82–83, 88, 170
Daniel W. McMorris, 88–89
R. E. Meath Company (Portland), 37
Frank Milward, 59
Milwaukee Bridge Company, 69
Leon Moisseiff, x, 61, 86, 234–44
Robert Montell, 169
Sid Morrison, 111
Morrison-Knudsen Company, 132
Munson Construction and Engineering Company (Seattle), 112
Andreas W. Munster, 89
T. J. Murphy, 217
Ray M. Murray, 82–83, 88, 89–90, 157, 160–61, 170
Lacey V. Murrow, 43, 82–83, 86, 87, 90–91, 100, 170–72

N

Leonard K. Narod, 91

O

James O'Farrell, 204
William Oliver Bridge Company (Spokane), 53, 80
Olson and Johnson Company (Missoula), 129
R. M. Ordish, 142
Oregon State Department of Highways, Bridge Department, 104
Oregon-Washington Bridge Company, 97

PQ

Parker-Schram Company (Portland), 101
Henry Petroski, 5, 52
Alton V. Phillips Company (Seattle), 116
Captain George E. Pickett, 30–31
J. H. Pomeroy and Company, 100, 102–3
Portland Bridge Company, 37, 215
Portland Cement Association, 39, 49, 73, 87, 169, 170, 205, 220
Harry R. Powell, 47, 91, 194–95, 209
Harry R. Powell and Associates, 55, 143, 209
Thomas and Caleb Pratt, 6, 8–9
Puget Construction Company (Seattle), 141, 191–92
Puget Sound Bridge and Dredging Company (Seattle), 97, 105, 146, 147, 221, 222
Puget Sound Energy, 78

R

J. C. Ralston, 77, 124, 126
Ramsey and Company, 145
Raymond International (New York), 100
Red Mountain Railroad Company, 119
Oliver S. Reed, 71
Caro Reese, 82–83
Larry Robertson, 189

S

San Francisco Bridge Company, 129
Sargent Engineers Inc. (Olympia), 74
Scherzer Rolling Lift Bridge Company (Chicago), 155
City of Seattle Engineering Department, 48, 90, 154, 161
City of Seattle Lighting Department, 193
Charles G. Sheely Construction Company (Denver), 10, 35, 68
F. D. Sheffield, 69
Dexter R. Smith, x, 203
Raymond G. Smith (Seattle), 192, 193
Tom Smith, 82–83
Spokane Engineering Department, 130
St. Paul Foundry Company, 80
George Stevens, 112, 224
J. B. Strauss Company, 19
Strauss-Bascule Bridge Company (Chicago), 79, 154, 221, 222
Strauss Engineering Corporation (Chicago), 102
Superior Portland Cement Company, 40

T

Jack Taylor, 82–83
Te Wanna Toll Bridge Company, 97, 105
"R. H." Thompson, 82–83, 85, 88, 91–92, 151, 153, 170
J. H. Tillman Company, 118
Tudor Engineering Company (San Francisco), 109, 131, 132, 225

UV

United States Steel Corporation, American Bridge Division, 100
United States Steel Corporation, American Bridge Division (Gary, Indiana), 112, 146, 193
United States Steel Corporation, American Bridge Division (Portland, Oregon), 132
United States Steel (Memphis, Tennessee), 192
U.S. Army Corps of Engineers, 76, 116
U.S. Bureau of Public Roads (San Francisco), 37, 197, 210–12
U.S. Forest Service, 232
U.S. Forest Service Region One (Missoula), 72

WXYZ

J. A. L. Waddell, 38, 70, 189, 199, 236
Waddell and Harrington (Kansas City), 128, 198–99
Al Walley, 177
Ward and Ward Inc., 71
James Warren, 6, 11
Washington Bridge Corporation (Wenatchee), 112–14
Washington Emergency Recovery Administration, 216–17
Washington Paving Company (Seattle), 130
Washington Portland Cement Company, 40
Washington State Transportation Improvement Board, 78
Washington Toll Bridge Authority, x, 42, 61, 83, 87, 92, 103, 107, 132, 170–71, 174, 176, 203, 225
Carl West, 189, 190
West Coast Steel Works (Portland), 143
P. H. Winston, 82–83
A. M. Young, 217

The Authors

Craig Holstine, M.A., is a Cultural Resources Specialist with WSDOT who studies historic bridges and other historic resources. He is the author of *Forgotten Corner: A History of the Colville National Forest, Washington* (1987), and coauthor with Fred C. Bohm of *The People's History of Stevens County, Washington* (1983). He lives in Tumwater with his wife Marsha, sons Robert and Zachary, and dog Tipper.

Richard Hobbs, Ph.D., is an independent historian-consultant and has written extensively on historic bridges in Washington. He is the author of *The Cayton Legacy: An African American Family* (WSU Press, 2002) and a forthcoming book on the history of the Tacoma Narrows Bridges (WSU Press, ca. 2006). He lives on Whidbey Island with his wife Lynette, twins Rhianna and Ryder, and several horses.